Principles of Chemical Engineering Processes

MATERIAL AND ENERGY BALANCES

SECOND EDITION

Principles of Chemical Engineering Processes

MATERIAL AND ENERGY BALANCES

SECOND EDITION

NAYEF GHASEM

REDHOUANE HENDA

CRC Press
Taylor & Francis Group
Boca Raton London New York

CRC Press is an imprint of the
Taylor & Francis Group, an **informa** business

CRC Press
Taylor & Francis Group
6000 Broken Sound Parkway NW, Suite 300
Boca Raton, FL 33487-2742

© 2015 by Taylor & Francis Group, LLC
CRC Press is an imprint of Taylor & Francis Group, an Informa business

No claim to original U.S. Government works

Printed on acid-free paper
Version Date: 20141031

International Standard Book Number-13: 978-1-4822-2228-9 (Pack - Book and Ebook)

Library of Congress Cataloging-in-Publication Data

Ghasem, Nayef.
    Principles of chemical engineering processes : material and energy balances / authors, Nayef Ghasem, Redhouane Henda. -- Second edition.
        pages cm
    Includes bibliographical references and index.
    ISBN 978-1-4822-2228-9 (hardback)
    1. Chemical engineering. I. Henda, Redhouane. II. Title.

TP155.G47 2015
660--dc23   2014039170

Visit the Taylor & Francis Web site at
http://www.taylorandfrancis.com

and the CRC Press Web site at
http://www.crcpress.com

Contents

Preface

Purpose of the Book

The objective of this book is to introduce chemical engineering students to the basic principles and calculation techniques used in the field and to acquaint them with the fundamentals of the application of material and energy balances in chemical engineering. The book is mainly intended for junior chemical engineers. It exposes them to problems in material and energy balances that arise in relation to problems involving chemical reactors. The focus is on single and multiphase systems related to material and energy balances. This material is introduced in a manner that there will be little lecture content. Most of the material presented covers work on problem solving skills. The material intrusion theory is explained briefly and simply to make this book more user-friendly. Many examples are included and solved in a simple and understandable format that explains the related theory. A few examples related to unsteady state material and energy balance are solved using MATLAB®/Simulink®, and this enables students to use such software in solving complicated problems of material and energy balances. The book can be covered in one semester if based on a four-hour contact course or two semesters on a two-hour contact course. The new edition contains additional solved examples and homework problems. A new chapter related to single and multiphase systems has been added.

Course Objectives

At the completion of this course, students will

1. Acquire skills in quantitative problem solving, specifically the ability to think quantitatively (including numbers and units), the ability to translate words into diagrams and mathematical expressions, the ability to use common sense to interpret vague and ambiguous language in problem statements, the ability to make judicious use of approximations and reasonable assumptions to simplify problems

2. Understand processes typical of those found in the chemical and bio process industries

3. Know general principles used in process engineering, particularly stoichiometry, material and energy balances, gas laws, and phase equilibria

Course Outcomes

Every course that a student takes should further his or her knowledge, building on what was previously learned. By the end of this course, each student should be able to

1. Explain the concepts of dimensions, units, psychrometry, steam properties, and conservation of mass and energy
2. Solve steady-state and transient mass and energy balance problems involving multiple-unit processes and recycle, bypass, and purge streams
3. Solve and understand simple unsteady-state mass and energy balances
4. Solve more complicated problems using the software appropriate for the problem
5. Present the solutions to engineering problems in both oral and written form in a clear and concise manner

Relationship of Course to Program Outcomes (Accreditation Board for Engineering and Technology [ABET] Criteria)

At the completion of this course, students will have demonstrated the following skills, as described by the ABET:

1. An ability to identify, formulate, and solve engineering problems
2. An ability to communicate effectively
3. An ability to use the techniques, skills, and modern engineering tools necessary for engineering practice

A Note about the Web Downloads

The material available via the CRC Web site contains a summary of all chapters in the book, exercises, property estimation software, and psychometric chart parameter estimation software. This material is available from the CRC Web site: http://www.crcpress.com/product/isbn/9781482222289.

PowerPoint presentations that explain the simulations and property estimation using HYSYS are also available to instructors.

The website also contains POLYMATH—a software that the user will be allowed to run 200 times free of charge. For the last 50 uses, a warning will be issued where the software will indicate the number of trials remaining and gives information as to where the software can be purchased. The web link for this software is

http://www.polymath-software.com/ghasem/

More information about POLYMATH can be obtained at the following link:

http://www.polymath-software.com

MATLAB® is a registered trademark of The MathWorks, Inc. For product information, please contact:

The MathWorks, Inc.
3 Apple Hill Drive
Natick, MA 01760-2098 USA
Tel: 508-647-7000
Fax: 508-647-7001
E-mail: info@mathworks.com
Web: www.mathworks.com

Acknowledgments

We are thankful to Allison Shatkin, acquiring editor for this book, and Jessica Vakili, the project coordinator, for their help and cooperation. We also thank those professors who have given us the chance to see their course syllabi and contents. They include Professor Richard Zollars from Washington State University; Professors Prausnitz, Brown, Jessica, Bartling, Koros, and Timothy from the Georgia Institute of Technology; and Professors Palecek, Murphy, Root, and Rawling from the University of Wisconsin–Madison. We also extend our thanks to all professors who are teaching or contributing to this subject. The feedback from Professor Douglas R. Lloyd and David Himmelblau and others is highly appreciated.

Nayef Ghasem
Redhouane Henda

Authors

Nayef Ghasem is an associate professor of chemical engineering at the United Arab Emirates University, Al Ain, United Arab Emirates. He earned both BSc and MSc degrees from the Middle East Technical University, Ankara (Turkey), and earned his PhD from the University of Salford, Greater Manchester (UK). He teaches chemical process principles, natural gas processing, and process modeling and simulation as undergraduate courses along with other courses in chemical engineering. Previously, he taught these courses at the University of Malaya, Kuala Lumpur, Malaysia. He has published more than 40 journal papers, primarily in the areas of modeling and simulation, bifurcation theory, gas–liquid separation using membrane contactor, and fabrication of polymeric hollow fiber membranes. He has also authored *Computer Methods in Chemical Engineering*, published by CRC Press. Dr. Ghasem is a member of the Institute of Chemical Engineers (IChemE) and a member of the American Chemical Society (ACS).

Redhouane Henda is a professor of chemical engineering at the Laurentian University, Sudbury, Ontario, Canada. He earned both MSc and PhD from the Institut National Polytechnique de Toulouse (France) and spent a postdoctoral fellowship at the Universität Heidelberg (Germany). He has taught numerous core courses in chemical engineering at the undergraduate and graduate levels. Dr. Henda's research interests lie at the intersection of chemical engineering and materials science, with a focus on nanoscience and technology of thin films and on process engineering of complex systems. He has written a number of journal articles and book chapters in both focus areas and has developed some computer modules for research and education purposes. Among the recognitions he has received are fellowships from the French Ministry of Higher Education and the German Alexander von Humboldt Foundation, as well a scholarship from the Research Council of Norway. Dr. Henda is a licensed professional engineer in Ontario and has served on the editorial boards of numerous journals.

Systems of Units

Systems of units are defined with reference to Newton's second law for a system of constant mass: $F = ma$ (mass–length/time^2)

where F is the force required to accelerate a body of mass, m, at a rate a (length/time^2)

System	Length	Time	Mass	Force	g_c
SI	Meter m	Second s	Kilogram kg	Newton N	$1.0\ (\text{kg.m})/(\text{N} \cdot \text{s}^2)$
CGS	Centimeter cm	Second s	Gram g	dyne	$1.0\ (\text{g.cm})/(\text{dyne.s}^2)$
AES	Foot ft	Second s	Pound mass lb_m	Pound force lb_f	$32.17\ (lb_m.\text{ft})/(lb_f.\text{s}^2)$

1.0 N = Force that will accelerate a mass of 1.0 kg by 1.0 m/s^2
1.0 dyne = Force that will accelerate a mass of 1.0 g by 1.0 cm/s^2
1.0 lb_f = Force that will accelerate a mass of 1.0 lb_m by 32.174 ft/s^2

Metric prefixes

10^{12}	T	Tera	Trillion	10^{-1}	d	Deci	Tenths
10^{9}	G	Giga	Billion	10^{-2}	c	Centi	Hundredths
10^{6}	M	Mega	Million	10^{-3}	m	Milli	Thousandths
10^{3}	k	Kilo	Thousand	10^{-6}	μ	Micro	Millionths
10^{2}	h	Hector	Hundred	10^{-9}	n	Nano	Billionths
10^{1}	da	Deca	Ten	10^{-12}	p	Pico	Trillionths

Acceleration of gravity $g = 9.8066$ m/s^2 (sea level, 45° latitude)
$g = 32.174$ ft/s^2

Gas constant $R = 10.731$ psia-ft^3/lbmol = 0.7302 atm-ft^3/lbmol

$R = 0.082056$ atm-L/mol-K = 8.3143 Pa-m^3/mol-K

$R = 0.08314$ L-bar/mol-K = 1.987 Btu/lb mol = 8314.3 J/kg mol-K

$R = 8.3143$ J/mol-K = 62.36 L-mmHg/mol-K = 1.987 cal/mol-K

Density of water at 4°C ρ (H_2O, 4°C) = 1.0 g/cm^3 = 1.0 kg/L = 10^3 kg/m^3

$\rho(H_2O$, 4°C) = 8.34 lb_m/gal = 62.43 lb_m/ft^3

Specific gravity of water = 1.0

Specific gravity of Hg = 13.6

Conversion Factors

Mass	$1\ lb_m = 5 \times 10^{-4}\ t = 0.453593\ kg = 453.593\ g = 16\ oz$
	$1\ kg = 1000\ g = 2.20462\ lb_m = 0.001\ t$
	$1\ t = 2000\ lb_m;\ 1\ t = 1000\ kg$
Length	$1\ ft = 12\ in.;\ 1\ ft = 0.3048\ m = 30.48\ cm;\ 1\ in. = 2.54\ cm;\ 1\ mile = 5280\ ft$
	$1\ m = 10^{10}\ Å = 39.37\ in. = 3.2808\ ft = 1.0936\ yd = 0.0006214\ mi$
Volume	$1\ ft^3 = 7.481\ gal = 1728\ in.^3 = 28.317\ L = 28{,}317\ cm^3$
	$1\ gal = 231\ in.^3;\ 1\ in.^3 = 16.387\ cm^3$
	$1\ cc = 1\ cm^3 = 1\ mL;\ 1000\ mL = L$
	$1000\ L = 1\ m^3 = 35.3145\ ft^3 = 220.83\ imperial\ gallons = 264.17\ gal = 1056.68\ qt$
	$8\ fl\ oz = 1\ cup;\ 4\ cup = 1\ quart;\ 4\ quart = 1\ gal = 128\ fl\ oz$
Density	$1\ g/cm^3 = 1\ kg/L = 1000\ kg/m^3 = 62.428\ lb/ft^3 = 8.345\ lb_m/gal$
Force	$1\ lb_f = 32.174\ lb_m\text{-}ft/s^2 = 4.448222\ N = 4.4482 \times 10^5\ dynes$
	$1\ N = 1\ kg\text{-}m/s^2 = 10^5\ dynes = 10^5\ g\text{-}cm/s^2 = 0.22481\ lb_f$
Pressure	$1\ bar = 10^5\ Pa = 100\ kPa = 10^5\ N/m^2$
	Pascal (Pa) is defined as $1\ N/m^2 = 1\ kg/m\text{-}s^2$
	$1\ atm = 1.01325\ bar = 14.696\ lb_f/in.^2 = 760\ mmHg\ at\ 0°C\ (torr) = 29.92$ in Hg at 0°C
	$1\ psi = 1\ lb_f/in.^2;$ psia (absolute) = psig (gauge) + 14.696
Temperature	$1\ K = 1.8°R$ (absolute temperature)
	$T\ (°C) = T\ (K) - 273.15$
	$T\ (°F) = T\ (°R) - 459.67$
	$T\ (°F) = 1.8T\ (°C) + 32$
Energy	$1\ J = 1\ N\text{-}m = 1\ kg\text{-}m^2/s^2 = 10^7\ ergs = 10^7\ dyne\text{-}cm = 2.778 \times 10^{-7}\ kW\text{-}h$
	$= 0.23901\ cal = 0.7376\ ft\text{-}lb_f = 9.486 \times 10^{-4}\ Btu$
	$1\ cal = 4.1868\ J;\ 1\ Btu = 778.17\ ft\text{-}lb_f = 252.0\ cal$
	$1\ Btu/lb_m\text{-}F = 1\ cal/g\text{-}°C$
Power	$1\ hp = 550\ ft\text{-}lb/s = 0.74570\ kW$
	$1\ W = 1\ J/s = 0.23901\ cal/s = 0.7376\ ft\text{-}lb_f/s = 9.486 \times 10^{-4}\ Btu/s$
	$1\ kW = 1000\ J/s = 3412.1\ Btu/h = 1.341\ hp$

1

Introduction

This chapter introduces the chemical engineering profession and reviews the main systems of units and related features. The chapter is organized to provide newcomers with a view of what chemical engineering is. Perhaps the first question a student who considers enrolling in a chemical engineering major has in mind is: What is chemical engineering about? This chapter begins with addressing this question. Next, the system of units and how to carry out conversions between units are reviewed. This is followed by an explanation of the use of an appropriate number of significant figures in calculations. The concept of dimensional homogeneity and how it can be used to identify valid and invalid equations are explained. Next, the major features of a typical process, including process streams and their identification though a variety of process variables (properties) units, are defined. Then, types of manometers and how to calculate pressure based on the type of manometer used are explained. Finally, process classification and introduction to the basic forms of material and energy balances are provided. The following outlines the major learning objectives of this chapter.

Learning Objectives

1. Understand what chemical engineering is about (Section 1.1).
2. Have a working knowledge of units and conversion between units (Section 1.2).
3. Determine and use the appropriate number of significant figures (Section 1.3).
4. Identify an invalid equation based on dimensional arguments (Section 1.4).
5. Calculate process stream flow rates in a variety of units (Section 1.5).
6. Calculate stream compositions in a variety of units (Section 1.6).
7. Calculate pressure through the application of the manometer equation (Section 1.7).
8. Be aware of the importance of material and energy balances (Section 1.8).

1.1 Definitions of Chemical Engineering

There is no universally accepted definition of the field of chemical engineering to date. A definition found in a standard dictionary states, "Chemical engineering is a branch of engineering that involves the design and operation of large-scale chemical plants, petrochemical refineries, and the like." Another definition from the information highway goes like this, "Chemical engineering is concerned with processes that cause substances to undergo required changes in their chemical or physical composition, structure, energy content, or physical state." A more appropriate definition [1] of chemical engineering states,

> Chemical engineering is the profession in which a knowledge of mathematics, chemistry, biology and other natural sciences gained by study, experience, and practice is applied with judgment to develop economic ways of using materials and energy for the benefit of mankind. The profession encompasses the spectrum from products, to the processes and equipment for making them, and to their applications.

Despite the lack of consensus among chemical engineers on the definition of their discipline, which may be attributed to the broad nature of the discipline itself, there is no disagreement that chemical engineers turn low-value materials into high-value products are involved in product design and development; design processes to manufacture products; are involved in process scale-up, development, and optimization; perform economic analysis of the production process; operate and control the processes to ensure that product quality satisfies the required specification; and are involved in the management of the processes and in product sales and technical service.

Because of the reliance of chemical engineering on principles and laws that are versatile in the natural world, the possibilities offered to by chemical engineering as well as the challenges faced by chemical engineers are quite formidable and constantly changing. Without intending to be exhaustive, the list of chemical engineering applications includes

- Traditional areas, for example, mining, pulp and paper, oil refining, materials (rubber, plastics, etc.), and environment
- Nontraditional areas such as microelectronics (semiconductor manufacturing), biotechnology (pharmaceutical production processes, genetic engineering, etc.), and nanotechnology
- Other areas, for example, medicine, consulting, law, and business

A similarity among all chemical engineering systems is that they involve processes designed to transform raw materials (generally, low value) into desired products (with added value). A typical problem in the design of a new process or the modification of an existing process is as follows: "Given the amount and the properties of the raw materials, calculate the amount of the products and determine their properties or vice versa." In order to solve this problem, chemical engineers have powerful tools at their disposal that consist of the principles of material and energy balances (and momentum balance, which is not covered in this text), and a broad array of fundamental sciences.

1.2 Sets of Units and Unit Conversion

Chemical engineers, like many other engineers, use values, units, and dimensions all the time. For example, in a statement like "2.0 L of drinking water…," the "value" is the numerical figure (2.0), the "unit" (L) indicates what the measured quantity represents, and the "dimension" is the measurable quantity that the unit represents. In this case, liter is a unit of volume. A measured or counted quantity has a numerical value (e.g., 2.0) and a unit (whatever there are 2.0 of). It is necessary in engineering calculations to report both the value and the unit. A value without its unit is meaningless. A dimension is a property that can be measured, for example, length, mass, time, and temperature, or calculated by multiplying or dividing other dimensions, for example, length/time = velocity, length^3 = volume, and mass/length^3 = density. Measurable units are specific values of dimensions that have been defined by convention, such as grams for mass, seconds for time, and centimeters for length. Units can be treated like algebraic variables when quantities are added, subtracted, multiplied, or divided. Derived units can be obtained by multiplying or dividing base units (i.e., units of length, mass, time, and temperature) (see Tables 1.1 and 1.2, [2–4]).

There are several systems of units, but two primary systems that engineers use are the International System of Units (SI system) and the American Engineering System of Units (AES). Other systems are the centimeter–gram–second (CGS), foot–pound–second (FPS), and the British System of Units (British). Table 1.3 shows the units associated with these systems of units [5].

1.2.1 Conversion of Units

A measured quantity can be expressed in terms of units having the appropriate dimension. For example, velocity may be expressed in terms of feet per second, miles per hour, centimeters per year, or any other ratio of length and time. The numerical value of velocity depends on the units chosen, for example, 20 m/s = 66 ft/s. To convert a quantity expressed in terms of one unit to its equivalent in terms of another unit, you will need to multiply the given

TABLE 1.1

Quantities, Units, and Symbols of the SI System

Quantity	Unit (Base Unit)	Symbol
Length	Meter	m
	Centimeter	cm
Mass	Kilogram	kg
	Gram	g
Time	Second	s
	Day	day
Temperature	Celsius	°C
	Kelvin	K

Source: Reklaitis, G.V., *Introduction to Material and Energy Balances*, John Wiley & Sons, New York, 1983.

TABLE 1.2

Units for Mass, Length, and Time

Quantity	Unit	Symbol	In Terms of Base Units
Volume	Liter	L	$0.001\ m^3$
Force	Newton	N	$1\ (kg \cdot m)/s^2$
Energy	Joule	J	$1\ N \cdot m = 1\ kg \cdot m^2/s^2$
Pressure	Pascal	Pa	N/m^2
Density			g/cm^3
Molecular weight			g/mol

Sources: Felder, R.M. and Rousseau, R.W., *Elementary Principles of Chemical Processes*, 3rd edn., John Wiley & Sons, New York, 1999; Himmelblau, D.M., *Basic Principles and Calculations in Chemical Engineering*, 3rd edn., Prentice-Hall, Englewood Cliffs, NJ, 1974.

TABLE 1.3

Units Associated with Systems of Units

System	Mass (m)	Length (l)	Time (t)	Temperature (T)
SI	Kilogram (kg)	Meter (m)	Second (s)	Kelvin (K)
AES	Pound mass (lb_m)	Foot (ft)	Second (s)	Degree Fahrenheit (°F)
CGS	Gram (g)	Centimeter (cm)	Second (s)	Kelvin (K)
FPS[a]	Pound mass (lb_m)	Foot (ft)	Second (s)	Degree Fahrenheit (°F)
British	Slug	Foot (ft)	Second (s)	Degree Celsius (°C)

[a] Imperial system units are sometimes referred to as FPS.

quantity by the conversion factor (new unit/old unit). On pages xix and xxi of the text, you will find a list of commonly used systems of units and conversion factors.

Example 1.1 Conversion of Units

Problem

 a. Convert 30 mg/s to its equivalent in kg/h.
 b. Convert 30 lb_m/s to its equivalent in kg/min.
 c. Convert a volume of 5 ft³ to its equivalent in m³.

Solution

Known quantities: 30 mg/s, 30 lb_m/s, 40 ft³.

Find: The equivalent of 30 mg/s in kg/h, 30 lb_m/s in kg/min, and 5 ft³ in m³.

Analysis: Use the table of conversion factors.

 a. Conversion of 30 mg/s to its equivalent in kg/h:

$$\frac{30\,\text{mg}}{\text{s}} \times \left(\frac{1\,\text{g}}{1000\,\text{mg}}\right)\left(\frac{1\,\text{kg}}{1000\,\text{g}}\right)\left(\frac{3600\,\text{s}}{\text{h}}\right) = 0.11\frac{\text{kg}}{\text{h}}$$

 b. Conversion of 30 lb_m/s to its equivalent in kg/min:

$$\frac{30\,\text{lbm}}{\text{s}} \times \left(\frac{1\,\text{kg}}{2.205\,\text{lbm}}\right)\left(\frac{60\,\text{s}}{\text{min}}\right) = 816.3\frac{\text{kg}}{\text{min}}$$

 c. Conversion of a volume of 5 ft³ to its equivalent in m³:

$$5\,\text{ft}^3 \times \left(\frac{1\,\text{m}}{3.28\,\text{ft}}\right)^3 = 0.14\,\text{m}^3$$

1.2.2 Temperature Measurement

Temperature is a measure of the average kinetic energy of a substance. We can measure temperature using physical properties of a substance, which change as a function of temperature. Such properties include volume of a fluid (thermometer), resistance of a metal (resistance thermometer), voltage at the junction of two dissimilar metals (thermocouple), and spectra of emitted radiation (pyrometer).

Common temperature scales are listed as follows [3]:

 1. *Fahrenheit*: On this scale, the freezing point of water is 32 and boiling point is 212.

 2. *Celsius*: The freezing point of water is 0 and boiling point is 100.

3. *Rankine*: On this scale, the absolute zero (when all kinetic energy vanishes) is 0, 1 Rankine increment = 1 Fahrenheit increment (459.67°R = 0°F).

4. *Kelvin*: The absolute zero (when all kinetic energy vanishes) is 0, 1 Kelvin increment = 1 Celsius increment (273.15 K = 0°C).

1.2.3 Temperature Conversion

The following relations are used to convert between °C, °F, °R, and K [6].

$$T(°F) = 1.8 \times T(°C) + 32 \tag{1.1}$$

$$T(K) = T(°C) + 273.15 \tag{1.2}$$

$$T(°R) = T(°F) + 459.67 \tag{1.3}$$

$$T(°R) = 1.8 \times T(K) \tag{1.4}$$

Example 1.2 Temperature Conversion

Problem
Perform the following temperature unit conversions:

(a) Convert a temperature of 32°C to °F, °R, and K.
(b) Convert a temperature of −40°F to °C, K, and °R.
(c) The temperature difference between the inside and outside of a process unit is 100°F. What is the temperature difference (ΔT) in °F, °R, °C, and K?

Solution

Known quantities: Temperature of 32°C, temperature of −40°F, and temperature difference of 100°F.

Find: The temperature difference (ΔT) in °R, °C, and K.

Analysis: Use temperature conversion relations (1.1–1.4).

(a) The temperature outside is 32°C, the temperature in °F, °R, and K is calculated as follows:

A temperature of 32°C in °F:

$$32°C \frac{1.8°F}{1°C} + 32°F = 89.6°F = 90°F$$

A temperature of 32°C equivalent to 90°F in °R:

$$\left(90°F + 460°F\right)\frac{°R}{°F} = 550°R$$

A temperature of 32°C in K:

$$\left(32°C + 273.15°C\right)\frac{K}{°C} = 305 \text{ K}$$

(b) Conversion of a temperature of −40°F to °C, K, and °R:
The temperature of −40°F in °C:

$$-40°F = T(°C)\frac{1.8°F}{1°C} + 32°F$$

Rearranging, $T(°C) = \dfrac{\left(-40°F - 32°F\right)}{\left(1.8°F/1°C\right)} = -40°C$

The temperature of −40°F in K:

$$\left(-40°C + 273.15°C\right)\frac{K}{°C} = 233 \text{ K}$$

The temperature of −40°F in R:

$$\left(-40°F + 460°F\right)\frac{1°R}{1°F} = 420°R$$

(c) The temperature change (ΔT) in °R, °C, and K is calculated as follows:
The temperature difference of 100°F in °R:

$$\Delta T = 100°F\frac{1°R}{1°F} = 100°R$$

The temperature difference of 100°F in °C:

$$\Delta T = 100°F\frac{1°C}{1.8°F} = 55.56°C$$

The temperature difference of 100°F in K:

$$\Delta T = 100°F\frac{1 K}{1.8°F} = 55.56 \text{ K}$$

1.3 Significant Figures

The significant figures of a number are the digits from the first nonzero digit on the left to either the last digit (zero or nonzero) on the right, if there is a decimal point, or the last nonzero digit of the number, if there is no decimal point. The number of significant figures is readily seen if scientific notation is used. Scientific notation is generally a more convenient way of representing large numbers (Tables 1.4 and 1.5).

Numbers can be written in either standard notation or scientific notation to show significant figures.

1. *With decimal point*: Count from nonzero on the left to the last digit on the right:

   0.0012 (1.2×10^{-3}): 2 significant figures

   0.001200 (1.200×10^{-3}): 4 significant figures

2. *Without decimal point*: Count from nonzero on the left to the last nonzero on the right:

   40500 (4.05×10^4): 3 significant figures

3. *Scientific notation*: Typically only significant figures are shown:

   1.47×10^8: 3 significant figures

   4.0500×10^4: 5 significant figures

TABLE 1.4

Significant Figures for Numbers without Decimal Point

Numbers	Significant Figures
2300	2
230001	6
02301	4
023010	4

TABLE 1.5

Significant Figures for Numbers with Decimal Points

Numbers	Significant Figures
2300	4
230.001	6
0230.10	5
0.0103	3
0.01030	4

There is an accepted convention regarding the reporting of computed results. It is simply that a final result should be listed to no more significant figures than the least accurate factor in the expression evaluated. For example, the distance traveled by a vehicle moving at a velocity of 16.223 m/s in 2.0 s is 32 m. The fact that your calculator may list 32.446000 m as the result of the computation is an electronic artifact. It is your job to round the result to an appropriate number of significant figures. Intermediate calculations should be performed with the full precision of the tool you are using (calculator or computer), but the result must reflect the true precision. The rules that apply to the result of multiplication/division and addition/subtraction operations are explained next.

1.3.1 Multiplication, Division, Addition, and Subtraction of Significant Numbers

The number of significant figures in your answer should equal the lowest number of significant figures of any of the numbers being multiplied or divided. In the case of addition and subtraction, compare the positions of the last significant figure of each number: the position farthest to the left is the position of the last significant figure in your answer.

Example 1.3 Multiplication of Significant Figures

Problem
Use the appropriate significant figure in calculating the weight of a 10 kg sample.

Solution

Known quantities: Sample mass of 10 kg.

Find: The sample weight.

Analysis: The weight of a sample is obtained by multiplying its mass by gravitational acceleration:

$$W = mg = (10 \text{ kg})(9.81 \text{ m/s}^2) = 98.1 \text{ N}$$

In multiplication, the lowest number of significant figures of either numbers is considered. In this case, 10 has one significant figure and 9.81 has three significant figures. Therefore, the weight of the sample is 100 N.

Example 1.4 Addition of Significant Figures

Problem
A large tank weighs 1510 N; a fitting of 3.55 N is added to the tank. What is the final tank weight?

Solution

Known quantities: Weight of the tank 1510 N, weight of the fitting 3.55 N.

Find: The weight of the tank.

Analysis: The total weight is the addition of weight of the tank and the fitting.

The final tank weight:

$$1510\,N + 3.55\,N = 1513.55\,N$$

For addition and subtraction, the result should have as many decimal places as the number with the smallest number of decimal places, the first number (1510) has zero decimal places and the second number (3.55) has two decimal places, accordingly, the answer should have zero decimal places, $1514\,N$.

1.4 Dimensional Homogeneity

The dimensions on both sides of the "equals" sign in an equation must be the same. Another way to say this is that equations must be dimensionally homogeneous. Consequently, the units of each term in the equation must be the same (via conversion) as the units of other terms it is added to or subtracted from. It is also good practice to identify an invalid equation based on dimensional arguments.

Example 1.5 Dimensional Homogeneity

Problem
Which of the following equations are dimensionally homogeneous?

1. $x(m) = x_0(m) + 0.3048\left(\dfrac{m}{ft}\right)v\left(\dfrac{ft}{s}\right)t(s) + 0.5a\left(\dfrac{m}{s^2}\right)t(s^2)$

2. $P\left(\dfrac{kg}{ms^2}\right) = 101{,}325.0\left(\dfrac{Pa}{atm}\right)1\left(\dfrac{kg/ms^2}{Pa}\right)P_0(atm) + \rho\left(\dfrac{kg}{m^3}\right)v\left(\dfrac{m}{s}\right)$

Solution

Known quantities: Two possible equations are for distance and pressure.

Find: The dimensional homogeneity of each equation.

Analysis: An equation is dimensionally homogeneous if both sides of the equation have the same units.

1. Write the units in vertical fractions to make easy cancellations:

$$x(m) = x_0(m) + 0.3048\left(\dfrac{m}{\cancel{ft}}\right)v\left(\dfrac{\cancel{ft}}{\cancel{s}}\right)t(\cancel{s}) + 0.5a\left(\dfrac{m}{\cancel{s^2}}\right)[t(\cancel{s^2})]$$

$$m\ [=]\ m\ +\qquad\qquad m\qquad\quad +\qquad m$$

This equation is dimensionally homogeneous because each term has the unit of length (m).

2. Use the same approach as in the earlier point 1:

$$P\left(\frac{kg}{ms^2}\right) = 101,325.0\left(\frac{\cancel{Pa}}{\cancel{atm}}\right)1\left(\frac{kg}{ms^2\,\cancel{Pa}}\right)P_0(\cancel{atm}) + \rho\left(\frac{kg}{m^3}\right)\upsilon\left(\frac{m}{s}\right)$$

$$\frac{kg}{ms^2} \quad = \quad \frac{kg}{ms^2} \quad + \frac{kg}{m^2s}$$

As can be seen, this equation is not dimensionally homogeneous. A dimensionally homogeneous equation does not mean that it is valid. Dimensional considerations act as a first test for validity. Because all equations must be dimensionally consistent (i.e., the units on both sides of the equal sign must be the same), engineers can exploit that constraint to put things in different units and to find the units of variables.

Example 1.6 Maintaining Dimensional Consistency

Problem
The following relation is dimensionally homogenous and consistent. Suppose that it is known that the composition C varies with time, t, in the following manner:

$$C = 0.03e^{(-2.00t)}$$

where
C has the units kg/L
t has the units seconds

What are the units associated with 0.03 and 2.00?

Solution

Known quantities: The relation is dimensionally homogeneous and consistent.

Find: The units associated with the constants 0.03 and 2.0.

Analysis: Dimensionally homogeneous and consistent equations mean that both sides of the equation have the same units.

The unit of 0.03 is the same as the unit of C (i.e., kg/L). Because the argument of an exponential function must be dimensionless, the figure 2.00 must have a unit s^{-1}.

1.4.1 Dimensionless Quantities

An example of dimensionless quantities is the Reynolds number (Re):

$$Re = \frac{\rho \upsilon D}{\mu} \tag{1.5}$$

where
  ρ is the fluid density
  v is the fluid velocity
  D is the pipe diameter
  μ is the fluid viscosity

After substitution of units in the equation, it becomes clear that the Reynolds number is dimensionless:

$$\text{Re} = \frac{\left(\dfrac{\text{kg}}{\text{m}^3}\right)\left(\dfrac{\text{m}}{\text{s}}\right)(\text{m})}{\dfrac{\text{kg}}{\text{m}\cdot\text{s}}} \tag{1.6}$$

The units cancel each other and the Re number is left without units.

Example 1.7 Dimensionless Groups

Problem
Engineers use dimensionless groups. What are the units of h, knowing that $h/(C_p G)$ is unitless?

$$\left(\frac{h}{C_p G}\right)\left(\frac{C_p \mu}{k}\right)^{2/3} = \frac{0.23}{(DG/\mu)}$$

where
  C_p is in units Btu/(lb$_m$°F)
  μ is in lb$_m$/(h ft)
  k is in Btu/(h ft°F)
  D is in ft
  G is in lb$_m$/(h ft^2)

Solution

Known quantities: Empirical correlation.

Find: The units of h, knowing that $h/(C_p G)$ is unitless.

Analysis: Since each individual group must be unitless or dimensionless, all we need to do is work from a single group containing h. Thus, $h/C_p G$ is dimensionless:

$$[h] = [C_p G] = \frac{\text{Btu}}{\text{lb}_m \cdot {}^{\circ}\text{F}} \times \frac{\text{lb}_m}{\text{h}\cdot\text{ft}^2} = \frac{\text{Btu}}{(\text{h}\cdot\text{ft}^2 \cdot {}^{\circ}\text{F})}$$

1.5 Process and Process Variables

A process is any operation or series of operations that causes physical or chemical changes in a substance or a mixture of substances.

1.5.1 Process Flow Sheet

A process flow sheet (PFD) is a sequence of process units connected by process streams. It shows the flow of materials and energy through the process units (Figure 1.1).

1.5.2 Process Unit

A process unit is an apparatus/equipment in which one of the operations that constitutes a process is carried out. Each process unit has associated with it a set of input and output "process streams," which consists of materials that enter and leave the unit.

1.5.3 Process Streams

A process stream is a line that represents the movement of material to or from process units. Typically, these streams are labeled with information regarding the amount, composition, temperature, and pressure of the components.

1.5.4 Density, Mass, and Volume

Density (ρ) is defined as mass (m) per unit volume (V) of a substance. The proper units reflect mass/length^3 (e.g., kg/m^3, g/cm^3, and lb_m/ft^3):

$$\rho = \frac{m}{V} = \frac{\text{mass}}{\text{volume}} \tag{1.7}$$

Density can be used as a conversion factor to relate the mass and volume of a substance. Densities of pure solids and liquids are essentially independent

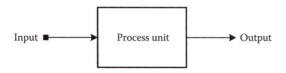

FIGURE 1.1
Schematic of a process unit with single input and single output stream.

of pressure and vary slightly with temperature. For most compounds, density decreases with temperature (volume expansion). Solids and liquids are incompressible, which means that density is constant with change in pressure. Gases (vapors) are compressible, which means that density changes as pressure changes. Specific volume ($1/\rho$) is defined as volume occupied by a unit mass of a substance. The proper units reflect length³/mass (reciprocal of density). Specific gravity (SG) is the ratio of the density (ρ) of a substance to the density (ρ_{ref}) of a reference substance at a specific condition:

$$SG = \frac{\rho}{\rho_{ref}} = \frac{\text{density of substance}}{\text{density of a reference substance}} \tag{1.8}$$

The reference commonly used for solids and liquids is water at 4°C and air for gas, whose density $\rho_{ref} = \rho_{H_2O}\big|_{@4°C} = 1.0\ \text{g/cm}^3 = 1000\ \text{kg/m}^3 = 62.43\ \text{lb}_m/\text{ft}^3$.
The following notation signifies the specific gravity of a substance at 20°:

$$SG = \frac{\rho_{sub} @ 20°C}{\rho_{ref} @ 4°C} \tag{1.9}$$

1.5.5 Mass and Volumetric Flow Rates

A flow rate is the rate (on a time basis) at which a material is transported through a process line (Figure 1.2):

$$\text{Mass flow rate } (\dot{m}) = \text{mass/time}$$

$$\text{Volume flow rate } (\dot{V}) = \text{volume/time}$$

where the "dot" above m and V refers to a flow rate, which is relative to time. The density of a fluid can be used to convert a known volumetric flow rate of a process stream to the mass flow rate of that particular stream or vice versa:

$$\text{Density} = \rho = \frac{m}{V} = \frac{\dot{m}}{\dot{V}} \tag{1.10}$$

$\dot{m}$ (kg fluid/s) or
$\dot{V}$ (m³ fluid/s)

FIGURE 1.2
Material transports through a pipe.

Example 1.8 Volumetric Flow Rate

Problem

Nitrogen from a tank flows at a rate of 6 lb_m/min at $-350°F$, then enters a heater where it is heated, and leaves the heater as a gas at 150°F and 600 psia. Calculate the volumetric flow rate and the specific volume of the gas leaving the heater. Molecular weight of N_2 is 28 g/mol or 28 lb_m/lbmol. Assume the ideal gas law.

Solution

Known quantities: Nitrogen mass flow rate, inlet temperature, exit temperature, and pressure are known.

Find: The volumetric flow rate of gas leaving the heater, specific volume.

Analysis: Use the ideal gas law, $P\dot{V} = \dot{n}RT$.

Volumetric flow rate $(\dot{V})$ of gas leaving the heater

$$\dot{V} = \frac{\dot{n}RT}{P}$$

Molar flow rate

$$\dot{n} = \left(\frac{6\,lb_m}{min} \times \frac{lbmol}{28\,lb_m} \right) = 0.214 \frac{lbmol}{min}$$

Absolute gas exit temperature $T = 150 + 459.7 = 609.7R$

Gas constant

$$R = \frac{10.73\,ft^3\,psia}{lbmol \cdot R}$$

Gas volumetric flow rate

$$\dot{V} = \frac{\left(0.214 \frac{lbmol}{min} \right)\left(\frac{10.73\,ft^3\,psia}{lbmol.R} \right)(609.7R)}{600\,psia} = 2.33 \frac{ft^3}{min}$$

The specific volume is obtained by dividing the volumetric flow rate by the mass flow rate:

$$v = \frac{\dot{V}}{\dot{m}} = \frac{2.33\,ft^3/min}{6\,lb_m/min} = 0.39\,ft^3/lb_m$$

The specific molar volume is obtained by dividing the volumetric flow rate by the molar flow rate:

$$\bar{v} = \frac{\dot{V}}{\dot{n}} = \frac{2.33\,ft^3/min}{0.214\,lbmol/min} = 10.9\,ft^3/lbmol$$

1.5.6 Moles and Molecular Weight

Atomic weight is the mass of an atom of an element. *Mole* is the amount of the species whose mass in grams is numerically equal to its molecular weight. One mole of any species contains approximately 6.023×10^{23} (Avogadro's number) molecules of that species. *Molecular weight* is the sum of the atomic weights of the atoms that constitute a molecule of the compound (same as molar mass); units are of the form kg/kmol, g/mol, or lb/lbmol. Molecular weight is the conversion factor that relates the mass and the number of moles of a quantity of a substance.

The average molecular weight based on mole fraction is

$$\overline{M}w = y_A M_{w,A} + y_B M_{w,B} \tag{1.11}$$

The average molecular weight based on mass fraction is

$$\frac{1}{\overline{M}w} = \frac{x_A}{M_{w,A}} + \frac{x_B}{M_{w,B}} \tag{1.12}$$

where
$\overline{M}w$ is the average molecular weight
x_i is the mass fraction of component i in a mixture
y_i is the mole fraction of component i in the mixture

Example 1.9 Average Molecular Weight

Problem
A liquid stream flowing at 100 g/min contains 0.3 mole fraction benzene (Mw = 78) and the balance 0.7 mole fraction toluene (Mw = 92). Calculate the molar flow rate of each component in the stream.

Solution

Known quantities: Stream mass flow rate and component mole fraction.

Find: The molar flow rate of benzene and toluene.

Analysis: Use the average molecular weight ($\overline{M}w$) based on mole fraction to convert the mass flow rate to mole flow rate:

$$\overline{M}w = y_A M_{w,A} + y_B M_{w,B} = 0.3 \times 78 + 0.7 \times 92 = 87.8$$

Convert the stream total mass flow rate to mole flow since the stream composition is in mole fraction:

$$\text{Total molar flow rate of the stream} = \frac{\text{mass}}{\overline{M}w} = 100 \cdot \frac{g}{\min} \times \frac{\text{mol}}{87.8\,g} = 1.14 \frac{\text{mol}}{\min}$$

$$\text{Molar flow rate of benzene} = 0.3 \times 1.14 \frac{\text{mol}}{\text{min}} = 0.342 \frac{\text{mol}}{\text{min}}$$

$$\text{Molar flow rate of toluene} = 0.7 \times 1.14 \frac{\text{mol}}{\text{min}} = 0.80 \frac{\text{mol}}{\text{min}}$$

1.6 Compositions of Streams

1.6.1 Mass Fraction and Mole Fraction

Process streams occasionally contain one substance; more often they consist of mixtures of liquids or gases, or solutions of one or more solutes in a liquid solvent. The following terms may be used to define the composition of a mixture of substances, including a species A:

$$\text{Mass fraction} : x_A = \frac{\text{mass of A}}{\text{total mass}} \qquad (1.13)$$

$$\text{Mole fraction} : y_A = \frac{\text{moles of A}}{\text{total number of moles}} \qquad (1.14)$$

Mass fractions can be converted to mole fractions or vice versa by assuming a basis of calculation. Remember mass and mole fractions are unitless.

Example 1.10 Mole Fractions

Problem
A mixture of methanol (CH_3OH, $Mw = 32.04$) and ethanol (C_2H_5OH, $Mw = 46.07$) is flowing through a circular pipe at a flow rate of 3.0 m/s. The mixture contains 30.0 wt% methanol and 70.0 wt% ethanol. The specific gravity of the mixture is 0.80. If the inside diameter of the pipe is 0.10 m, what is the flow rate of the mixture in kg/s, kmol/s?

Solution

Known quantities: Mixer velocity, specific gravity, and inner pipe diameter are known.

Find: The molar flow rate of the mixture.

Analysis: The mass flow rate is obtained by multiplying its density by the volumetric flow rate:

$$\dot{m} = \rho \upsilon A = \left(0.80 \times \frac{1000\,\text{kg}}{\text{m}^3} \right) \left(3.0 \frac{\text{m}}{\text{s}} \right) \left[\pi \frac{(0.10\,\text{m})^2}{4} \right] = 18.85 \frac{\text{kg}}{\text{s}}$$

Since we know the total mass flow rate, we can compute the molar flow rate of methanol and ethanol:

$$\dot{n}_{CH_3OH} = \frac{0.3\left(18.85\dfrac{kg}{s}\right)}{32.04\dfrac{kg}{kmol}} = 0.176\frac{kmol}{s}$$

Ethanol molar flow rate

$$\dot{n}_{C_2H_5OH} = \frac{0.7\left(18.85\dfrac{kg}{s}\right)}{46.07\dfrac{kg}{kmol}} = 0.282\frac{kmol}{s}$$

The total molar flow rate then is

$$\dot{n}_{Total} = \dot{n}_{CH_3OH} + \dot{n}_{C_2H_5OH}$$

$$\dot{n}_{Total} = 0.176\frac{kmol}{s} + 0.282\frac{kmol}{s} = 0.458\frac{kmol}{s}$$

Example 1.11 Mass Fractions

Problem
A liquid stream containing only benzene, toluene, and p-xylene flows through a conduit that has a square cross section. The total flow rate of the stream is 10.0 mol/s. If the mole fraction of benzene (Mw = 78) is 0.300, that of toluene (Mw = 92) is 0.500, and that of p-xylene (Mw = 106) is 0.200, what is the mass fraction of each of the three components? If the density of the liquid is 0.87 g/cm^3 and the conduit is 0.10 m on a side, what is the mass fraction and volumetric flow rate of the liquid inside the conduit?

Solution

Known quantities: Mixture mole fraction of benzene, toluene, and p-xylene, density of mixture and conduit diameter, and molar flow rate are known.

Find: The mass fraction of each component.

Analysis: The mass flow rate for the three components is equal to their respective molar flow rate times their molecular weight:

$$\dot{m}_i = y_i \times \dot{n}_{tot} \times Mw_i$$

where
    $\dot{m}_i$ is the component mass flow rate
    y_i is the component mole fraction
    $\dot{n}_{tot}$ is the total molar flow rate
    Mw_i is the component molecular weight

Mass flow rate of benzene:

$$\dot{m}_B = 0.3\left(10\frac{mol}{s}\right)\left(\frac{78g}{mol}\right) = 234\frac{g}{s}$$

Mass flow rate of toluene:

$$\dot{m}_T = 0.5\left(10.0\frac{mol}{s}\right)\left(\frac{92g}{mol}\right) = 460\frac{g}{s}$$

Mass flow rate of xylene:

$$\dot{m}_X = 0.2\left(10.0\frac{mol}{s}\right)\left(\frac{106g}{mol}\right) = 212\frac{g}{s}$$

The total mass flow rate is the sum of these three:

$$\dot{m}_{Total} = \dot{m}_B + \dot{m}_T + \dot{m}_X = 234 + 460 + 212 = 906\frac{g}{s}$$

The mass fractions are the individual mass flow rates divided by the total mass flow rate.

$$\text{Mass fraction of benzene: } x_B = \frac{234\,g/s}{906\,g/s} = 0.258$$

$$\text{Mass fraction of toluene: } x_T = \frac{460\,g/s}{906\,g/s} = 0.508$$

$$\text{Mass fraction of xylene: } x_X = \frac{212\,g/s}{906\,g/s} = 0.234$$

The volumetric flow rate obtained by dividing the mass flow rate by the density is

$$\dot{V} = \frac{\dot{m}_{Total}}{\rho} = \frac{906\,g/s}{0.87\;g/cm^3} = 1041\frac{cm^3}{s}$$

1.6.2 Concentration

Concentrations can be expressed in many ways: weight/weight fraction (w/w), weight/volume fraction (w/v), molar concentration (M), and mole fraction. The weight/weight concentration is the weight of the solute divided by the total weight of the solution, and this is the fractional form of the percentage composition by weight. The weight/volume concentration is the weight of solute divided by the total volume of the solution. The molar concentration is the number of moles of the solute, expressed in moles, divided by the volume of the solution. The mole fraction is the ratio of the number of moles of the solute to the total number of moles of all species present in the solution [7].

Example 1.12 Air Compositions

Problem

If 100 g of air consists of 77% by weight of nitrogen (N_2, molecular weight $= 28$), and 23% by weight of oxygen (O_2, molecular weight $= 32$), calculate (a) the mean molecular weight of air, (b) mole fraction of oxygen, (c) concentration of oxygen in mol/m^3 and kg/m^3, if the total pressure is 1.5 atm and the temperature is 25°C.

Solution

Known quantities: Mass of air and its composition in weight fraction are known.

Find: The air composition.

Analysis: The number of moles is obtained by dividing the mass by molecular weight.

a. The mass of N_2 and O_2 that 100 g of air contains:

$$\text{Mass of } N_2 = 0.77 \times 100 \text{ g} = 77 \text{ g}$$

$$\text{Mass of } O_2 = 0.23 \times 100 \text{ g} = 23 \text{ g}$$

$$\text{Number of moles of } N_2 = \frac{77 \text{ g}}{28 \dfrac{\text{g}}{\text{mol}}} = 2.75 \text{ mol}$$

$$\text{Number of moles of } O_2 = \frac{23 \text{ g}}{32 \dfrac{\text{g}}{\text{mol}}} = 0.72 \text{ mol}$$

$$\text{Total number of moles} = 2.75 + 0.72 = 3.47 \text{ mol}$$

The mean molecular weight of air is the total mass/total moles. Therefore, the mean molecular weight of air $= 100$ g/3.47 mol $= 28.8$ g/mol.

b. Mole fraction of oxygen (O_2) $= 0.72/3.47 = 0.21$, and this is also the volume fraction.

c. In the ideal gas equation, n is the number of moles present, P is the pressure (in atm, in this instance), and T is the temperature (in K, in this case). The corresponding value of R is 0.08206 m^3 atm/(mol K).

$$PV = nRT \text{ and molar concentration} = n/V$$

O_2 molar concentration

$$\frac{n}{V} = \frac{y_{O_2}P}{RT} = \frac{0.21 \times 1.5\,\text{atm}}{0.08206\frac{\text{m}^3\text{atm}}{\text{mol K}}(25+273)} = 0.013\frac{\text{mol}}{\text{m}^3}$$

The O_2 mass concentration is the molar concentration multiplied by the molecular weight:

$$n \times Mw = 0.013\frac{\text{mol}}{\text{m}^3}\frac{32\text{g}}{\text{mol}}\frac{1\text{kg}}{1000\text{g}} = 4.16 \times 10^{-4}\frac{\text{kg}}{\text{m}^3}$$

Example 1.13 Concentration of Salt in Water

Problem
Salty water is prepared by mixing salt (NaCl, $Mw = 58.5$) in water ($Mw = 18$). A solution is prepared by adding 20 kg of salt to 100 kg of water, to make a liquid of density 1323 kg/m³. Calculate the concentration of salt in this solution as a (a) weight/weight fraction, (b) weight/ volume in kg/L, and (c) mole fraction.

Solution

Known quantities: Mass of salt and mass of water, mixture density.

Find: (a) Weight/weight fraction, (b) weight/volume in kg/L, and (c) mole fraction.

Analysis: Weight fraction is obtained by dividing the mass of each component by the total mass.

(a) Weight fraction of salt $= \dfrac{20\,\text{kg}}{(20\,\text{kg} + 100\,\text{kg})} = 0.167$

(b) Salt concentration is the weight of salt/total volume. The total volume is the mass of the mixture divided by mixture density.

A density of 1323 kg/m³ means that 1 m³ of solution weighs 1323 kg.

$$\text{Volume of mixture} = \frac{100\,\text{kg} + 20\,\text{kg}}{1323\,\text{kg/m}^3} = 0.091\,\text{m}^3$$

$$\text{Weight of salt/volume} = \frac{20\,\text{kg}}{(0.091)} = 220\frac{\text{kg}}{\text{m}^3}$$

$$\text{Weight/volume in kg/L} = 220\frac{\text{kg}}{\text{m}^3}\frac{\text{m}^3}{1000\text{L}} = 0.22\frac{\text{kg}}{\text{L}}$$

(c) Moles of water and salts are obtained by dividing the mass of each component by its respective molecular weight:

$$\text{Moles of water} = \frac{100 \text{ kg}}{18 \text{ kg/kmol}} = 5.56$$

$$\text{Moles of salt} = \frac{20 \text{ kg salt}}{58.5 \text{ kg/kmol}} = 0.34$$

Mole fraction is obtained by dividing moles of each component by the total number of moles:

$$\text{Mole fraction of salt} = \frac{5.56}{(5.56 + 0.34)} = 0.942$$

1.7 Pressure Measurement

Pressure is the ratio of force perpendicular to the area on which the force acts. Pressure units are force units divided by area units (e.g., N/m^2 or Pascal (Pa), dyn/cm^2, and $lb_f/in.^2$ or psi). The pressure at the base of a vertical column of fluid (nonmoving) of density ρ and height h, called the hydrostatic pressure, is given by

$$P = P_0 + \rho g h \tag{1.15}$$

where
P_0 is the pressure exerted on the top of the column
g is the acceleration of gravity

The earth atmosphere can be considered as a column of fluid with zero pressure at the top. The fluid pressure at the base of this column (e.g., at sea level) is the atmospheric pressure, P_{atm}.

The absolute pressure, P_{abs}, of a fluid is the pressure relative to a perfect vacuum ($P = 0$). Many pressure measurement devices report the gauge pressure of the fluid, which is the pressure of the fluid relative to atmospheric pressure (reference pressure). A gauge pressure of 0 indicates that the absolute pressure of the fluid is equal to the atmospheric pressure. The relationship for converting between absolute and gauge pressure is given by

$$P_{absolute} = P_{gauge} + P_{atmospheric} \tag{1.16}$$

Pressure is defined as force per unit area. Pressure can be expressed as a relative pressure or absolute pressure. The units of pressure are

$$\text{SI unit}: \frac{N}{m^2} (\text{or Pa})$$

$$\text{CGS unit}: \frac{dyn}{cm^2}$$

$$\text{AES unit}: \frac{lb_f}{in.^2} (\text{or psi})$$

Chemical engineers are most often interested in pressures that are caused by a fluid. If a fluid is flowing through a horizontal pipe and a leak develops, a force must be applied over the area of the hole that causes the leak to stop the fluid from leaking. This pressure is called the fluid pressure (the force applied must be divided by the area of the hole). This is schematically shown in Figure 1.3.

If a vertical container contains a fluid, the mass of the fluid will exert a force on the base of the container. This pressure is called the hydrostatic pressure. Hydrostatic pressure is the pressure caused by the mass of a fluid, as shown in Figure 1.4.

1.7.1 Types of Pressures

1. Atmospheric pressure, P_{atm}, is the pressure caused by the weight of the earth's atmosphere. Often atmospheric pressure is called barometric pressure.

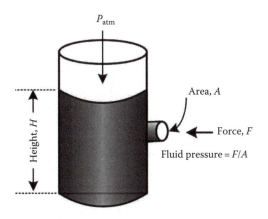

FIGURE 1.3
Force on a side hole in an open tank.

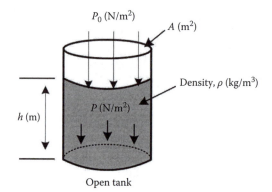

FIGURE 1.4
Pressure at the bottom of an open tank.

2. Absolute pressure, P_{abs}, is the total pressure. An absolute pressure of 0.0 is a perfect vacuum. Absolute pressure must be used in all calculations, unless a pressure difference is used.
3. Gauge pressure, P_{gauge}, is pressure relative to atmospheric pressure.
4. Vacuum pressure, P_{vac}, is a gauge pressure that is below atmospheric pressure. It is used so that a positive number can be reported.

Standard atmosphere is defined as the pressure equivalent to 760 mmHg or 1 atm at sea level and at 0°C. Other units used for the standard atmosphere (atm) are

$$1 \text{ atm} = 760 \text{ mmHg} = 76 \text{ cmHg}$$

$$= 1.013 \times 10^5 \frac{N}{m^2} \text{ (or Pa)} = 101.3 \text{ kPa} = 1.013 \text{ bar}$$

$$= 14.696 \text{ psi} \left(\text{or} \frac{lb_f}{in.^2} \right) = 29.92 \text{ in Hg} = 33.91 \text{ ft H}_2\text{O}$$

The units psi and atm often carry a suffix "a" or "g" to indicate absolute or gauge (relative) pressure, respectively. Thus, by psig, we mean gauge pressure in psi, and by psia, we mean absolute pressure in psi. So is the meaning of atma or atmg, if nothing is noted otherwise.

1.7.2 Standard Temperature and Pressure

Standard temperature and pressure are used widely as a standard reference point for expression of the properties and processes of ideal gases:

$$\text{Standard temperature} = T_s = 0°C \Rightarrow 273 \text{ K}$$

$$\text{Standard pressure} = P_s = 1 \text{ atm} = 760 \text{ mmHg} = 101.3 \text{ kPa}$$

$$\text{Standard molar volume} = \hat{V}_s = 22.4 \frac{\text{m}^3(\text{STP})}{\text{kmol}} \Leftrightarrow 22.4 \frac{\text{L(STP)}}{\text{mol}} \Leftrightarrow 359 \frac{\text{ft}^3(\text{STP})}{\text{lb mol}}$$

One molecule of an ideal gas at 0°C and 1 atm occupies 22.4 L.

1.7.3 Pressure-Sensing Devices

The types of pressure-sensing devices include the Bourdon gauge, diaphragm capsule, and capacitance sensor, column of fluid, manometer, barometer, silicon diaphragm, and semiconductor strain gauges. The latter are used for better response and increased sensitivity. A manometer is a U-shaped device that uses a fluid having greater density than other fluids in the process unit. Manometer operation is based on the fact that hydrostatic pressure at the same level in the same fluid must be the same in each leg. To understand how a manometer works, we must understand how to determine the hydrostatic pressure caused by a mass of a column of fluid.

By definition of pressure,

$$P = \frac{F}{A} = \frac{mg}{A} = \frac{\rho V g}{A} = \frac{\rho A h g}{A} = \rho g h$$

Whenever we need to determine the hydrostatic pressure caused by a mass of fluid, it is simply given by

$$P = \rho g h$$

Gauge pressure is the pressure relative to atmospheric pressure. Gauge pressure is positive for pressures above atmospheric pressure, and negative for pressures below it. Atmospheric pressure does add to the pressure in any fluid not enclosed in a rigid container. The total pressure, or absolute pressure, is thus the sum of gauge pressure and atmospheric pressure:

$$P_{abs} = P_{gauge} + P_{atm}$$

where
P_{abs} is absolute pressure
P_{gauge} is gauge pressure
P_{atm} is atmospheric pressure

Example 1.14 Fundamentals of Pressure

Problem

Consider the manometer in Example Figure 1.14.1. If $h = 10$ in. and the manometer fluid is mercury ($\rho = 13.6$ g/cm³), calculate the gauge pressure and absolute pressure (14.7 psia).

Solution

Known quantities: Manometer fluid is mercury, density of mercury, and change in mercury height, h.

Find: The gauge and absolute pressure.

Analysis: Absolute pressure is obtained by adding the gauge pressure to atmospheric pressure.

$$\text{Gauge pressure} = \text{absolute pressure} - \text{atmospheric pressure}$$

$$\rho = 13.6 \text{ g/cm}^3 \text{ (or 13.6 times greater than } H_2O\text{)}$$

Thus, gauge pressure $= P_{\text{gauge}} = \rho \dfrac{g}{g_c} h$

$$= \left(13.6 \times \frac{62.4 \text{ lb}}{\text{ft}^3} \left| \frac{(1 \text{ ft})^3}{(12 \text{ in.})^3} \right| \left(\frac{32.174 \text{ ft/s}^2}{32.174 \text{ lb.ft/lb}_f\text{s}^2} \right) \right) (10 \text{ in.})$$

$$= 4.9 \frac{\text{lb}_f}{\text{in.}^2} \left| \frac{\text{psig}}{\text{lb}_f/\text{in.}^2} \right. = 4.9 \text{ psig}$$

The absolute pressure, $P_{\text{absolute}} = P_{\text{gauge}} + P_{\text{atm}} = 4.9 + 14.7 = 19.6$ psia.

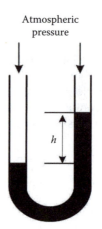

EXAMPLE FIGURE 1.14.1
Manometer filled with mercury.

Example 1.15 Pressure Gauges in a Tank

Problem
A pressure gauge on a tank reads 25 psi. What is the absolute pressure, considering the atmospheric, $P_{atm} = 14.7$ psi?

Solution

Known quantities: Pressure gauge.

Find: The gauge pressure.

Analysis: Absolute pressure is obtained by adding the gauge pressure and local atmospheric pressure.

The gauge reads the gauge pressure directly. Therefore, the absolute pressure

$$P_{abs} = 25 \text{ psig} + 14.7 \text{ psi} = 39.7 \text{ psia} = 40 \text{ psia}$$

Example 1.16 Absolute Pressures from Vacuum Pressure

Problem
The pressure gauge on a tank reads 20 cmHg vacuum. The atmospheric pressure in that tank area is 760 mmHg. What is the absolute pressure in the tank?

Solution

Known quantities: Vacuum pressure.

Find: The absolute pressure.

Analysis: The gauge reads a vacuum gauge pressure directly. The pressure is measured below atmospheric pressure relative to the latter and in the reverse direction toward zero absolute pressure, that is, $P_{vac} = P_{atm} - P_{abs}$. Thus, the absolute pressure is

$$P_{abs} = P_{atm} - P_{vac} = 760 - 200 = 560 \text{ mmHg}$$

Example 1.17 Pressure Caused by a Column of Fluid

Problem
The fluid used in a manometer has a specific gravity of 0.80, and the column manometer height is 0.50 m. The manometer is used to measure the pressure in a pipe. The local atmospheric pressure is 100 kPa. Determine the pressure in the column.

Solution

Known quantities: Fluid specific gravity and column height are known.

Find: The pressure in the manometer column.

Analysis: The density of the fluid is obtained by multiplying its specific gravity by the density of water, which is 1000 kg/m³.

$$\rho = \rho_s \times \rho_{H_2O} = 0.8 \times 1000 \text{ kg/m}^3 = 800 \text{ kg/m}^3$$

From Equation 1.14 and substituting known quantities,

$$P = 100 \text{ kPa} + \left(800 \text{ kg/m}^3\right)\left(9.81 \text{ m/s}^2\right)\left(0.50 \text{ m}\right)$$

$$\times \left(\frac{1 \text{ N}}{1 \text{ kg} \cdot \text{m/s}^2}\right)\left(\frac{1 \text{ kPa}}{1000 \text{ N/m}^2}\right) = 103.924 \text{ kPa}$$

Example 1.18 Pressure Calculations in American Engineering System

Problem

What is the pressure (in psi) caused by a 10 ft column of fluid with a specific gravity of 0.8 at ambient temperature? The gravitational acceleration is 32.174 ft/s².

Solution

Known quantities: Fluid specific gravity, height, and acceleration are known.

Find: The column pressure.

Analysis: Use $P = \rho g h$.

Density of water (reference fluid)

$$\rho_{ref} = 62.4 \frac{\text{lb}_m}{\text{ft}^3}$$

$$P = \rho g h = \left(\rho_{ref} \times \rho_s\right) \times g \times h$$

Substitute known quantities:

$$P = \left(0.8 \times 62.4 \text{ lb}_m/\text{ft}^3\right)\left(32.174 \text{ ft/s}^2\right)\left(10 \text{ ft}\right)$$

$$\times \left(\frac{1 \text{ lb}_f}{32.174 \text{ lb}_m \cdot \text{ft/s}^2}\right)\left(\frac{1 \text{ ft}}{12 \text{ in.}}\right)^2 = 3.47 \text{ psi}$$

Example 1.19 Pressure Calculations in SI Units

Problem

What is the pressure (in kPa) caused by a 10 m long column filled with water at ambient temperature? Assume the gravitational acceleration to be 9.81 m/s².

Solution

Known quantities: Fluid specific gravity, height, and acceleration are known.

Find: The column pressure.

Analysis: Use $P = \rho g h$.

Density of water

$$\rho = 1000\frac{kg}{m^3}$$

$$P = \rho g h$$

Substitute known quantities:

$$P = \left(1000\frac{kg}{m^3}\right)\left(9.81\frac{m}{s^2}\right)(10\,m)\left(\frac{1\,N}{1\,kg\cdot m/s^2}\right)\frac{kPa}{1000\,N/m^2} = 98.1\,kPa$$

Calculation of pressure based on the type of manometer used

Three different arrangements of manometers are shown in Figure 1.5. These manometers can be used to measure pressure using a column of a dense liquid.

1. An open-end manometer can give the gauge pressure.
2. A differential manometer gives ΔP between two points. Note that the pressure decreases in the direction of flow.
3. A closed-end manometer gives absolute pressure. A manometer that has one leg sealed and the other leg open measures atmospheric pressure. It is called a barometer.

The basic variables that need to be considered in a manometer are those that measure pressure or pressure difference as shown in Figure 1.6. The line 1–2 is at the interface between the manometer fluid and the higher-pressure fluid. The hydrostatic pressure on each leg is the same along the line within the same fluid. It becomes our reference point. We do a pressure balance by equating the pressures on each leg. Applying this allows us to develop the general manometer equation as

$$\sum \text{Pressure on leg }1 = \sum \text{Pressure on leg }2$$

$$\text{i.e., } P_1 + \rho_1 g d_1 = P_2 + \rho_2 g d_2 + \rho_3 g h$$

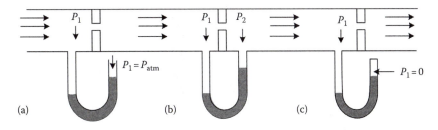

FIGURE 1.5
(a) Open-end, (b) differential, and (c) sealed-end manometers.

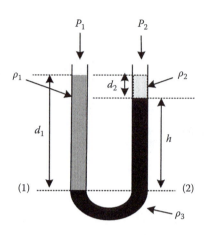

FIGURE 1.6
Open-end manometer.

Example 1.20 Pressure Drop across an Orifice

Problem
Determine the pressure drop across the orifice meter as shown in Example Figure 1.20.1.

Solution

Known quantities: Water density and manometer fluid height difference are known.

Find: The pressure drop across the orifice meter P_1-P_2.

Analysis: This is a differential manometer. Note that the hydrostatic pressure above the x mark is the same on both sides. Manometers are simple, inexpensive, and accurate devices used to measure fluid pressure.

$$P_1 + \rho_1 g d = P_2 + \rho_f g d$$

EXAMPLE FIGURE 1.20.1
Pressure drop across an orifice.

We can now write the pressure difference across the orifice from the earlier equation as

$$P_1 - P_2 = gd(\rho_f - \rho_1)$$

Substituting the appropriate known quantities, we get

$$P_1 - P_2 = \left(9.81\frac{m}{s^2}\right)(0.022\ m)(1100 - 1000)\frac{kg}{m^3}\left(\frac{1\ N}{1\ kg \cdot m/s^2}\right)$$

$$= 21.6\frac{N}{m^2} = 21.6\ Pa$$

Example 1.21 Pressure in a Manometer

Problem

Consider the tank filled with natural gas as shown in Example Figure 1.21.1. Assume that the manometer fluid is a liquid with a specific gravity of 0.87 and the elevation difference is of 0.01 m ($h = 0.01$ m). Calculate the pressure inside the tank (P_B).

Solution

Known quantities: Elevation difference and manometer fluid specific gravity.

Find: The pressure in the tank.

Analysis: Since the manometer used is a closed-end one, the pressure inside the tank is calculated using the following equation:

$$P_{gauge} = \rho g h$$

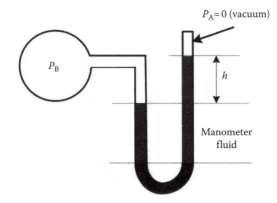

EXAMPLE FIGURE 1.21.1
Pressure in a closed tank measured using a closed-end manometer.

Substitute known quantities:

$$P_{tank} = \left(0.87 \times 1000 \frac{kg}{m^3}\right) \times \left(\frac{9.81\,m}{s^2}\right) \times (0.01\,m)\left(\frac{N}{kg \cdot m/s^2}\right)\left(\frac{Pa}{N/m^2}\right) = 85.3\,Pa$$

Example 1.22 Open-End Manometers

Problem

An open-end manometer is connected to a pipe in which a gas is flowing (Example Figure 1.22.1). The manometer fluid density is 2.00 g/cm³. A gas bubble is trapped in the left leg of the manometer as shown in Example Figure 1.22.1. What is the pressure of the flowing gas inside the pipe (in atm)? Is this pressure a gauge pressure or an absolute pressure? How do you know?

Solution

Known quantities: As per schematic diagram.

Find: The pressure of the flowing gas.

Analysis: At the level marked with the letter x, the pressure in the more dense fluid (density = 2.00 g/cm³) must be the same since the fluid is not moving. On the right-hand side of the manometer, this pressure must be the pressure exerted on its free surface plus the pressure from the 12.00 cm of this fluid. Since the tube is open to atmospheric pressure, the pressure at level x in the manometer fluid is

$$P_x = \rho g h + P_{atm} = \frac{2000\,kg}{m^3} \times 9.81 \frac{m}{s^2} \times 0.12\,m + P_{atm}$$

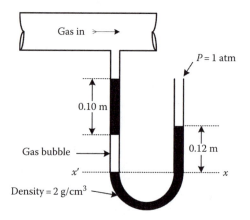

EXAMPLE FIGURE 1.22.1
Pressure in a gas pipe measured using an open-end manometer.

$$P_x = 2354.4 \frac{kg}{ms^2} \left(\frac{1\,Pa}{1\frac{kg \cdot m}{m^2 \cdot s^2}} \right) \left[\frac{1\,atm}{1.01325 \times 10^5\,Pa} \right] + P_{atm} = 0.0232\,atm + P_{atm}$$

Adding up all of the hydrostatic pressures and the pressure from the flowing gas in this leg gives

$$P_x = 0.0232\,atm + P_{atm}$$

$$P_{x'} = P_{gas} + \rho_{gas} g_{gas} h_{gas} + \rho_{fluid} g_{fluid} h_{fluid}$$

We can ignore the contribution of the gas bubble to the hydrostatic head, since gas densities are much smaller than liquid densities. Thus the earlier equation gives

$$P_{x'} = P_{gas} + 0 + 10\,cm \left(2\frac{g}{cm^3} \right) \left(981\frac{cm}{s^2} \right) \frac{1\,atm}{1.01325 \times 10^6\,g/cm^2}$$

$$= P_{gas} + 0.0194\,atm$$

Rearranging,

$$P_x = P_{x'}$$

$$0.0232\,atm + P_{atm} = P_{gas} + 0.0194\,atm$$

$$P_{gas} - P_{atm} = 0.0232\,atm - 0.0194\,atm = 0.0038\,atm$$

The difference between gas pressure and atmospheric pressure is the gauge pressure:

$$P_{guage} = P_{gas} - P_{atm} = 0.0037\,atm$$

Had the right leg of the manometer been open to the atmosphere, the *manometer reading* would have been a gauge pressure.

Example 1.23 Pressure in a Gas Pipe

Problem
Gas is flowing in a circular pipe. A closed-end manometer is connected to the gas pipe. The manometer fluid density is 2000 kg/m³. A gas bubble has been trapped in the left leg of the manometer as shown in Example Figure 1.23.1. The closed end of the manometer is in a vacuum.

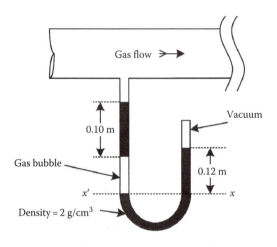

EXAMPLE FIGURE 1.23.1
Measuring pressure in a gas pipe using a closed-end manometer.

What is the pressure of the flowing gas inside the pipe (in cmHg)? Is the manometer reading the gauge or the absolute pressure of the pipe?

Solution

Known quantities: See the schematic diagram.

Find: The pressure of the flowing gas.

Analysis: At the level marked with the letter x the pressure in the more dense fluid (specific gravity = 2.00) must be the same since the fluid is not moving. On the right-hand side of the manometer, this pressure must be the pressure exerted on its free surface plus the pressure from the 12.00 cm of this fluid. Since there is a vacuum above the fluid, the pressure exerted on its free surface is zero. Thus, the pressure at level x in the manometer fluid is given by

$$P_x = 0.12 \text{ m} \times 9.81 \frac{\text{m}}{\text{s}^2} \times 2000 \frac{\text{kg}}{\text{m}^3} \frac{1 \text{ atm}}{1.013 \times 10^5 \text{ Pa}} \left| \frac{76 \text{ cmHg}}{1 \text{ atm}} \right| + 0 = 1.77 \text{ cmHg}$$

At level x in the left leg of the manometer, the pressure must be the same as the afore-calculated pressure (1.77 cmHg). Adding up all of the hydrostatic pressures and the pressure from the flowing gas in this leg gives

$$P_{x'} = P_{gas} + \rho_{gas} g h_{gas} + \rho_{fluid} g h_{fluid}$$

The contribution to the hydrostatic head from the gas bubble can be ignored since gas densities can be neglected relatively to liquid densities. Thus the earlier equation gives

$$P_x = P_{x'} = 1.77 \text{ cmHg} = P_{gas} + 0.10 \text{ m} \times 9.81 \frac{\text{m}}{\text{s}^2}$$

$$\times 2000 \frac{\text{kg}}{\text{m}^3} \left. \frac{1 \text{ atm}}{1.013 \times 10^5 \text{ Pa}} \frac{76 \text{ cmHg}}{1 \text{ atm}} \right| = P_{gas} + 1.47 \text{ cmHg}$$

$$P_{gas} = 1.77 \text{ cmHg} - 1.47 \text{ cmHg} = 0.3 \text{ cmHg}$$

The manometer reading is the absolute pressure. We know this because the pressure exerted on the fluid surface in the right leg of the manometer is zero (vacuum). Thus the *manometer reading* is the absolute pressure. Had the right leg of the manometer been open to the atmosphere, the manometer reading would have been a gauge pressure.

Example 1.24 Pressure in Closed and Open Tanks

Problem
Consider the equipment connected to a closed-end manometer, as shown in Example Figure 1.23.1. When the equipment is fully open to the atmosphere, the difference in the manometer liquid height in the closed leg of the manometer is h_1. When the equipment is in operation, a pressure gauge on the equipment reads 50.0 mmHg, and the difference between the liquid level in the closed leg of the manometer and the liquid level in the leg connected to the equipment is now 30.0 cm more than it was when the equipment was open. Assume that there is a partial vacuum pressure above the liquid in the closed end of the manometer. Assume that the ambient pressure is 710 mmHg. What is the density of the liquid in the manometer?

Solution

Known quantities: See the schematic diagram.

Find: The pressure of the flowing gas.

Analysis: In this problem, the two scenarios are schematically represented in Example Figure 1.24.1.

At the position marked "×," the pressure in the manometer fluid must be the same in both legs. The pressure above the left leg is an ambient pressure (710 mmHg). The pressure on the other side is the hydrostatic head from the manometer fluid plus any pressure resulting from the gas in the closed end. Let us call this pressure P_v. Thus,

$$710 \text{ mmHg} = \rho g h_1 + P_v$$

The system will look essentially the same when the equipment is in operation, except that the manometer fluid in the right leg will be 30 cm above the fluid in the left leg. The pressure gauge on the equipment will read 50 mmHg, but remember that this is a gauge pressure. Thus the

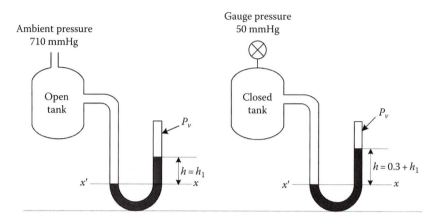

EXAMPLE FIGURE 1.24.1
Pressure inside open tank and closed tank.

actual pressure is 710 + 50 mmHg. Proceeding as mentioned earlier, you would get

$$710 \text{ mmHg} + 50 \text{ mmHg} = \rho g(h_1 + 30 \text{ cm}) + P_v$$

Subtracting the first equation from the second equation gives

$$50 \text{ mmHg} = \rho g(0.30 \ m)$$

Solving for ρ gives

$$\rho = \frac{50 \text{ mmHg}\left(\dfrac{1.01325 \times 10^5 \text{ Pa}}{760 \text{ mmHg}}\right)\left(\dfrac{1\dfrac{\text{kg} \cdot \text{m}}{\text{m}^2 \cdot \text{s}^2}}{\text{Pa}}\right)}{0.3 \times 9.81\dfrac{\text{m}}{\text{s}^2}} = 2.27\frac{\text{g}}{\text{cm}^3}$$

Example 1.25 Inclined Manometers

Problem

The manometer shown here contains three liquids with densities $\rho_1 = 2000 \text{ kg/m}^3$, $\rho_2 = 1000 \text{ kg/m}^3$, and $\rho_3 = 1600 \text{ g/cm}^3$. One end of the manometer is connected to a vessel filled with a gas, while the other end is open to the atmosphere (P_B). The ambient pressure is 100 kPa. The inclined leg of the manometer is at an angle of 30° from the horizontal. What is the pressure (P_A) inside the vessel?

Solution

Known quantities: See schematic diagram (Example Figure 1.25.1).

Find: The pressure in the tank, P_A.

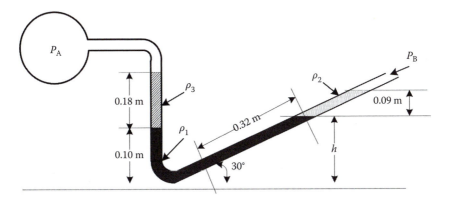

EXAMPLE FIGURE 1.25.1
Pressure in a closed tank measured using an inclined manometer.

Analysis: Starting at the lowest point in the manometer and going up each leg, we can balance the pressure as

$$P_A + \rho_3 g(0.18 \text{ m}) + \rho_1 g(0.10 \text{ m}) = P_B + \rho_2 g(0.09 \text{ m}) + \rho_1 g(h)$$

where $h = 0.32 \sin(30) = 0.16\,m$.
Substitute known quantities:

$$P_A + \left(1600\,\frac{\text{kg}}{\text{m}^3}\right)\left(9.81\,\frac{\text{m}}{\text{s}^2}\right)(0.18 \text{ m}) + \left(2000\,\frac{\text{kg}}{\text{m}^3}\right)\left(9.81\,\frac{\text{m}}{\text{s}^2}\right)(0.10 \text{ m})$$

$$= P_B + \left(1000\,\frac{\text{kg}}{\text{m}^3}\right)\left(9.81\,\frac{\text{m}}{\text{s}^2}\right)(0.09 \text{ m}) + \left(2000\,\frac{\text{kg}}{\text{m}^3}\right)\left(9.81\,\frac{\text{m}}{\text{s}^2}\right)(0.16 \text{ m})$$

Note the units of N and Pa:

$$N = \frac{\text{kg} \cdot \text{m}}{\text{s}^2} \quad \text{and} \quad Pa = \frac{\text{N}}{\text{m}^2} \quad \text{and}$$

Using this to converse units gives

$$P_A + 2825\,\frac{\text{N}}{\text{m}^2}\left(\frac{\text{kPa}}{1000 \text{ N/m}^2}\right) + 1962\,\frac{\text{N}}{\text{m}^2}\left(\frac{\text{kPa}}{1000 \text{ N/m}^2}\right)$$

$$= 100 \text{ kPa} + 883\,\frac{\text{N}}{\text{m}^2}\left(\frac{\text{kPa}}{1000 \text{ N/m}^2}\right) + 3139\,\frac{\text{N}}{\text{m}^2}\left(\frac{\text{kPa}}{1000 \text{ N/m}^2}\right)$$

Simplifying,

$$P_A + 2.825 \text{ kPa} + 1.962 \text{ kPa} = 100 \text{ kPa} + 0.882 \text{ kPa} + 3.139 \text{ kPa}$$

The pressure in the tank $P_A = 99$ kPa.

1.8 Process Classification and Material Balance

Before writing a material balance you must first identify the type of process in question.

Batch: In batch processes no material is transferred into or out of the system over the period of time of interest (e.g., heating a sealed bottle of milk in a water bath).

Continuous: A material is transferred into and out of the system continuously (e.g., pumping liquid at a constant rate into a distillation column and removing the product streams from the top and bottom of the column).

Semibatch: Any process that is neither batch nor continuous (e.g., slowly blending two liquids in a tank).

Steady state: Process variables (i.e., T, P, V, flow rates) do not change with time.

Transient: Process variables change with time.

1.8.1 Material and Energy Balances

Material and energy balances are very important in the chemical process industries. Material balances can be used to describe material quantities as they pass through a process operation (system). Such balances are statements on the conservation of mass. If no accumulation and generation occur in the system, what goes into a process is equal to what comes out. Material and energy balances can be simple; however, sometimes they can be very complicated. In all cases, the basic approach is the same. A balance on a conserved quantity (i.e., mass or energy) in a system may be written as

$$\text{Accumulation} = (\text{in} - \text{out}) + (\text{generation} - \text{consumption}) \qquad (1.17)$$

Working with simpler systems, such as individual unit operations, is a practical approach to apply the balance to more complicated situations. The increasing availability of computers has meant that very complex mass and energy balances can be set up and manipulated quite readily and, therefore, used in everyday process management to maximize product yields and minimize costs.

Proper understanding and mastering of material and energy balances is critical to the chemical engineering profession. The formulation of the

1.13 An open-end manometer connected to a closed tank is used to measure the tank pressure. The manometer/tank system is shown in Problem Figure 1.13.1. What is the pressure inside the tank (P_{tank}) in kPa? (101.521 kPa)

1.14 The manometer shown in Problem Figure 1.14.1 contains three liquids with densities $\rho_1 = 2000$ kg/m³, $\rho_2 = 1000$ kg/m³, and $\rho_3 = 1600$ kg/m³. One end of the manometer is connected to a vessel filled with a gas, and the other end is open to the atmosphere (P_B). The ambient pressure is 100 kPa. The inclined leg of the manometer is at an angle of 30° from the horizontal. What is the pressure inside the vessel in kPa? (99 kPa)

1.15 The tank shown in Problem Figure 1.15.1 is filled with water and is 2.0 m long. The pressure gauge on the tank reads 0.5 bars. Determine the height, h, in the open water tube. The level of the open tube is located at half the length of the water in the tank. (6.10 m)

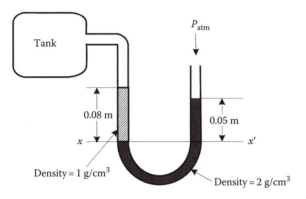

PROBLEM FIGURE 1.13.1
Pressure in closed tank using open-end manometer.

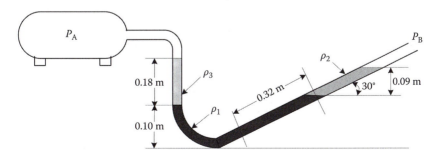

PROBLEM FIGURE 1.14.1
Pressure in closed tank using inclined manometer.

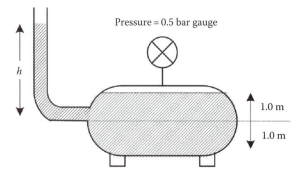

PROBLEM FIGURE 1.15.1
Height of water in an open tube connected with closed tank.

References

1. AIChE. (2008) The long-range strategy for the AIChE. *Report of the strategy project steering committee.* April, 2008.
2. Reklaitis, G.V. (1983) *Introduction to Material and Energy Balances*, John Wiley & Sons, New York.
3. Felder, R.M. and R.W. Rousseau (1999) *Elementary Principles of Chemical Processes*, 3rd edn., John Wiley & Sons, New York.
4. Himmelblau, D.M. (1974) *Basic Principles and Calculations in Chemical Engineering*, 3rd edn., Prentice-Hall, Englewood Cliffs, NJ.
5. Whirwell, J.C. and R.K. Toner (1969) *Conservation of Mass and Energy*, Blaisdell, Waltham, MA.
6. Cordier, J.-L., B.M. Butsch, B. Birou, and U. von Stockar (1987) The relationship between elemental composition and heat of combustion of microbial biomass. *Appl. Microbiol. Biotechnol. 25*, 305–312.
7. Atkinson, B. and F. Mavituna (1991) *Biochemical Engineering and Biotechnology Handbook*, 2nd edn., Macmillan, Basingstoke, U.K.
8. Scott Fogler, H. (1999) *Elements of Chemical Reaction Engineering*, 3rd edn., Prentice Hall Inc., Upper Saddle River, NJ.

2

Process Units and Degrees of Freedom Analysis

This chapter starts with a brief clarification of the essential unit operations in chemical processes. Introduction to process flow diagrams (PFDs) is provided, and the procedure to draw these for a single process from a given written or oral description of a process is illustrated. The next section elucidates the concept of degrees of freedom analysis (DFA) for a single unit system and shows how it can be used to determine whether the problem at hand is solvable. Finally, the approach used for a single process is extended to a multiunit process. The principal learning objectives of this chapter are outlined in the following section.

Learning Objectives

1. Understand the function of common unit operations encountered in chemical processes (Section 2.1).

2. Draw a flowchart, given a written or verbal description of a system (Section 2.2).

3. Properly label all known qualities and unknown quantities on a flowchart (Section 2.3).

4. Conduct a DFA for a single unit system (Section 2.4).

5. Determine whether a set of equations is independent or not (Section 2.5).

6. Properly construct a process flowchart and label all known qualities and unknown quantities for a multiunit process with recycles and purge (Section 2.6).

7. Conduct a DFA for a multiunit process (Section 2.7).

2.1 Process Units: Basic Functions

In this section, a brief description is given of the most frequently used unit operations in chemical engineering processes. The explanation is focused on typical operations involving the transfer of mass through physical or chemical routes [1].

2.1.1 Divider (Splitter)

A splitter is used to divide the flow rate in a certain stream into two or more streams with different flow rates. The schematic diagram of a splitter is shown in Figure 2.1. In this case, the composition of streams F_1, F_2, and F_3 is the same since no operation is taking place between inlet and exit streams. There is only one independent material balance even in the case of a multi-component system, since all compositions are equal. Mass flow rates F_1, F_2, and F_3 may be different.

2.1.2 Mixer (Blender)

The mixing process has the following characteristics: There are two or more entering streams, and only one exit stream resulting from the blending of the incoming streams. The streams can be in any phase, that is, gas, liquid, or solid. The mixing process flow sheet is shown in Figure 2.2.

2.1.3 Dryer (Direct Heating)

Drying is a mass transfer process resulting in the removal of moisture by evaporation from a solid, semisolid, or liquid to produce a solid state. To achieve this operation, the dryer is supplied with a source of heat. Vapor is produced in the process. The flow sheet of the drying process is shown in Figure 2.3. Solvent stream leaves as a pure vapor and is free of solids.

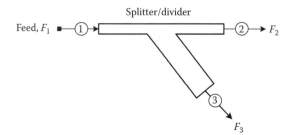

FIGURE 2.1
Schematic of a splitter.

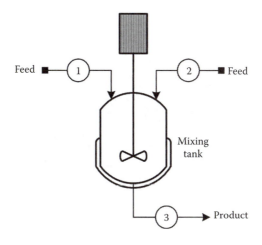

FIGURE 2.2
Schematic of a mixer.

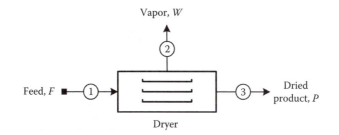

FIGURE 2.3
Schematic of a dryer.

Resulting dried products are in solid phase. Dried solids may not be solvent free. Feed can be solid, slurry, or solution.

2.1.4 Filter

Filtration is a technique used either to remove impurities from a liquid or to isolate a solid from a fluid. Filtration is commonly a mechanical or a physical operation that is used for the separation of solids from fluids (liquids or gases) by interposing a medium through which only the fluid can pass. A cross section of the filter press is shown in Figure 2.4. Filtration can also be used to separate particles that are suspended in a fluid, where the latter can be a liquid, a gas, or a supercritical fluid. Depending on the application, either one or both of the components may be isolated.

In the filtration process, filtrate, the exit liquid, is free of solids. The filtrate is saturated with soluble components. The filter cake remains with

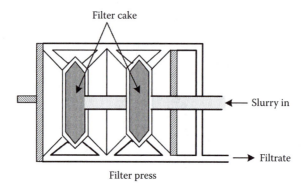

Filter cake

Slurry in

Filtrate

Filter press

FIGURE 2.4
Schematic of a filter press.

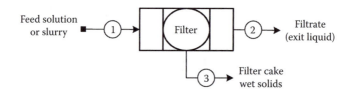

Feed solution or slurry — 1 → Filter → 2 → Filtrate (exit liquid)

3 → Filter cake wet solids

FIGURE 2.5
Schematic of a filter.

some liquid left out (see Figure 2.5). Concentration of stream 2 and the liquid remaining with the filter cake is the same.

2.1.5 Distillation Column

Figure 2.6 is a schematic diagram of a distillation column. Distillation is a method of separating chemical substances based on differences in their volatilities. Distillation usually forms part of a larger chemical process. In the distillation column, more volatile components are in the distillate, while less volatile components are in the bottoms. Separation is accomplished by boiling. However, perfect separation is not possible [2].

Each tray accomplishes a fraction of the separation task by transferring the more volatile species to the gas phase and the less volatile species to the liquid phase. Material and energy balances can be performed on an individual tray, the column, bottom reboiler, or top condenser, or the entire system.

2.1.6 Multieffect Evaporator

Figure 2.7 is a schematic diagram of a three-effect evaporator. The process of evaporation is used in the different branches of the industry for

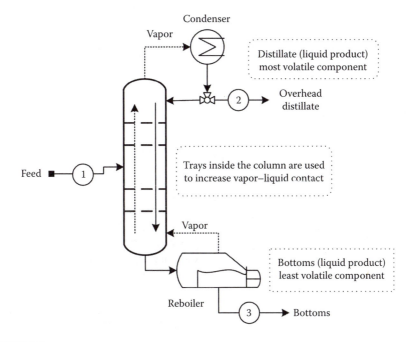

FIGURE 2.6
Schematic of a distillation column.

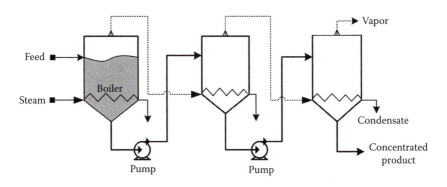

FIGURE 2.7
Schematic of a multieffect evaporator.

food or chemicals processes, in which the concentration of the solutions is required. Theoretically, multiple-effect evaporators allow decreased consumption of energy for a concentration almost proportionally equal to the number of effects (evaporators). However, being expensive, evaporators require the reduction in the number of effects, in order to be cost-effective. The optimal number of effects is generally determined via calculations.

The specifications of an evaporator are similar to those of a dryer, except that both process streams (feed and condensate) are liquids in the case of an evaporator [2,3].

2.1.7 Dehumidification

A dehumidifier with internal cooling or heating coils is shown in Figure 2.8. It is a device that reduces the level of humidity in air or a gas stream. A dehumidification process has the following characteristics: Feed stream contains a condensable component and a noncondensable component, and the condensate is a liquid with the condensable component only, such as water in air [4].

2.1.8 Humidifier

Humidifier (Figure 2.9) is a device that increases the amount of moisture in indoor air or a stream of air. It operates by allowing water to evaporate from a pan or a wetted surface, or by circulating air through an air-washer compartment that contains moisture. Humidifier processes have the following characteristics: Feed gas is not saturated, liquid is evaporated in the process unit, and exit product may or may not be saturated [4].

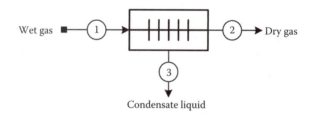

FIGURE 2.8
Schematic of a dehumidifier.

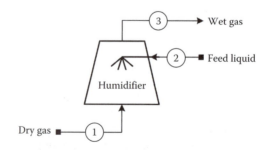

FIGURE 2.9
Schematic of a gas humidifier.

2.1.9 Leaching and Extraction

Leaching is the removal of materials from solids by dissolving them. The chemical process industries use leaching, but the process is usually called extraction. Leaching of toxic materials into groundwater is a major health concern. Extraction processes have the following characteristics: Two liquid solvents must be immiscible and have different specific gravities, and at least one component is transferred from one solvent to the other by a difference in solubility. The process is often called liquid–liquid extraction. If one of the feed streams is a solid, the process is called leaching or liquid–solid extraction (Figure 2.10). In leaching, the liquid to which materials are extracted from a carrier is not always a solvent.

2.1.10 Absorber (Stripper)

In gas absorption, a soluble component is absorbed by contact with a liquid phase in which the component is soluble. An absorber is often called a scrubber (Figure 2.11). This system is used for absorbing impurities from a gas stream of certain components such as hydrogen sulfide, carbon dioxide, and ammonia, using a suitable solvent. Absorption processes have the following characteristics: The purpose of the unit is to have the liquid absorb a component from the feed gas. The liquid stream flows down through the tower due to gravity, while the gas stream is pumped upward through the tower. No carrier gas is transferred to the liquid. Generally, no liquid solvent is transferred to the gas stream. Desorption is the same process as gas absorption except that the component transferred leaves the liquid phase and enters the gas phase.

In general, in an absorption tower (absorber), a gas is contacted with a liquid such that one or more components in the gas are transferred into the

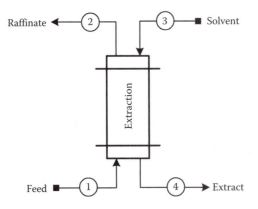

FIGURE 2.10
Schematic of an extraction column.

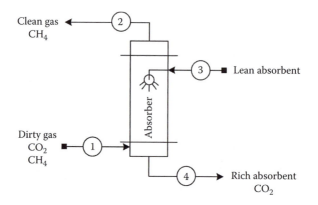

FIGURE 2.11
Schematic of CO_2 absorption from CO_2/CH_4 gas mixture.

liquid. A stripping tower (stripper) also involves a gas contacting a liquid, but components are transferred from the liquid into the gas.

2.1.11 Partial Condenser and Flash Separator

A partial condenser partly condenses a vapor stream. Partial condensers have the following characteristics: Feed stream contains only condensable vapor components, and exit streams contain liquid, L, and vapor, V, which are in equilibrium. Condensation is caused by cooling or increasing pressure (Figure 2.12). Liquid and vapor emerging from the partial condenser are separated using a flash separator.

2.1.12 Flash Separator

Flash separator splits a liquid feed into vapor- and liquid-phase products (Figure 2.13). Flash units have the following characteristics: The process is the same as that of a partial condenser except that the feed is a liquid, and

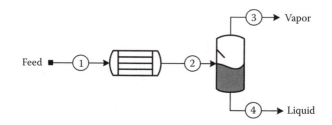

FIGURE 2.12
Schematic of a partial condenser followed by a flash unit.

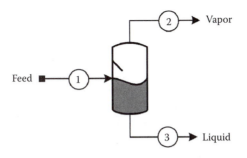

FIGURE 2.13
Schematic of a flash unit.

vaporization is caused by reducing the pressure or by heating. Vapor and liquid streams are in equilibrium [6].

2.1.13 Crystallizer

Crystallizers are used in industry to achieve liquid–solid separation. The process for a crystallizer involves a crystallizer–filter combination so as to separate solid crystals from a solution (Figure 2.14). Solid crystals are formed in the unit by a change in temperature. Crystallization is capable of generating high purity products with a relatively low energy input [5].

2.1.14 Reactors

A chemical reactor carries out a chemical reaction that converts molecular species in the input (whereby a species loses its identity) to different molecular species in the output. Figure 2.15 shows a typical reactor that has two reactant feed streams and a recycle stream. A reactor is often named by the

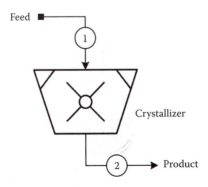

FIGURE 2.14
Schematic of a crystallization unit.

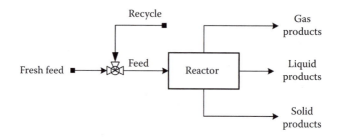

FIGURE 2.15
Schematic of a chemical reactor.

chemical reaction taking place within it. A reactor is sometimes preceded by a fictitious mixer, if the combined reactor feed is specified or must be determined. Multiple exit streams are shown to remind you to watch for streams that separate because of their different phases [5].

There are various types of reactors used in industry. The most common ones are the batch reactor, plug flow reactor (PFR), packed bed reactor (PBR), continuous stirred tank reactor (CSTR), and fluidized bed reactor.

2.1.15 Batch Reactor

A simplified schematic of a batch reactor is shown in Figure 2.16. The reactor content is loaded all at once and continuously mixed. The key characteristics of a batch reactor are unsteady-state operation (by definition) and spatial uniformity of concentration and temperature (perfectly mixed); that is, the reactor is a lumped parameter system. Batch operation is mainly used for small-scale production and is suitable for slow reactions. A batch reactor is mainly (not exclusively) for liquid-phase reactions with large charge-in/cleanup times [5].

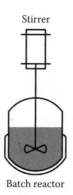

Batch reactor

FIGURE 2.16
Schematic of a batch reactor.

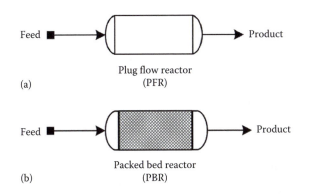

FIGURE 2.17
Schematic of a PFR (a) and a PBR (b).

2.1.16 PFRs and PBRs

The key characteristics of a PFR are steady-state operation [6], variation of concentration and temperature on space, no mixing along the reactor but complete mixing in the radial direction of the reactor; PBR is suitable for fast reactions and mainly used for gas-phase reactions with difficult temperature control, and no moving parts (Figure 2.17).

2.1.17 CSTR and Fluidized Bed Reactor

Figure 2.18a shows the diagram of a CSTR where inlet and outlet streams are continuously fed and removed, respectively. Fluidized bed reactors are sometimes treated as stirred tank reactors (Figure 2.18b). The key characteristics of a CSTR are steady-state operation, good mixing leading to spatially uniform concentration and temperature, and the condition of the outlet stream being the same as the condition in the reactor. CSTR is used for liquid-phase reactions and is suitable for viscous liquids.

2.2 Process Flow Diagram

A PFD is a diagram commonly used in chemical and process engineering to describe the general flow of plant processes and equipment. PFD displays the relationship between major equipment of a plant facility and does not show minor details such as piping and control designations. Another commonly used term for a PFD is a flow sheet. In this section, a few examples are presented to illustrate how to draw a flowchart, given a description of a system. In drawing the flowchart, one must know (or be able to determine) the total amount of the flow within the stream and composition of the stream.

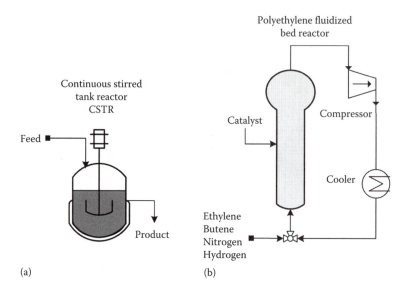

FIGURE 2.18
(a) Schematic of a CSTR and (b) a fluidized bed reactor.

Label what you do not know with variables. PFDs are considered as prelimi-
nary drawings and are used to develop initial project estimates. A piping
and instrumentation diagram, sometimes called process and instrumentation diagram (P&ID), is a diagram that shows the interconnection of process
equipment and the instrumentation used to control the process. In the chemical process industry, a standard set of symbols is used to prepare drawings
of processes.

2.3 Labeling a PFD

As mentioned before, a PFD identifies all processes at a facility. Each piece of
equipment is labeled with a name and a number assigned by the engineer or
client and must be shown on the PFD. Major flow streams are represented by
arrow lines directed from left to right in a diagram. Each stream line should
have a specification indication as a minimum unit number and line number.

Example 2.1 Separator PFD

Problem
An amount of 100 kg/h of a mixture of 50% benzene and 50% toluene is
separated in a distillation column. The distillate contains 90% benzene

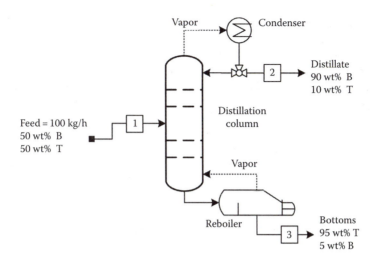

EXAMPLE FIGURE 2.1.1
Binary component separation column.

and the bottom stream composition is 95% toluene (compositions are in weight percent). Draw and label the process flowchart, and specify vapor and liquid streams.

Solution

Known quantities: Inlet and exit stream compositions are known.

Find: Draw the process flowchart.

Analysis: Read the problem statement carefully, start with feed stream, and then connect the distillation column block. Two streams leave the distillation column, that is, top product (distillate) and bottom product. The following process flow sheet can be constructed (Example Figure 2.1.1). Dashed lines indicate vapor streams.

2.4 Degrees of Freedom Analysis

When attempting to solve a material balance problem, typical questions that may arise are: How many equations do I need, and where do these potentially come from? The DFA is used to address these questions. DFA is a highly useful tool for a systematic analysis of block flow diagrams. It provides a rapid means for assessing if a specific problem is "solvable," that is, if the information available is sufficient, and provides a structured approach to decide on the order the equations must be solved. Basically, one simply counts the number of independent variables and the number of equations. To carry the analysis, you need to draw a flow diagram, label each stream with the components that are present in that stream, and make a list of additional information such as known flow rates, compositions, ratios, and conversions. There are two main points here. The first has to do with drawing "balance boundaries," that is, the number of systems where you can write the material balance equation. There are three rules for drawing system boundaries: draw a boundary around each process unit, draw a boundary around junction points, and draw a boundary around the entire process (unless there is only one boundary). The second point has to do with how many equations you can write for each drawn boundary. You can write as many equations as there are unique components passing through the boundary. For a reacting system, the number of degrees of freedom (NDF) is defined as

NDF = number of unknowns + number of independent reactions
     – number of independent material balance equations – number
     of useful auxiliary relations.

2.4.1 Possible Outcomes of DFA

The NDF can have three possible values, that is, if

1. $NDF = 0$, the system is completely defined. You get a unique solution.
2. $NDF > 0$, the system is under-defined (under-specified). There are an infinite number of solutions. More independent equations are needed.
3. $NDF < 0$, the system is over-defined (over-specified). There are too many restrictions. Check if you have too many equations or too many restrictions. Over-defined problems cannot be solved to be consistent with all equations.

2.5 Independent Equations

A set of equations are said to be independent, if you cannot derive one by adding and subtracting combinations of the others. Sources of equations that relate unknown process variables include

1. Material balances for a nonreactive process. Usually, but not always, the maximum number of independent equations that can be written equals the number of chemical species in the process.
2. Energy balances.

3. Process specifications given in the problem statement.

4. Physical properties and laws, for example, density relation, gas law.

5. Physical constraints: mass or mole fractions must add to unity.

6. Stoichiometric relations for systems with chemical reactions.

For example, the following set of equations derived from a material balance of a unit process is independent because we cannot derive any one by adding and/or subtracting combinations of the others:

$$m_1 + 2m_2 + m_3 = 100$$

$$2m_1 + m_2 - m_3 = 200$$

$$m_1 + m_2 + 2m_3 = 500$$

While the following set is not independent because we can obtain the second equation by dividing the third equation by a value of 2:

$$n_1 + 2n_2 + n_3 = 100$$

$$2n_1 + 4n_3 = 100$$

$$4n_1 + 8n_3 = 200$$

Example 2.2 Binary Separation Process

Problem
Feed stream to a distillation column flows at a rate of 300 mol/h and contains 50 mol% of component A and 50 mol% of component B. The distillate flow is at a rate of 200 mol/h and contains 60 mol% of component A. Draw and label the process flowchart. Perform a DFA.

Solution

Known quantities: Feed and distillate stream flow rates and compositions.

Find: Draw the PFD and perform a DFA.

Analysis: The schematic of the process flowchart is shown in Example Figure 2.2.1.

The DFA is demonstrated in the following table:

#	DFA	No.	Justification
1	Number of unknowns	2	B, $x_{A,B}$ only, $x_{B,B} = 1 - x_{A,B}$
2	Number of independent equations	2	Overall + one of the components A or B
3	Number of relations	0	No auxiliary relations
4	NDF	0	$NDF = 2 - 2 - 0 = 0$

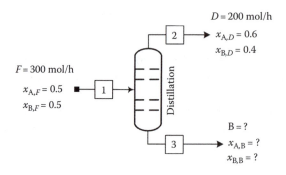

EXAMPLE FIGURE 2.2.1
Schematic of a binary distillation process.

The number of unknowns is equal to the number of components. Since we have two components, two independent equations can be written: one is the overall material balance and the second is the component balance for one of either component.

Example 2.3 Binary Component Separation Process

Problem
A feed stream flowing at a rate of 300 mol/h contains 20 mol% of components 1 and 80 mol% component 2. The distillate flow rate is 200 mol/h. Draw and label the process flowchart and perform DFA.

Solution

Known quantities: Feed flow rate and composition, distillate flow rate.

Find: Draw and label the process flowchart, and perform DFA.

Analysis: The process flow sheet is shown in Example Figure 2.3.1.

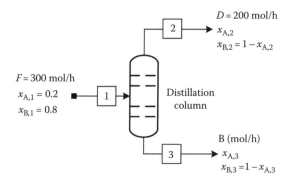

EXAMPLE FIGURE 2.3.1
Schematic of a distillation column.

The DFA is demonstrated in the following table:

#		No.	Justification
1	Number of unknowns	3	$B, x_{A,2}, x_{A,3}$
2	Number of independent equations	2	One overall + one of the two components
3	Number of relations	0	No auxiliary relations
4	NDF	1	$NDF = 3 - 2 - 0 = 1$

Comments: NDF is greater than 0 and, accordingly, the problem is under-specified. One extra piece of information is needed for the problem to be solvable.

Example 2.4 Multicomponent Separation Process

Problem
A feed stream to a distillation column contains three components (A, B, and C). Component A's mass flow rate is 100 kg/s, and the flow rates of components B and C are unknown. The distillate flow rate is 100 kg/s and contains 60 kg/s of component A and 40 kg/s of component B. It has been found that 40% of component A in the feed stream ends up in the bottom stream. The distillate and bottoms flow rates are equal. Draw and label the process flowchart, and perform DFA.

Solution

Known quantities: Distillate components mass flow rates.

Find: Draw and label the process flowchart and perform DFA.

Analysis: The process flow sheet is shown in Example Figure 2.4.1.

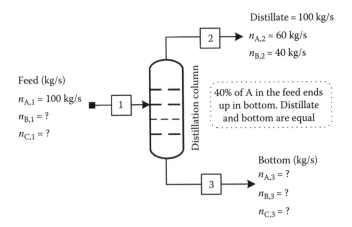

EXAMPLE FIGURE 2.4.1
Ternary component separation process.

The DFA is demonstrated in the following table:

#		No.	Justification
1	Number of unknowns	5	$F, n_{B,1}, B, n_{A,3}, n_{B,3}$
2	Number of independent equations	3	One overall + two of the three components
3	Number of relations	2	40% of A in the feed ends up in the bottom, D and B are equal
4	NDF	0	$NDF = 5 - 3 - 2 = 0$

In the feed stream, two unknowns are considered because the component flow rates are related; that is, their sum is equal to the feed flow rate, F. The same is applied for the bottom stream.

Example 2.5 Tertiary Component Separation Process

Problem

A feed to a distillation column contains 60 kg/s of benzene (B) and 10 kg/s of toluene (T), and a small amount of xylene (X). The distillate contains pure benzene. The bottom stream flow rate is 100 kg/s. Hundred percent of toluene in the feed ends up in the bottom. Draw and label the process flowchart, and perform DFA.

Solution

Known quantities: Benzene and toluene flow rates, distillate contains pure B, bottom flow rate.

Find: Draw and label the process flowchart, and perform DFA.

Analysis: The process flow sheet is shown in Example Figure 2.5.1.

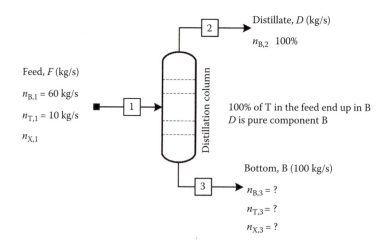

Distillate, D (kg/s)

$n_{B,2}$ 100%

Feed, F (kg/s)

$n_{B,1}$ = 60 kg/s

$n_{T,1}$ = 10 kg/s

$n_{X,1}$

100% of T in the feed end up in B

D is pure component B

Bottom, B (100 kg/s)

$n_{B,3}$ = ?

$n_{T,3}$ = ?

$n_{X,3}$ = ?

EXAMPLE FIGURE 2.5.1
Tertiary separation process.

The DFA is demonstrated in the following table:

#		No.	Justification
1	Number of unknowns	4	$F, D, n_{B,3}, n_{T,3}$
2	Number of independent equations	3	One overall + two of the three components
3	Number of relations	1	Two auxiliary relations but one is already used (distillate contains pure B)
4	NDF	0	$NDF = 4 - 3 - 1 = 0$

The first relationships have useful information, so it is considered as one relationship. The second relationship (distillate contains pure B) is already specified in the chart as 100% component B.

Example 2.6 Distillation Column

Problem
A feed stream flows at a molar flow rate of 100 mol/h and contains three components (20% component A, 30% component B, and the balance, component C). Note that 80% of A in the feed and 50% of feed rate end up in the distillate. The bottom stream contains 10% A, 70% B, and 20% C (by moles). Draw and label the process flowchart and perform DFA.

Solution

Known quantities: Feed flow rate and compositions, bottom stream composition.

Find: Draw and label the process flowchart and perform DFA.

Analysis: The process flow sheet is shown in Example Figure 2.6.1.

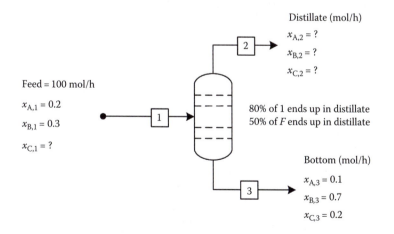

EXAMPLE FIGURE 2.6.1
Separation of a three component system.

#		No.	Justification
1	Number of unknowns	4	$D, x_{A,2}, x_{B,2}, B$
2	Number of independent equations	3	Overall + two of the three components
3	Number of relations	2	Auxiliary relations shown in the diagram
4	NDF	0	$NDF = 4 - 3 - 2 = -1$

This problem is over-specified since the number of pieces of information given is more than the number of unknowns.

Example 2.7 Binary Component Distillation Column

Problem

An ethanol (E)–methanol (M) stream is fed at a rate of 1000 kg/h to be separated in a distillation column. The feed has 40% ethanol and the distillate has 90% methanol. The flow rate of the bottom product is 400 kg/h. Draw and label the PFD and perform DFA.

Solution

Known quantities: Feed flow rate and composition, distillate composition, bottoms flow rate.

Find: Draw and label process flowchart, and perform DFA.

Analysis: The process flow sheet is shown in Example Figure 2.7.1.

DFA of the process is as follows:

#		No.	Justification
1	Number of unknowns	2	D, x
2	Number of independent equations	2	Overall + one of components E or M
3	Number of relations	0	No auxiliary relations
4	NDF	0	$NDF = 2 - 2 - 0 = 0$

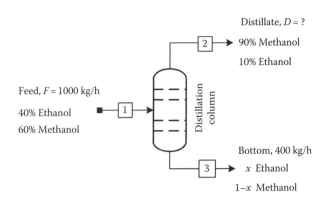

EXAMPLE FIGURE 2.7.1
Methanol–ethanol separation process.

Since the NDF is zero, the problem is solvable and has one unique solution.

Example 2.8 Drying of Wet Solid Material

Problem

Two hundred kilograms of wet leather is to be dried by heating in a dryer. The wet leather enters the drier with 1.5 g H_2O per gram bone dry leather (BDL). The leather is to be dried to residual 20% moisture. Draw and label the process flowchart, and perform DFA.

Solution

Known quantities: Feed rate and water to BDL ratio, and exit product concentration are known.

Find: Draw and label the process flowchart, and perform DFA.

Analysis: The process flow sheet is shown in Example Figure 2.8.1.

DFA is shown in the following table:

#		No.	Justification
1	Number of unknowns	2	W, P
2	Number of independent equations	2	Overall + one of components W or BDL
3	Number of relations	0	No auxiliary relations
4	NDF	0	$NDF = 2 - 2 - 0 = 0$

Since $NDF = 0$, the problem has a unique solution.

Example 2.9 Binary Component Separation Process

Problem

A stream of ethanol–methanol mixture (40 wt% ethanol, 60 wt% methanol) is fed at a rate of 100 kg/h to a distillation column. The distillate has 90% methanol and the balance is ethanol. Eighty percent of methanol fed to the distillation column is to be recovered in the distillate. Draw and label the process flowchart, and perform DFA.

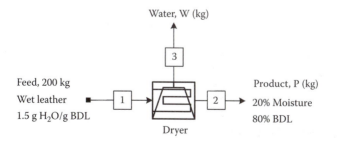

EXAMPLE FIGURE 2.8.1
Drying of a wet leather system.

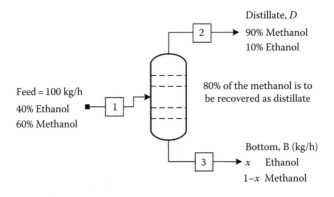

EXAMPLE FIGURE 2.9.1
Schematic of the methanol–ethanol process.

Solution

Known quantities: Feed stream flow rate and composition, distillate composition, 80% of methanol is to be recovered in the distillate.

Find: Draw and label the process flowchart and perform DFA.

Analysis: The process flow sheet is shown in Example Figure 2.9.1. DFA is shown here:

#		No.	Justification
1	Number of unknowns	3	D, B, x
2	Number of independent equations	2	Overall + one of components E or M
3	Number of relations	1	80% of methanol to distillate
4	NDF	0	$NDF = 3 - 2 - 1 = 0$

Since $NDF = 0$, the problem has a unique solution.

Example 2.10 Drying Process

Problem
One hundred kilograms of wet slurry is to be dried by heating in a furnace. The wet slurry is placed in the furnace with 60% moisture and 40% dry solid (S). Note that 90% of the water is removed. Draw and label the process flowchart, and perform DFA.

Solution

Known quantities: Wet slurry flow rate and composition, 90% of water in the wet leather is to be removed.

Find: Draw and label the process flowchart, and perform DFA.

Analysis: The process flow sheet is shown in Example Figure 2.10.1.

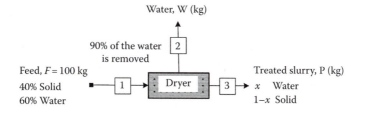

EXAMPLE FIGURE 2.10.1
Schematic of a slurry drying process.

The DFA is demonstrated in the following table:

#		No.	Justification
1	Number of unknowns	3	W, P, x
2	Number of independent equations	2	Overall + one of components W or S
3	Number of relations	1	90% of water is to be removed
4	NDF	0	$NDF = 3 - 2 - 1 = 0$

Since $NDF = 0$, the problem has a unique solution.

Example 2.11 Hydrocarbon Mixtures

Problem

A hydrocarbon feed consisting of a mixture of 20 wt% propane (C3), 30 wt% isobutene (iC4), 20 wt% isopentane (iC5), and 30 wt% n-pentane (nC5) is fractionated at a rate of 100 kg/h into a distillate. The latter contains all the propane and 78% of the isopentane in the feed. The mole fraction of isobutane in the distillate is 0.378. The bottom stream contains all the n-pentane fed to the unit. Draw and label the process flowchart and perform DFA.

Solution

Known quantities: Feed flow rate and composition.

Find: Draw and label the process flowchart, and perform DFA.

Analysis: The PFD is shown in Example Figure 2.11.1.

The DFA is demonstrated in the following table:

#		No.	Justification
1	Number of unknowns	7	$D, B, x_1, x_2, x_3, x_4, x_5$
2	Number of independent equations	4	One overall + three of the four components
3	Number of relations	3	1. Distillate contains all the C3. 2. 78% of i-C5 in the feed is in distillate. 3. The bottom contains all the n-C5 fed to the unit.
4	NDF	0	$NDF = 7 - 4 - 3 = 0$

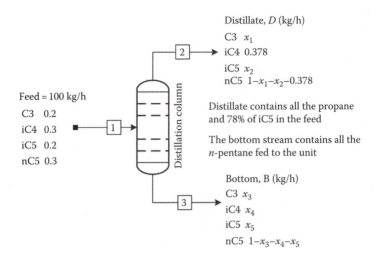

EXAMPLE FIGURE 2.11.1
Schematic of a multicomponent separation process.

2.6 Multiunit PFD

PFDs for single unit or multiple units differ in their structure and implementation. PFDs for multiple units do not include detailed information and are known as the schematic flow diagrams or block flow diagrams. The main reason for using process flowchart is to show the relation between major parts of the system. Process flowchart is used primarily in process engineering and chemical industry, where there is a requirement of depicting the relationship between major components only.

Example 2.12 Extraction Process

Problem

A stream containing 10.0 wt% acetone and the balance water enters a liquid/liquid extraction column where acetone is extracted with methyl isobutyl ketone (MIBK). The solvent is a fresh feed of MIBK and a recycled stream consisting of MIBK with a small amount of acetone. There are two streams leaving the liquid–liquid extraction process: a raffinate stream that contains only acetone and water, and an extract stream that contains only acetone and MIBK. The extract stream goes to a distillation process. The overhead stream from the distillation process is mainly acetone with some MIBK. The bottoms stream contains mainly MIBK with some acetone and is recycled to the liquid–liquid extraction process. The overhead stream from the distillation tower contains 90% of the acetone in stream 1. The overhead stream from the distillation tower also contains four times as much acetone as it contains MIBK. Draw a flow diagram for the process labeling all processing units and all streams. Give the information of the flow rate and/or composition. For each stream in this process, indicate as much information as you know about that stream as given in the process description.

Solution

Known quantities: See the process flow sheet, Example Figure 2.12.1.

Find: Draw and label the PFD.

Analysis: The PFD would look like Example Figure 2.12.1.

Example 2.13 Process Flow Sheet

Problem
The process for producing pure ethanol starts with a fresh feed stream that contains 96% ethanol and 4% water. The stream is fed to a distillation tower along with a stream of benzene plus a small amount of ethanol (this stream is being recycled from another portion of the system). This distillation tower produces two outgoing streams: the overhead stream (all vapor) is a mixture of water, benzene, and ethanol, and the bottom stream (all liquid) is 100% ethanol. The overhead stream is condensed to an all-liquid stream. The liquid stream is sent to a decanter where it separates into two distinct liquid phases. The top layer in the decanter consists primarily of benzene with a small amount of ethanol and is sent back to the first distillation tower to be mixed with the incoming ethanol/water mixture being fed to the system. The bottom layer from the decanter contains ethanol, water, and benzene. This stream is fed to a second distillation tower. The overhead stream from this second distillation tower is also all vapors and consists of ethanol, water, and benzene. This vapor stream is mixed with the overhead vapor stream from the

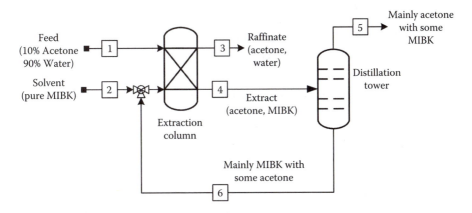

EXAMPLE FIGURE 2.12.1
Schematic of an extraction/regeneration column.

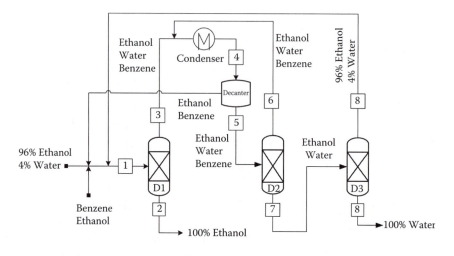

EXAMPLE FIGURE 2.13.1
Process flow sheet of the ethanol–water–benzene separation process.

first distillation tower before both streams are condensed prior to the decanter. The bottoms stream from this second tower (again all liquid) consists of water and ethanol. This bottoms stream is fed to a third distillation tower. The overhead product from this third distillation tower is a 96 wt% ethanol/4.0 wt% water stream that is sent to the start of the entire process to be mixed with the incoming feed of ethanol and water. The bottoms stream from the third distillation tower is all water. Draw a flow diagram for the process described earlier clearly labeling all processing units and all streams. For each stream in this process, you should indicate as much information as you know about that stream as given in the process description. If you know the flow rate and/or composition, give this information. If all you know about a stream are its components, indicate those.

Solution

Known quantities: See the diagram.

Find: Draw and label the PFD.

Analysis: The PFD would look like Example Figure 2.13.1.

2.7 DFA, Multiunit Process

For more complex processes, those consisting of a single operation, it is of utmost importance that you have a method of determining if a problem is solvable at all, given the information that you have at hand. The DFA for a single-unit process can be easily extended to multiunit processes. There are three ways to describe a problem in terms of its solvability:

1. If the problem has a unique set of solutions then it is called well-defined.
2. The problem is over-specified; that is, you have too much information and it is either redundant or inconsistent. It could be fixed by removing an assumption about the system that one had made.
3. The problem is under-specified; that is, you do not have enough information to solve for all your unknowns. There are several ways to deal with this. The most obvious is to gather additional information, such as measuring additional process variables (e.g., temperatures, flow rates, etc.), until you have a well-defined problem. Another way is to use additional equations or information about what we want to achieve out of a process (e.g., conversion level of a reaction, efficiency of a separation unit, etc.). Finally, we can make assumptions in order to simplify the equations, and perhaps they will simplify enough that they become solvable.

Multiunit process systems are more involved in their analysis than single-unit systems, but they can be analyzed, provided a structured approach is followed. The next few steps can prove to be extremely useful:

1. Label a flowchart completely with all the relevant unknowns for all units and streams.
2. Perform a DFA on each unit operation.
3. Determine the NDF for each unit and the overall system.

Example 2.14 DFA for a Multiunit Process

Problem

An absorber–stripper system is used to remove carbon dioxide and hydrogen sulfide from a feed consisting of 30% CO_2 and 10% H_2S in nitrogen. In the absorber, a solvent selectively absorbs hydrogen sulfide and carbon dioxide. The absorber overhead contains only 1% CO_2 and no H_2S. N_2 is insoluble in the solvent. The rich solvent stream leaving the absorber is flashed, and the overhead stream consists of 20% solvent, and contains 25% of the CO_2 and 15% of the H_2S in the raw feed to the absorber. The liquid stream leaving the flash unit is split into equal portions, one being returned to the absorber. The other portion, which contains 5% CO_2, is fed to the stripper. The liquid stream leaving the stripper consisting of pure solvent is returned to the absorber along with makeup solvent. The stripper overhead contains 30% solvent. Draw and completely label a flow sheet of the process and perform a DFA.

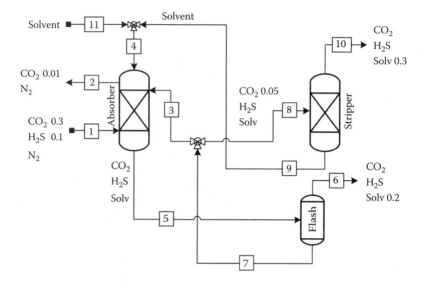

EXAMPLE FIGURE 2.14.1
Process flow sheet of the acid gas absorption process.

Solution

The labeled process flow sheet is shown here (Example Figure 2.14.1).

The DFA is demonstrated in the following table:

	System					
	Absorber	Flash	Stripper	Splitter	Mixer	Overall
Number of unknowns	8	7	5	4	3	7
Number of independent equations	4	3	3	1	1	4
Number of relations	0	2	0	1	0	2
NDF	4	2	2	2	2	1

The lowest degrees of freedom value is for the overall process. Specifying a basis will reduce its NDF to zero. There are two common situations where you will find fewer independent equations than species. They balance around a divider, since the single input and two or more output streams have the same composition, thus resulting in only one independent equation (mass balance). If two species are in the same ratio to each other wherever they appear in a process and this ratio is incorporated in the flowchart labeling, balances on those species will not be independent equations. This situation occurs frequently when air is present in a nonreactive process (21 mol% O_2; 79 mol% N_2).

Homework Problems

2.1 An ethanol–methanol–propanol stream is fed at a rate of 2000 kg/h to a distillation column. Feed is 20% methanol. Ninety percent of the methanol is to be recovered in the distillate along with 60% of ethanol. All of the 400 kg/h of propanol fed to the process must be sent to the bottom. Draw and label the PFD and perform a DFA.

2.2 An ethanol–methanol stream fed at a rate of 100 kg/h is to be separated in a distillation column. The feed has 40 wt% ethanol and the distillate has 80% methanol by weight. Note that 80 wt% of methanol in the feed is to be recovered as distillate. Perform a DFA (Problem Figure 2.2.1).

2.3 A stream of aqueous hydrochloric acid, 57.3 wt% HCl, is mixed with pure water to produce a stream of 16.5% acid. Perform a DFA (Problem Figure 2.3.1).

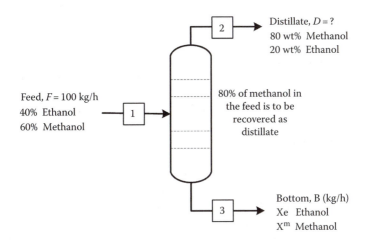

PROBLEM FIGURE 2.2.1
Ethanol–methanol separation process.

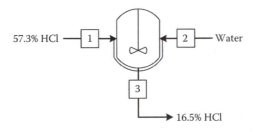

PROBLEM FIGURE 2.3.1
Mixing process.

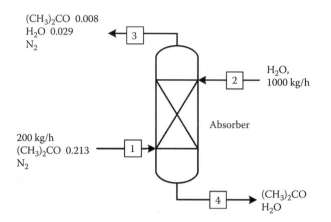

PROBLEM FIGURE 2.4.1
Absorption process.

2.4 An absorber is used to remove acetone from a nitrogen carrier gas. The feed, with acetone weight fraction 0.213, enters at a rate of 200 kg/h. The absorbing liquid is water, which enters at a rate of 1000 kg/h. The exit gas stream is 0.8 wt% acetone and 2.9% water vapor. Perform a DFA (Problem Figure 2.4.1).

2.5 Air containing 3% acetone and 2% water is fed to an absorber column. The mass flow rate of air is 1000 kg/h. Pure water is used as absorbent to absorb acetone from air. The air leaving the absorber should be free of acetone. The air leaving the absorber was found to contain 0.5% water. The bottom product of the absorber is sent to a distillation column to separate acetone from water. The bottom of the distillation column was found to contain 4% acetone, and the balance is water. The vapor from the head of the absorber is condensed. The concentration of the condensate is 99% acetone and the balance is water. Draw and label a process flowchart, and perform a DFA.

2.6 A liquid mixture containing 38 mol% benzene (B), 35.0 mol% toluene (T), and 27.0 mol% xylene (X) is fed to a distillation column. The bottom product contains 97.0 mol% X and the balance is T, bearing in mind that 93.0% of X in the feed is recovered in this stream. The overhead product is fed to a second column. The overhead product from the second column contains 95 mol% B and 5.0 mol% T, bearing in mind that 96.0 mol% of the benzene fed to the system is recovered in this stream. Draw and completely label a flow sheet of the process, and perform a DFA.

2.7 A copper ore contains 1.00 wt% copper and 99.0 wt% rock. The crushed ore is mixed with a stream of fresh acid plus a recycled stream. The rock is allowed to settle out of the copper/acid solution completely. The rock leaving this stage contains all of the rock in the ore as well as some of the copper/acid solution produced in this stage. Every 3.00 lb of rock that leaves this stage contains 1.00 lb of the copper/acid solution. It is desired that only 10% of the copper originally contained in the ore be lost in this stream. The remaining copper/acid solution, containing no rock, is sent to an electrical recovery process where pure copper is separated and a solution containing copper and acid is generated. In the electrical recovery process, 90% of the copper that enters the electrical recovery process is recovered as the pure copper product. The copper/acid solution generated in the electrical recovery process is recycled and mixed with the fresh acid stream entering the mixing stage. For every 2000 lb of ore that enters this process, draw and label the PFD.

2.8 A chemical A is to be removed from its ore. Hundred kilograms per hour of ore is fed to a dissolution tank where it is mixed with a stream of pure water (W) and a recycle stream. The tank is heated to 90°C so that all of A (but none of the other junk, J) dissolves, forming a saturated solution. The material exits the tank and is sent to a filter (separator) where all of the junk and a small portion of the (still saturated) solution are removed. Finally, the remaining solution is fed to a crystallizer where it is cooled to 25°C in order to form some solid A, which is shipped off to be packaged and sold (some filtrate also leaves with the solid). The remaining filtrate, now saturated at 25°C, is recycled to the dissolution tank. Upon analyzing the products from the filter and crystallizer, it is determined that the waste stream from the filter contains 40.0 kg/h of J, 12 kg/h of A, and 4 kg/h of water (W), and that the final product stream from the crystallizer contains 42 kg/h, 6 kg/h A, and 6 kg/h W. From the process description, draw a flowchart that concisely summarizes each stage of the process (properly labeling all streams and species within the streams).

2.9 An absorber–stripper system is used to remove carbon dioxide and hydrogen sulfide from a feed consisting of 30% CO_2 and 10% H_2S in nitrogen. In the absorber, a solvent selectively absorbs hydrogen sulfide and carbon dioxide. The absorber overhead contains only 1% CO_2 and no H_2S. N_2 is insoluble in the solvent. The rich solvent stream leaving the absorber is flashed, and the overhead stream consists of 20% solvent, and contains 25% of the CO_2 and 15% of the H_2S in the raw feed to the absorber. The liquid stream leaving the flash unit is split into equal portions, one being returned to the absorber. The other portion, which contains 5% CO_2, is fed to the stripper. The liquid stream leaving the stripper consists of pure solvent and is returned to the absorber along with makeup solvent. The stripper overhead contains 30% solvent. Draw and completely label a PFD of the process and perform a DFA.

2.10 Fresh feed containing 20% by weight KNO_3 (K) in H_2O (W) is combined with a recycle stream and fed to an evaporator. The concentrated solution leaving the evaporator, containing 50% KNO_3, is fed to a crystallizer. The crystals obtained from the crystallizer are 96% KNO_3 and 4% water. The supernatant liquid from the crystallizer constitutes the recycle stream and contains 0.6 kg KNO_3 per 1.0 kg of H_2O. Draw and label the process flowchart and perform a DFA.

References

1. Reklaitis, G.V. (1983) *Introduction to Material and Energy Balances*, John Wiley & Sons, New York.
2. Felder, R.M. and R.W. Rousseau (1999) *Elementary Principles of Chemical Processes*, 3rd edn., John Wiley & Sons, New York.
3. Himmelblau, D.M. (1974) *Basic Principles and Calculations in Chemical Engineering*, 3rd edn., Prentice-Hall, Englewood Cliffs, NJ.
4. Whirwell, J.C. and R.K. Toner (1969) *Conservation of Mass and Energy*, Blaisdell, Waltham, MA.
5. Scott Fogler, H. (1999) *Elements of Chemical Reaction Engineering*, 3rd edn., Prentice-Hall Inc., Upper Saddle River, NJ.
6. Mersmann, A. (1995) *Crystallization Technology Handbook*, Marcel Dekker, Inc., New York.

3

Material Balance on Single-Unit Process

Material balances are important when designing a new process or analyzing an existing one. They are approximately at all times prerequisite to all other calculations in the solution of process engineering problems. They are used in the industry to calculate mass flow rates of different streams entering or leaving chemical or physical processes. In this chapter, the concept of material balance as it is related to chemical processes is introduced. The material balance for a general system and highlights on its importance in the analysis of chemical engineering processes are illustrated. Further, the difference between systems in terms of their operation mode, that is, steady or unsteady state, and open or closed, is defined. How to solve the steady-state material balance in a single-unit process with no chemical reactions is explained. The following highlights the focal learning objectives that should be achieved toward the end of this chapter.

Learning Objectives

1. Understand material balance and its practical importance (Section 3.1).

2. Develop a conceptual understanding of the different types of systems and their implications on material balance (Section 3.2).

3. Write the mass balance on single-unit processes without reactions (Section 3.3).

4. Choose a suitable basis in mass balance calculations (Section 3.4).

3.1 Introduction to Material Balance

Material balances are the application of the law of conservation of mass, which states that mass can neither be created nor destroyed. Thus, you cannot, for example, specify an input to a reactor of 1 ton of naphtha and an output of 2 tons of gasoline or gases or anything else. One ton of total material input will only give 1 ton of total output, that is,

$$\text{Total mass of input} = \text{total mass of output}$$

It is one (in addition to energy and momentum) of the fundamental laws of conservation, namely, mass conservation. The simplest form of a mass balance is for systems wherein there is no accumulation and generation (consumption) of matter. For such systems, mass balance problems have a standard theme: Given the mass flow rates of some input and output streams, calculate the mass flow rates of remaining streams [1]. (Usually, it is not feasible to measure the masses and compositions of all streams entering and leaving a system.) Unknown quantities can then be calculated using mass balance principles.

The concept of mass balance is a powerful tool in chemical engineering analysis. This chapter explains how the law of conservation of mass is applied to atoms, molecular species, and total mass, and describes formal techniques for solving material balance problems without reaction. For a given system, a material balance can be written in terms of the following conserved quantities: total mass (or moles), mass (or moles) of a chemical compound, mass (or moles) of an atomic species. It is to be mentioned that the number of moles is not always conserved, and hence may not be appropriate to represent an amount in a material balance. To apply the material balance, you need to define the system and the quantities of interest (see Figure 3.1). A system is a region of space defined by a real or imaginary closed envelope (system boundary), and may be a single process unit, collection of process units, or an entire process [2,3]. The general expression of a mass balance takes the following form:

$$\text{Accumulation} = (\text{input} - \text{output}) + (\text{generation} - \text{consumption})$$

If the system is under steady state (defined next), then the accumulation term is removed from the balance. If the balanced quantity is total mass, the generation term will be canceled since total mass is neither created nor destroyed (with the exception of nuclear reactions). If the balanced substance is a

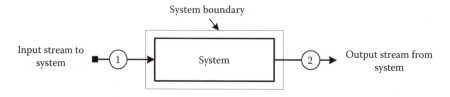

FIGURE 3.1
System over which material balance is made.

nonreactive species, neither a reactant nor a product, no generation or consumption takes place; these terms are removed from the balance, accordingly.

3.2 Material Balance Fundamentals

As explained previously, the concept of a material balance is based on the fundamental law of conservation of mass:

$$\text{Accumulation} = (\text{input} - \text{output}) + (\text{generation} - \text{consumption})$$

where
    Accumulation is the change in quantity of material inside the system
    Input is the material that enters the system by crossing system boundary
    Output is the material that exits the system by crossing system boundary
    Generation is the material that is produced by chemical reaction within
        the system
    Consumption is the material that is used up by chemical reaction within
        the system

In particular, chemical engineers are concerned with writing mass balances around chemical processes. Chemical engineers write mass balances to account for what happens to each of the chemicals participating in a chemical process. Processes can operate either under steady state or in a transient mode (unsteady state). A steady-state system is one whose properties do not change with time. Every time we take a snapshot of the process, all the variables have the same values as they did when measured the first time. A transient system is one whose properties change with time. Every time a snapshot is taken, process variables take on.

Processes can be further classified as continuous, batch, or semibatch. A continuous process refers to a process where feed and product streams move chemicals into and out of the process all the time. In a batch process, the material content is loaded in the process unit all at once. Products are withdrawn from the process at specific times only. A semibatch process is one that has some characteristics of continuous and batch processes. For instance, input streams in the process are handled batch wise, while output streams flow out of the process continuously, and vice versa.

The type of process has clear implications on the formulation of material balance. There are two types of material balances, that is, differential or integral, based on a system mode of operation. A differential mass balance is a form of the balance that represents a process continuous in time. This type of balance is generally applied to continuous steady-state processes, and is used comprehensively in this book. An integral balance is a form of the balance written over a time interval is generally applied to batch processes. For example, a water storage tank contains 20 L of water initially. After 30 min of filling up the tank with water, the water content of the tank is found to be 50 L. In this case, the accumulated amount of water after 30 min is 30 L of water:

$$\text{Accumulation} = \text{final amount (at 30 min)} - \text{initial amount (at some } t_o)$$
$$= 50 - 20 = 30 \text{ L}$$

3.3 Mass Balance on Steady-State Processes

A flowchart is a convenient way of organizing process information for subsequent calculations, as mentioned in the previous chapter. To obtain maximum benefit from the flowchart in material balance calculations, one must write the values and units of all known stream variables at the locations of the streams on the chart. Assign algebraic symbols to unknown stream variables and write these variable names and their associated units on the chart. The use of consistent notation is generally advantageous. The following notation is used. For example, m (mass), $\dot{m}$ (mass flow rate), n (moles), $\dot{n}$ (mole flow rate), V (volume), $\dot{V}$ (volumetric flow rate), x (component fractions (mass or mole) in liquid streams), and y (component fractions in gas streams).

A systematic procedure will be outlined for solving single-unit processes where there are no reactions (consumption = generation = 0), and when processes are continuous and under steady state (accumulation = 0). The procedure will form the foundation for more complex problems involving multiple units and processes with reactions (described in later chapters). For a stream to be fully specified, the flow rate and the composition of each component should be known. If any of these items are not given, then it will be considered as unknown. If the stream composition is unknown or if some of the component masses are known, represent the component masses directly and use a lower case letter for each chemical. If the stream composition is known from fractional compositions, represent the component masses directly and label them. If the stream composition is partially known with fractional compositions and the total is known, represent the component masses indirectly and use lowercase x, y, and z for each fractional composition [4,5].

3.3.1 Stream Specification

A material stream is fully specified if we can express the mass flow rates of each component in the stream. For example, the following stream is fully specified since the mass flow rate of components in the stream are known.

Stream F contains 30 kg/min of O_2 and 70 kg/min of CH_4. Note that the component masses must be added and be equal to the total mass. The total mass in F is 100 kg/min:

$$\underrightarrow{\quad \text{Stream } F(100 \text{ kg/min})\quad}$$
$$m_{O_2} = 30 \text{ kg/min}$$
$$m_{CH_4} = 70 \text{ kg/min}$$

Suppose that 100 kg/s of a mixture containing O_2, N_2, and CH_4 is fed to a process. The stream contains 20% O_2 by mass. The mass flow rate of component i in the stream, $m_i = F \times x_i$.

In this case, the stream compositions are partially known, accordingly the stream is not fully specified. Note that the fractional compositions in a specific stream must add to 1; thus, we can write two alternatives:

$$\underrightarrow{\quad \text{Stream } F \, (100 \text{ kg/s})\quad}$$
$$x_{O_2} = 0.2$$
$$x_{N_2} = ?$$
$$x_{CH_4} = 1.0 - 0.2 - x_{N_2} = 0.8 - x_{N_2}$$

or using component masses,

$$\underrightarrow{\quad \text{Stream } F(100 \text{ kg/s})\quad}$$
$$m_{O_2} = 20 \text{kg/s}$$
$$m_{N_2} = ?$$
$$m_{CH_4} = 100 - 20 - m_{N_2} = 80 - m_{N_2}$$

Example 3.1 Material Balance on Distillation Column

Problem
A mixture of three components (A, B, and C) enters a separation process. The three components appear in the distillate with variable composition. In contrast, only B and C appear in the bottom. Write a proper set of material balance equations.

Solution

Known quantities: No data is given.

Find: Possible material balance equation.

Analysis: The process-labeled flowchart is shown in Example Figure 3.1.1.

System: Distillation column
We can write four equations in total, but only three are independent, because the fourth one can be derived from the other three. So, in doing the math, we have our choice on which of the three equations we want to use.

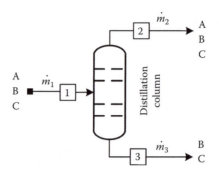

EXAMPLE FIGURE 3.1.1
Schematic of a distillation column.

Component balance ($\dot{m}$: mass flow rate):

$$A: \quad \dot{m}_{A,1} = \dot{m}_{A,2} + \dot{m}_{A,3}, \quad \dot{m}_{A,3} = 0 \text{ (no A exists in stream 3)}$$

$$B: \quad \dot{m}_{B,1} = \dot{m}_{B,2} + \dot{m}_{B,3}$$

$$C: \quad \dot{m}_{C,1} = \dot{m}_{C,2} + \dot{m}_{C,3}$$

Total mass balance:

$$\dot{m}_1 = \dot{m}_2 + \dot{m}_3$$

The number of independent material balance equations equals the number of components. When writing down the independent equations, use the overall material balance around the system plus all component balances less one, as shown here:

$$\left\{ \begin{array}{l} \text{Independent material} \\ \text{balance equations} \end{array} \right\} = \begin{array}{c} \text{Relations} + \text{overall total} + (n-1) \\ \text{component balances} \end{array}$$

where n is the number of components involved in the system.

Example 3.2 Ethanol–Water Separation Process

Problem
A mixture containing 10% ethanol (E) and 90% H_2O (W) by weight is fed into a distillation column at the rate of 100 kg/h. The distillate contains 60% ethanol and the distillate is produced at a rate of one tenth that of the feed. Draw and label a flowchart of the process. Calculate all unknown stream flow rates and compositions.

Solution

Known quantities: Feed and distillate flow rates and composition.
Find: Calculate all unknown stream flow rates and composition.

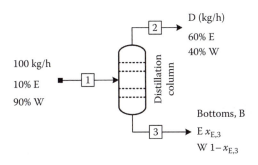

EXAMPLE FIGURE 3.2.1
Schematic of the ethanol–water separation process.

Analysis: The process-labeled flowchart is shown in Example Figure 3.2.1.

Assumptions: Continuous process, steady state, no reactions.

Basis: 100 kg/h of feed.

$$NDF = \text{number of unknowns} - \text{number of independent equations} - \text{number of relations}$$

$$NDF = 3 - 2 - 1 = 0 \text{ (solvable)}$$

Three independent material balance equations: Overall material balance and a component mass balance (one of the two components).

Relation (one relation only: distillate flow rate is one-tenth that of the feed).

$$D = 0.1 \times F = 0.1 \times 100 = 10 \text{ kg/h}$$

Overall total material in the system:

$$F = D + B$$

$$100 \text{ kg/h} = 10 + B; \ B = 90 \text{ kg/h}$$

Component balance (ethanol):

$$0.1 \times 100 \text{ kg/h} = 0.6 \times 10 \text{ kg/h} + x_{E,3} \times 90 \text{ kg/h} \Rightarrow x_{E,3} = 0.044$$

Check your answer:

Perform a mass balance on the components (water) not used in the earlier solution procedure.

$$\text{Mass of water in feed} = \text{mass of water in the distillate} + \text{mass of water in the bottom}$$

$$90 \text{ kg} = 4 \text{ kg} + 0.956 \times 90 \text{ kg} = 90 \text{ kg}$$

Example 3.3 Separation Process

Problem

A feed stream is flowing at a mass flow rate of 100 kg/min. The stream contains 20 kg/min NaOH and 80 kg/min of water. The distillate flows

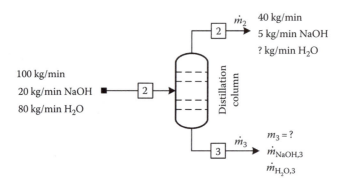

EXAMPLE FIGURE 3.3.1
Schematic of the sodium hydroxide–water separation process.

at 40 kg/min and contains 5 kg/min NaOH. Determine bottom stream mass flow rate and composition.

Solution

Known quantities: Feed and distillate streams are fully specified (i.e., mass flow rate and compositions are known).

Find: Bottom stream flow rate and compositions.

Analysis: The process-labeled flowchart is shown in Example Figure 3.3.1.

Basis: 100 kg/min of the feed stream.

$$NDF = 2 \text{ (unknowns)} - 2 \text{ (independent equations)} - 0 \text{ (relation)}$$
$$= 0 \text{ (problem is solvable)}$$

Total mass balance:

$$100 \text{ kg/min} = \dot{m}_2 + \dot{m}_3$$

$$100 \text{ kg/min} = 40 \text{ kg/min} + \dot{m}_3$$

Component mass balance (NaOH):

$$20 \text{ kg/min} = 5 \text{ kg/min} + \dot{m}_{\text{NaOH}}$$

There are two simple linear algebraic equations with two unknowns ($\dot{m}_3$, $\dot{m}_{\text{NaOH}}$), solving for $\dot{m}_3$ and $\dot{m}_{\text{NaOH}}$:

$$\dot{m}_3 = 60 \text{ kg/min}$$

$$\dot{m}_{\text{NaOH}} = 15 \text{ kg/min}$$

$$\dot{m}_{\text{H}_2\text{O}} = \dot{m}_3 - \dot{m}_{\text{NaOH}} = 60 - 15 = 45 \text{ kg/min}$$

3.4 Basis for Calculation

The amount or flow rate of one of the process streams can be used as a basis for calculation. It is recommended to bear the following in mind:

1. If a stream amount or flow rate is given in the problem statement, use this as the basis for calculation.
2. If no stream amounts or flow rates are known, assume one, preferably a stream of known composition.
3. If mass fractions are known, choose the total mass or mass flow rate of that stream (e.g., 100 kg or 100 kg/h) as the basis.
4. If mole fractions are known, choose the total number of moles or the molar flow rate.

Scaling from a basis amount or rate to certain values can be calculated by changing the values of all stream amounts or flow rates of a certain process by a proportional amount, while leaving the stream compositions unchanged. Figure 3.2a shows how a feed is scaled up by a factor of 2. The molar flow rates of A and B are doubled in the exit streams. The compositions are not affected by scaling up or down.

3.4.1 Procedure for Solving Material Balance Problems

1. Read the problem carefully and express what the problem statement asks you to determine. Analyze the information given in the problem.
2. Properly label all known quantities and unknown quantities on a flowchart.

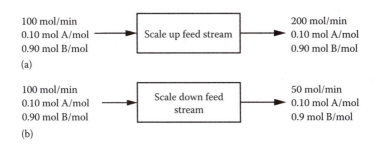

FIGURE 3.2
(a) Scale-up and (b) scale-down processes.

3. Choose a basis for calculation.

4. Select a system, draw its boundaries, and state your assumptions.

5. Determine the number of unknowns and the number of equations that can be written that relate to them.

6. Calculate the quantities requested in the problem statement. Check your solution and whether it makes sense.

Example 3.4 General Material Balance Equation

Write the material balance equation in simple words and briefly define each term using the word "system" in your definitions.

Solution

Known quantities: No quantities are given.

Find: Write material balance equation in words.

Analysis: The general material balance equation is

$$\text{Accumulation} = (\text{input} - \text{output}) + (\text{generation} - \text{consumption}),$$

where
Accumulation is the change in quantity of material inside the system
Input is the material that enters the system by crossing system boundary
Output is the material that exits the system by crossing system boundary
Generation and consumption are the material that is produced or used
up by chemical reaction within the system, respectively

Example 3.5 Concentrated Lemon Juice

Problem
The initial amount of pressed lemon juice contains 10% of total solids. It is desired to increase the figure to 20% of total solids by evaporation. The resulting concentrated juice consists of 20% total solids. Calculate the quantity of water that must be removed.

Solution

Known quantities: Pressed juice mass and concentration, concentrated juice solid concentration.

Find: The quantity of water to be removed.

Analysis: The process flow sheet is shown in Example Figure 3.5.1.

Basis: 100 kg of pressed juice.
Solid balance:

$$0.10 \times 100 \text{ kg} = 0.2 \times m_3$$

$$m_3 = 50 \text{ kg}$$

Amount of water evaporated $= 100 - 50 = 50$ kg

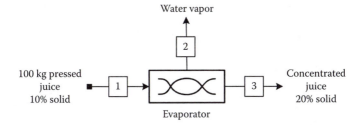

EXAMPLE FIGURE 3.5.1
Concentration of lemon juice.

Example 3.6 Separation of a Mixture of Benzene and Toluene

Problem
A 100 kg/h mixture of benzene (B) and toluene (T) containing 50% benzene by mass is separated by distillation into two fractions. The mass flow rate of benzene in the top stream is 45 kg B/h and that of toluene in the bottom stream is 47.5 kg T/h. The operation is under steady state. Write balances for benzene and toluene to calculate the unknown component flow rate in the output streams.

Solution

Known quantities: 100 kg/h of B and T; the feed contains 50 wt% B, the distillate contains 45 kg B/h, and the bottom contains 47.5 kg T/h.

Find: The unknown component flow rate in the output streams.

Analysis: The process-labeled flowchart is shown in Example Figure 3.6.1.

Basis: 100 kg/h of feed.

Assumptions: Steady state, no reaction.

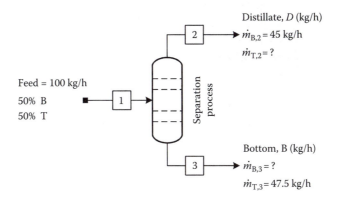

EXAMPLE FIGURE 3.6.1
Benzene–toluene separation process.

Component mass balance (benzene):

$$\text{B balance:} \quad 0.5(100) = 45\frac{\text{kg B}}{\text{h}} + \dot{m}_{B,3} \rightarrow \dot{m}_{B,3} = 5\frac{\text{kg}}{\text{h}}$$

Material balance (toluene):

$$\text{T balance:} \quad 0.5(100) = \dot{m}_{T,2} + 47.5\frac{\text{kg T}}{\text{h}} \rightarrow \dot{m}_{T,2} = 2.5\frac{\text{kg}}{\text{h}}$$

Example 3.7 Methanol–Water Mixtures

Problem

Two methanol–water mixtures are mixed in a stirred tank. The first mixture contains 40.0 wt% methanol and the second one contains 70.0 wt% methanol. If 200 g of the first mixture is combined with 150 g/s of the second one, what are the mass and composition of the product?

Solution

Known quantities: Two methanol/water mixtures with known compositions are mixed together.

Find: The mass and composition of the product.

Analysis: The process-labeled flowchart is shown in Example Figure 3.7.1.

Basis: 200 g/s of the 40 wt% methanol (i.e., stream 1).

Assumptions: Steady state, no reaction.

Material Balance

Total mass balance:

$$\dot{m}_1 + \dot{m}_2 = \dot{m}_3$$

$$200 + 150 = \dot{m}_3 \Rightarrow \dot{m}_3 = 350$$

Component balance (methanol):

$$0.4(200) + 0.7(150) = x_{M,3}(350) \Rightarrow x_{M,3} = 0.53$$

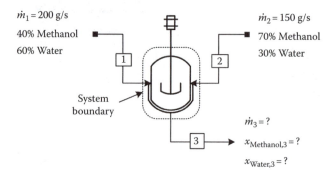

EXAMPLE FIGURE 3.7.1
Methanol–water mixture.

Example 3.8 Synthesis of Strawberry Jam

Problem

To make strawberry jam, strawberries containing 15 wt% solids and 85 wt% water are crushed. The crushed strawberries and sugar are mixed in a 4/5 mass ratio and the mixture is heated to evaporate water. The residue contains one-third water by mass. Calculate the amounts of strawberries needed to make 100 g of jam, and of evaporated water.

Solution

Known quantities: Strawberries contain 15 wt% solids and 85 wt% water.

Find: Amount of strawberries needed to make 100 g of jam.

Analysis: The process-labeled flowchart is shown in Example Figure 3.8.1.

Basis: 100 g of jam is to be produced; $m_3 = 100$ g.

Assumption: Steady state, no reaction. We have three components (solid, water, and sugar) so three independent equations can be written. A relationship is given, which can be expressed as

$$\frac{\text{strawberries}}{\text{sugar}} = \frac{m_1}{m_2} = \frac{4}{5}$$

Material Balance

Component balance (solid):

$$0.15m_1 = x_{s,3}(100 \text{ g})$$

Relation:

$$\frac{m_1}{m_2} = \frac{4}{5} \Rightarrow m_2 = m_1 \times \frac{5}{4} = 1.25m_1$$

Component balance (water):

$$0.85(m_1) = \frac{1}{3}(100 \text{ g}) + m_4$$

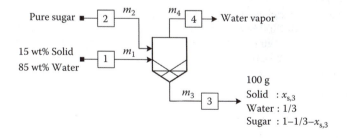

EXAMPLE FIGURE 3.8.1
Synthesis of strawberry jam.

Overall total mass balance:

$$m_1 + m_2 = m_4 + 100$$

Substituting m_2 in terms of m_1 using the relation,

$$m_1 + 1.25m_1 = m_4 + 100$$

Substituting m_4 in terms of m_1 using water mass balance equation,

$$m_1 + 1.25m_1 = \left\{0.85m_1 - \frac{1}{3}(100)\right\} + 100$$

Rearranging and solving for m_1,

$$1.4m_1 = \frac{2}{3}(100), \quad m_1 = 47.6 \text{ g}$$

Substitute $m_1 = 47.6$ g in the following equation:

$$m_1 + 1.25m_1 = m_4 + 100$$

$$47.6 + 1.25(47.6) = m_4 + 100 \Rightarrow m_4 = 7.1 \text{ g}$$

Substitute $m_1 = 47.6$ g and $m_4 = 7.1$ g in the overall material balance equation:

$$m_1 + m_2 = m_4 + 100$$

$$47.6 + m_2 = 7.1 + 100 \Rightarrow m_2 = 59.5 \text{ g}$$

$$m_1 = 47.6 \text{ g}, \quad m_2 = 59.5 \text{ g}, \quad m_4 = 7.1 \text{ g}, \quad x_{s,3} = 0.0714$$

Example 3.9 Roasting of Cement Raw Materials

Problem
Cement is produced by roasting raw materials at high temperature in a rotating kiln. 1000 kg/min of raw material enters the kiln, producing 700 kg/min of cement. It is known that gaseous by-products are produced during roasting. From the data, determine the gases emission mass flow rate.

Solution

Known quantities: 1000 kg/min of raw material.

Find: Gases emission rates.

Analysis: The process-labeled flowchart is shown in Example Figure 3.9.1.

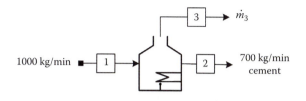

EXAMPLE FIGURE 3.9.1
Roasting of cement materials.

Basis: 1000 kg/min feed.

Assumption: Steady state, no reaction.

Material Balance

Total mass balance:

$$\dot{m}_1 = \dot{m}_2 + \dot{m}_3$$

$$1000\,\frac{kg}{min} = 700\,\frac{kg}{min} + \dot{m}_3 \Rightarrow \dot{m}_3 = 300\,\frac{kg}{min}$$

Example 3.10 Partial Vaporization

Problem

A liquid mixture of benzene and toluene contains 55.0% benzene by mass. The mixture is to be partially evaporated to yield a vapor containing 85.0% benzene and a residual liquid containing 10.6% benzene by mass. Suppose that the process is to be carried out continuously and under steady state, with a feed rate of 100.0 kg/h of the 55% mixture. Let V (kg/h) and L (kg/h) be the mass flow rates of the vapor and liquid product streams, respectively. Draw and label a process flowchart, and solve for unknown flow rates and compositions.

Solution

Known quantities: Feed rate and composition, vapor and liquid stream compositions.

Find: Draw and label a process flowchart and solve for unknown flow.

Analysis: The process-labeled flowchart is shown in Example Figure 3.10.1.

Basis: 100 kg/h of feed.

Assumption: Steady state, no reaction.

Material Balance

Total mass balance:

$$100 = \dot{V} + \dot{L}$$

Component balance (B):

$$0.55(100) = 0.85\dot{V} + 0.106\dot{L}$$

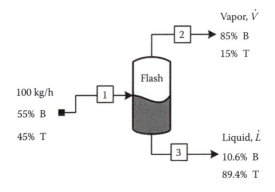

EXAMPLE FIGURE 3.10.1
Schematic of the benzene–toluene separation process.

Substitute L in terms of V from the total balance into the component balance equation and solve for V:

$$55 = 0.85\dot{V} + 0.106(100 - \dot{V})$$

$$\dot{V} = 59.68 \text{ kg/h}$$

$$\dot{L} = 40.32 \text{ kg/h}$$

Example 3.11 Cool Drying Process

Problem
A gas stream containing 40% O_2, 40% H_2, and 20 mol% H_2O is to be dried by cooling the stream and condensing out the water. If 100 mol/h of a gas stream is to be processed, what is the rate at which the water will be condensed out and what is the composition of dry gas?

Solution

Known quantities: Feed gas stream compositions and flow rate.

Find: Composition of dry gas and condensate rate.

Analysis: The process-labeled flowchart is shown in Example Figure 3.11.1.

Basis: 100 mol/h wet gas.

System: Dryer cooler.
We have three unknowns (D, C, x_2), which can be determined from the three available independent material balance equations (e.g., for O_2, H_2, and H_2O).
$DF = 3 - 3 = 0$ (problem is solvable). Consider the overall total material balance and tow component balances.

Total material balance:

$$100 = D + C$$

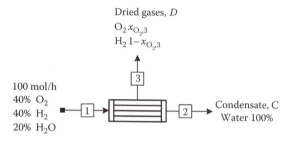

EXAMPLE FIGURE 3.11.1
Wet gas drying process.

Component balance (H_2O), all water in the feed is condensed:

$$0.2(100) = C$$

Component balance (O_2):

$$0.4(100) = x_{O_2,3}D$$

Solving the three material balance equations yields the result:

$$C = 20 \text{ mol/h}, \quad D = 80 \text{ mol/h}, \quad \text{and } x_{O_2,3} = 0.5$$

Example 3.12 Mixing of Binary Liquids

Problem

Pure streams of NaOH and water are mixed on a continuous basis to prepare an aqueous solution at 20 mol% NaOH. What is the rate of each stream required to prepare a 100 mol/h solution?

Solution

Known quantities: Pure NaOH and water are mixed to produce 100 mol/h of 20 mol% NaOH.

Find: The rate of each stream.

Analysis: The process-labeled flowchart is shown in Example Figure 3.12.1.

Basis: 100 mol/h of product stream (stream no. 3).

System: Mixer

Material Balance

There are two unknowns ($\dot{n}_1, \dot{n}_2$) and two independent equations; that is, $NDF = 0$ (problem is solvable).

Total overall material balance:

$$\dot{n}_1 + \dot{n}_2 = 100 \text{ mol/h}$$

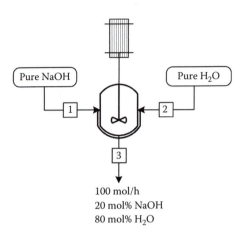

EXAMPLE FIGURE 3.12.1
Aqueous sodium hydroxide preparation process.

Component balance (NaOH):

$$\dot{n}_1 = 0.2(100) = 20 \text{ mol/h}$$

From the total balance equation,

$$20 \text{ mol/h} + \dot{n}_2 = 100 \text{ mol/h} \Rightarrow \dot{n}_2 = 80 \text{ mol/h}$$

Example 3.13 Extraction

Problem
Hexane is used to extract seed oil from cleaned cotton seed. Raw cotton seed consists of 15 wt% cellulose material, 35 wt% protein meal, and 50 wt% oil. Calculate the composition of oil extract obtained when 3 kg hexane is used per 1 kg mass raw seeds.

Solution

Known quantities: Raw seed composition, hexane/raw seed mass ratio.

Find: Composition of oil extract.

Analysis: The process-labeled flowchart is shown in Example Figure 3.13.1.

Basis: 100 kg of raw cotton seeds

Component balance:
  Protein meal: $0.35(100 \text{ kg}) = m_3$
  Cellulose: $0.15(100 \text{ kg}) = m_4$
  Oil: $0.50 (100 \text{ kg}) = x_{\text{oil},5} \, m_5$
  Hexane: $300 \text{ kg} = (1 - x_{\text{oil},5}) \, m_5$

Total mass balance:

$$100 \text{ kg} + 300 \text{ kg} = 35 \text{ kg} + 15 \text{ kg} + m_5$$

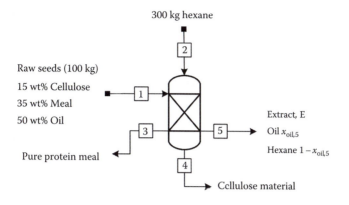

300 kg hexane

Raw seeds (100 kg)
15 wt% Cellulose
35 wt% Meal
50 wt% Oil

Pure protein meal

Extract, E
Oil $x_{oil,5}$
Hexane $1 - x_{oil,5}$

Cellulose material

EXAMPLE FIGURE 3.13.1
Schematic of the extraction process.

Mass of the extracted oil $= m_5 = 350$ kg

Substitute m_5 in the oil balance equation:

$$0.5(100 \text{ kg}) = x_{oil,5}\,350 \Rightarrow x_{oil,5} = 0.143$$

Example 3.14 Multicomponent Distillation Column

Problem

A feed rate of 100 mol/h of an equimolar mixture of ethanol (E), propanol (P), and butanol (B) is separated in a distillation column into two streams. The overhead stream (distillate) contains 2/3 ethanol and no butanol, and the bottom stream is free of ethanol. Calculate the rates and compositions of the overhead and bottom streams.

Solution

Known quantities: Feed flow rate and composition, distillate composition.

Find: Distillate rate and bottom flow rate and composition.

Analysis: The process-labeled flowchart is shown in Example Figure 3.14.1.

Basis: 100 mol/h of feed

Degrees of freedom analysis:
  Number of unknowns $= 3$ $(\dot{m}_d, \dot{m}_b, x_{P,3})$
  Number of independent equations $= 3$
  Number of relations $= 0$
  $DF = 3 - 3 = 0$

Material Balance

Total overall material balance:

$$100 \text{ mol/h} = \dot{m}_d + \dot{m}_b$$

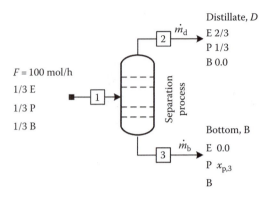

EXAMPLE FIGURE 3.14.1
Schematic of the tertiary-component separation process.

Component balance (E):

$$\frac{1}{3}(100\ \text{mol/h}) = \left(\frac{2}{3}\right)\dot{m}_d + 0.0 \Rightarrow \dot{m}_d = 50\ \text{mol/h}$$

From the total material balance,

$$100\ \text{mol/h} = 50\ \text{mol/h} + \dot{m}_b \Rightarrow \dot{m}_b = 50\ \text{mol/h}$$

Component balance (P):

$$\frac{1}{3}(100\ \text{mol/h}) = \left(\frac{2}{3}\right)\dot{m}_d + x_{P,3}\dot{m}_b$$

Substitute $\dot{m}_b$ and $\dot{m}_d$ in the P component balance:

$$\frac{1}{3}(100\ \text{mol/h}) = \frac{1}{3}(50\ \text{mol/h}) + x_{P,3}(50\ \text{mol/h})$$

Divide both sides of the equation by 50 to reduce to

$$\frac{1}{3}(2) = \frac{1}{3} + x_{P,3}$$

$$x_{P,3} = \frac{1}{3}$$

Example 3.15 Benzene–Toluene Separation Process

Problem

The feed to a distillation column contains 36% benzene (B) by weight, and the remainder toluene (T). The overhead distillate is to contain 52% benzene by weight, while the bottom is to contain 5% benzene by weight. Calculate the percentage of the benzene in the feed that is contained in the distillate, and the percentage of the total feed that leaves as distillate.

Solution

Known quantities: Compositions of feed, overhead, and bottom.

Find: Distillate and bottoms mass ratio.

Analysis: The process-labeled flowchart is shown in Example Figure 3.15.1.

Basis: 100 kg/h of feed to the column

Degrees of Freedom Analysis
Number of unknowns = 3 (flow rates of feed, overhead and bottom)
Number of independent equations = 2 (B, T)
Number of relations = 0 (no relations)
$NDF = 3 - 2 - 0 = 1$

Comment: Due to the problem being underspecified ($NDF = 1$), and since none of the streams flow rates are specified, a basis should be assumed.
Overall total material balance:

$$100 \text{ kg/h} = \dot{m}_d + \dot{m}_b$$

Component material balance:

$$B: 0.36(100 \text{ kg/h}) = 0.52\dot{m}_d + 0.05\dot{m}_b$$

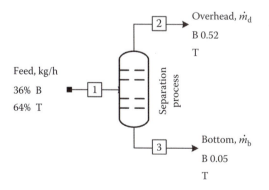

EXAMPLE FIGURE 3.15.1
Benzene–toluene separation process flow diagram.

Substitute $\dot{m}_b$:

$$0.36(100 \text{ kg/h}) = 0.52 \, \dot{m}_d + 0.05(100 - \dot{m}_d)$$

$$36 \text{ kg/h} = 0.52 \, \dot{m}_d + 5 - 0.05 \, \dot{m}_d \Rightarrow \dot{m}_d = 66 \text{ kg/h}$$

The bottom mass flow rate is obtained by substituting $\dot{m}_d$ in the total mass balance equation.

$$\text{Total balance}: 100 \text{ kg/h} = 66 \frac{\text{kg}}{\text{h}} + \dot{m}_b$$

The bottom mass flow rate, $\dot{m}_b = 34$ kg/h.

$$\text{The mass ratio of distillate to bottom mass flow rate} = \frac{\dot{m}_d}{\dot{m}_b} = \frac{66 \text{ kg/h}}{34 \text{ kg/h}}$$

Example 3.16 Evaporation Chamber

Problem

Three input streams are fed into an evaporation chamber to produce an output stream with the desired composition. Liquid water, fed at a rate of 20.0 cm³/min, air (21 mol% O_2, the balance N_2), and pure oxygen, fed at one-fifth of the molar flow rate of air stream. The output gas is analyzed and is found to contain 1.5 mol% water. Draw and label a flowchart of the process and calculate all unknown stream variables.

Solution

Known quantities: Feed stream composition and product water composition.

Find: All unknown stream variables.

Analysis: The process-labeled flowchart is shown in Example Figure 3.16.1.

Assumption: Steady state, no reaction

Basis: $20 \dfrac{\text{cm}^3}{\text{min}}$ of liquid water

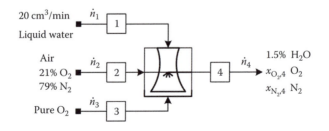

EXAMPLE FIGURE 3.16.1
Schematic of an evaporation chamber.

$$\text{Mass flow rate of water} = \dot{V} \times \rho = \left(20\ \frac{\text{cm}^3}{\text{min}}\right)\left(\frac{1\ \text{g}}{\text{cm}^3}\right) = 20\ \frac{\text{g}}{\text{min}}$$

$$\text{Mole flow rate of water} = 20\ \frac{\text{g}}{\text{min}} \div \frac{1}{M_w} = 20\ \frac{\text{g}}{\text{min}}\frac{1}{18\ \text{g/mol}} = 1.11\ \frac{\text{mol}}{\text{min}}$$

Relation:
Pure oxygen molar flow rate is one-fifth of the molar flow rate of air (stream 2):

$$n_3 = \frac{1}{5}(\dot{n}_2)$$

Material Balance

Component balance (water): $\dot{n}_1 = 0.015\dot{n}_4$

$$1.11\ \frac{\text{mol}}{\text{min}} = 0.015\dot{n}_4 \rightarrow \dot{n}_4 = 74.0\ \text{mol}$$

Total overall mole balance:

$$\dot{n}_1 + \dot{n}_2 + \dot{n}_3 = \dot{n}_4 \Rightarrow 1.11\ \frac{\text{mol}}{\text{min}} + \dot{n}_2 + \frac{1}{5}\dot{n}_2 = 74.0$$

$$\dot{n}_2 = 60.74$$

Component balance (N_2):

$$0.79\dot{n}_2 = x_{N_2,4}\dot{n}_4$$

$$0.79(60.74) = x_{N_2,4} \times 74.0$$

$$x_{N_2,4} = 0.648$$

Oxygen mole fraction in product stream:

$$x_{O_2,4} = 1 - 0.648 - 0.015 = 0.337$$

Check your answer against oxygen component balance:

$$0.21n_2 + \frac{1}{5}n_2 = 0.337n_4$$

After substitution: Substitute known quantities in the component balance equation for oxygen:

$$(0.21)(60.74) + \frac{1}{5}(60.74) = 0.337(74.0)$$

The values on both sides of the equation are identical:

$$25 = 25$$

Example 3.17 Ethanol–Methanol Separation Process

Problem
A 100 kg/h ethanol–methanol stream is to be separated in a distillation column. The feed has 40% ethanol and the distillate has 90% methanol. The flow rate of the bottom stream product is 40 kg/h. Determine the percentage of methanol in the bottom stream.

Solution

Known quantities: Feed flow rate and composition, distillate composition, bottom flow rate.

Find: The percentage of methanol in the bottom stream.

Analysis: The process-labeled flowchart is shown in Example Figure 3.17.1.

Basis: 100 kg/h of feed

Assumptions: Continuous process under steady state; accumulation $= 0$ No chemical reaction; generation $=$ consumption $= 0$

Degrees of Freedom Analysis
  Number of unknowns $= 2$ (D, x)
  Number of equations $= 2$, which equals the number of components

$$DF = \text{number of unknowns} - \text{number of equations} = 2 - 2 = 0$$

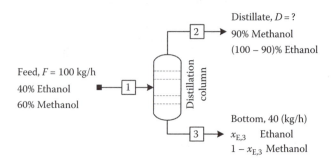

EXAMPLE FIGURE 3.17.1
Process flow sheet for methanol–ethanol.

Total mass balance:

$$100 \text{ kg/h} = D + 40 \text{ kg/h}$$

$$D = 60 \text{ kg/h}$$

Component balance (methanol):

$$0.6(100 \text{ kg/h}) = 0.9 (60 \text{ kg/h}) + (1 - x)\,40$$

$$x = 0.85$$

Example 3.18 Drying of Wet Polymeric Membrane

Problem

Two kilograms of wet polymeric hollow fiber membrane is to be dried by heating in a furnace. The wet fibers enter the drier with 1.5 kg H_2O per kilogram dry membrane (DM). The membrane is to be dried to 20% moisture. Determine the mass of water removed.

Solution

Known quantities: Feed rate, composition, product composition.

Find: The mass of water removed.

Analysis: The process-labeled flowchart is shown in Example Figure 3.18.1.

Material Balance

Total material balance:

$$2 \text{ kg} = m_3 + m_2$$

Component balance (M):

$$m_{M,1} = 0.8m_2$$

Feed content:

$$2 = m_{M,1} + m_{w,1}$$

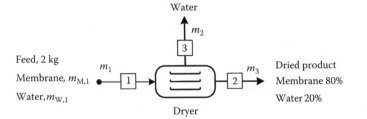

EXAMPLE FIGURE 3.18.1
Schematic of leather drying process.

The ratio of water to dry membrane in the feed stream is 1.5/1:

$$\frac{m_{w,1}}{m_{M,1}} = \frac{1.5}{1}$$

$$m_{w,1} = 1.5 \times m_{M,1}$$

Accordingly,

$$m_{M,1} = 0.8 \text{ kg}, \quad m_{w,1} = 1.2 \text{ kg}$$

Component balance dry membrane (M): 0.8 kg = 0.8 m_2
  The mass of dry product stream: $m_2 = 1.0$ kg
  Results: $m_2 = 1.0$ kg, $m_3 = 1.0$ kg

Homework Problems

3.1   Hundred kilograms per hour of an ethanol/methanol stream is to be separated in a distillation column. The feed consists of 40% ethanol and the distillate is made up of 90% methanol. Eighty percent of the methanol is to be recovered as distillate. Determine the percent methanol in the bottoms product. (26%)

3.2   Two hundred kilograms of wet leather is to be dried by heating. The wet leather enters the drier with 60% moisture and 90% of the water is removed. Determine the mass fraction of the moisture content of the dried leather. (0.13)

3.3   A feed stream to a distillation column, flowing at a rate of 1200 kg/h, contains a gas mixture of ethanol, methanol, and propanol. The feed contains 20% methanol. The distillate consists of 90% of the methanol and 60% of the ethanol contained in the feed. All of the 400 kg/h of propanol fed to the process ends up in the bottom of the column. Calculate bottom stream mass flow rate. (648 kg/h)

3.4   A 100 kg/h ethanol–water stream is to be separated in a distillation column. The feed stream contains 60 wt% ethanol and 40% water. The distillate contains 80% ethanol and the balance is water. Eighty percent of the ethanol in the feed stream is to be recovered as distillate. Determine the composition of ethanol in the bottom stream. (0.3)

3.5   A stream of aqueous hydrochloric acid, that is, 57.3 wt% HCl is mixed with pure water in a stirred vessel to produce a stream of 16.5% acid. What ratio of water to concentrated acid should be used? (2.47)

3.6   An absorber is used to remove acetone from a nitrogen carrier gas. The feed, with an acetone weight fraction of 0.213, enters at a rate of 200 kg/h. The absorbing liquid is water, which enters at a rate of 1000 kg/h. The exit gas stream is 0.8 wt% acetone and 2.9% water vapor. Determine the acetone mass fraction in the bottom stream. (0.04)

3.7 A 100 kg/h mixture containing equal parts by mass of benzene and toluene is distilled. The flow rate of the overhead product stream is 48.8 kg/h, and the bottom stream contains 7 wt% benzene. Draw and label a flowchart of the process. Calculate the mass fractions of benzene in the overhead product stream. (0.95)

3.8 A stream containing 25 wt% methanol in water is to be diluted with a second stream containing 10% methanol in water to form a product solution containing 17% methanol.

Draw and label a flowchart of this process, and calculate the ratio of the mass flow of the solution that consists of 17% methanol to that with 25% methanol. (0.875)

3.9 Liquid water and air flow into a humidification chamber where water evaporates completely. The air stream enters the chamber with a composition of 1.00 mol% H_2O and 20.8% O_2, the balance consists of N_2. The humidified air exits as humidifier with a composition of 10.0 mol% H_2O. Calculate the volumetric flow rate (cm^3/min) of water required to humidify 200 mol/min of the entering air. (360 cm^3/min)

3.10 A gas stream flowing at 100 mol/h contains 20 mol% oxygen, 70 mol% N_2, and 10 mol% H_2O is passed through a packed column with absorbent solids. The absorbents in the column absorb 97% of the water but none of the gases. Calculate the mole fraction of oxygen in the product gas. (0.22)

3.11 Wet sugar that contains 20% water is sent through a dryer where 75% of the water is removed. Taking l00 kg/h feed as basis, calculate the mass fraction of dry sugar in the wet sugar that leaves the dryer. Calculate the ratio of H_2O removed/kilogram wet sugar leaving the dryer. (0.18)

3.12 Hollow fiber membranes are a new technology used for separating gases. A porous membrane is used to concentrate the amount of oxygen in the product stream. The feed stream is air (21% O_2, 79% N_2). The concentrated stream is 30% O_2 and 70% N_2. If the total mole of waste stream is 75% of that of the input stream, the input stream is 100 mol. What is the molar composition of oxygen in the waste stream? (0.18)

3.13 A dilute 5 wt% caustic soda is to be concentrated 25 wt%. An evaporator is used for this purpose. Calculate the amount of water to be evaporated per 100 kg of feed. (80 kg)

3.14 A hydrocarbon feed consisting of a mixture of propane (20%), isobutene (30%), isopentane (20%), and *n*-pentane (30%) is fractionated at a rate of 100 kg/h into a distillate that contains all the propane and 78% of the isopentane in the feed. The mole fraction of isobutane in the distillate is 0.378. The bottom stream contains all the *n*-pentane fed to the unit. Determine the flow rate of the bottom stream. (42.765 kg/h)

3.15 Whole milk contains around 4.5% fat. Skimmed milk is prepared by the removal of some of the fat from whole milk. The skimmed milk is found to contain 0.1% fat. It is desired to prepare 100 kg of skimmed milk from original whole milk. Assuming that fat only was removed to make the skimmed milk, calculate the mass of the removed fat. (4.6 kg)

3.16 In the preparation of a soft drink, the total quantity of carbon dioxide required is equivalent to three volumes of gas to one volume of water at 0°C and atmospheric pressure. Ignoring all components other than CO_2 (assumed to be an ideal gas) and water, and based on 1000 kg of water (density of water 1000 kg/m^3), calculate the mole fraction of CO_2 in the drink. (2.41×10^{-3})

3.17 Raw potatoes are composed of about 18%–20% starch and approximately 75% water. Before frying, raw potatoes are peeled and then washed. Assume that 8% by weight of the raw potatoes is lost in the peeling. Potatoes are then dried to 7% total water content. What is the mass of dried potatoes produced from each 100 kg of raw potatoes? (24.73 kg)

3.18 Salted butter is produced in a continuous butter-making machine by adding slurry of salt to butter. The slurry contains 60% of salt and 40% of water by weight. If the final composition of the butter is to be 15.8% moisture and 1.4% salt, on a basis of 100 kg of produced butter, estimate the original percent moisture content of butter prior to salting, assuming original butter contains no salt. (15.2%)

3.19 In a flour mill, wheat is to be adjusted to a moisture content of 15%. If the whole grain received at the mill is found to contain 11.4% of water initially, how much water must the miller add, based on 100 kg of input grain as received, to produce the desired moisture content? (4.2 kg)

3.20 It is desired to prepare a sweetened concentrated orange juice. The initial pressed juice contains 5% of total solids, and it is desired to increase this amount to 10% of total solids by evaporation. Calculate the quantity of water that must be removed based on 100 kg of initial pressed juice. (50 kg)

References

1. Earle, R.L. (1983) *Unit Operations in Food Processing*, 2nd edn., Pergamon Common Wealth and International Library, Pergamon Press, Oxford, U.K.
2. Atkinson, B. and F. Mavituna (1991) *Biochemical Engineering and Biotechnology Handbook*, 2nd edn., Macmillan, Basingstoke, U.K.

3. Himmelblau, D.M. (1996) *Basic Principles and Calculations in Chemical Engineering*, 6th edn., Prentice-Hall, Upper Saddle River, NJ.
4. Reklaitis, G.V. (1983) *Introduction to Material and Energy Balances*, John Wiley & Sons, New York.
5. Felder, R.M. and R.W. Rousseau (1999) *Elementary Principles of Chemical Processes*, 3rd edn., John Wiley & Sons, New York.

4

Multiunit Process Calculations

The processes covered so far involved only single-unit systems; however, industrial processes rarely involve only one process unit. This chapter starts with describing the main features of a multiunit process. Next, the chapter extends the concepts of degrees of freedom analysis and material balance to a multiunit process without reaction. Finally, it defines some special streams commonly encountered in a multiunit process and explains how to properly choose subsystems so as to simplify the analysis of such processes. Many practical examples are discussed, which will help reinforce the concepts already learned.

Learning Objectives

1. Define a multiunit process and its main features (Section 4.1).
2. Apply the degrees of freedom analysis and material balance to solve multiunit process problems (Section 4.2).
3. Properly construct a process flowchart and label all known quantities and unknown quantities for a multiunit process involving recycle, bypass, makeup, and purge streams (Section 4.3).

4.1 Multiunit Process

A multiunit process is simply a process that contains more than one unit. A generic process is depicted in Figure 4.1. The process consists of one mixing point, two unit operations, and one splitter. In the process, one can define five subsystems delimited by a boundary (as explained in Chapter 3): the overall process system (S1), mixing point (S2), and unit one (S3), unit two (S4), and splitter (S5). The procedure for material balance calculations on single-unit processes can be readily extended to the solution of multiunit processes [1].

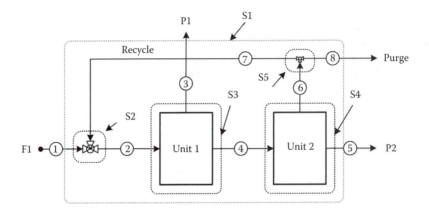

FIGURE 4.1
Multisystems process.

The material balance is then applied to each subsystem constituting the multiunit process being investigated and is expressed as

$$\text{Accumulation} = (\text{input} - \text{output}) + (\text{generation} - \text{consumption})$$

4.2 Degrees of Freedom Analysis

As explained in the previous section, the proper analysis of a multiunit process problem calls for the isolation of subsystems and application of the material balance to several subsystems of the process. The goal, again, is to determine all unknown process variables. For each system, it is strongly recommended that you perform a degrees of freedom analysis to find out if the problem at hand is solvable. Accordingly, you start by solving the material balance equations of the system that has the number of degrees of freedom equal to 0, that is, $NDF = 0$ [2].

Example 4.1 Degrees of Freedom Analysis for a Multiunit Process

Problem
Perform a degrees of freedom analysis for the process whose flowchart is shown in Example Figure 4.1.1.

Solution
The process consists of four subsystems: the overall process, column 1, mixing point, and column 2. For a stream to be fully specified, its flow rate and composition of all of its components must be known.

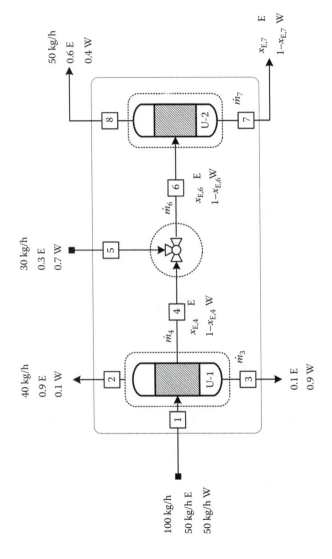

EXAMPLE FIGURE 4.1.1
Schematic diagram of multiunit process.

The degrees of freedom analysis is shown in the following table:

	Systems			
	Unit 1	Mixing	Unit 2	Process
Number of unknowns	2	4	4	3
Number of independent equations	2	2	2	2
Number of relations	—	—	—	—
NDF	0	2	2	1

The system with zero degree of freedom is the first system to start with, that is, either unit 1 or overall process, in this example.

4.3 Recycle, Bypass, Purge, and Makeup

Although we are not considering chemical reactions in this chapter, we will refer to them in this section to better explain the different types of special streams very common in multiunit processes. Consider the following chemical reaction taking place in a reactor [3,4]:

$$A \rightarrow B$$

Since it is rare for any chemical reaction to proceed to completion, some of A will remain in the product stream (Figure 4.2). This is not an ideal situation as some unreacted A is left in the product stream and the final product is not very pure in B.

4.3.1 Recycle

Recycling is common in chemical processes. There are several reasons for using a recycle stream, including recovery and reuse of unconsumed reactants, recovery of catalyst, dilution of a process stream, control of a process variable, and circulation of a working fluid. A process flow sheet with recycle stream is shown in Figure 4.3. Component A reacts to form a product B in an incomplete reaction. The remaining amount of A in the product stream is separated and recycled back to the reactor. When solving for the unknown process variables, write material balances (total material or components) around the entire process, the mixing point, the reactor, and the separator.

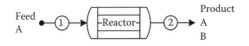

FIGURE 4.2
Reactor system.

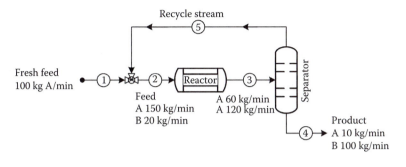

FIGURE 4.3
Process with recycle stream.

Pick the right subsystem ($NDF = 0$) to organize the problem for a sequential solution. In particular, the balance around the entire process system, terms describing the recycle (or bypass, defined next) stream do not appear in the equations; only the fresh feed and the product stream are considered [5].

4.3.2 Bypass

Bypass is defined as a fraction of the feed to a process unit that is diverted around the unit and combined with the output stream (Figure 4.4), thus affecting the composition and properties of the product. Processes involving bypass streams are treated in exactly the same manner as processes carrying recycle streams. The flow sheet is drawn and labeled, and, in this example, material balances are written down around the splitter, the process unit, or the mixing point downstream of the process unit.

4.3.3 Purge

A purge stream is a small stream bled off from a recycle loop to prevent buildup of inert or impurities in the system. Often, the purge flow rate is so much smaller than the recycle flow rate that it *can be neglected* in the steady-state overall material balance of the process [6]. The process flowchart with recycle and purge streams is shown in Figure 4.5.

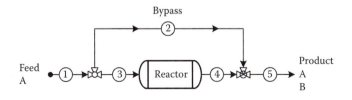

FIGURE 4.4
Process with bypass stream.

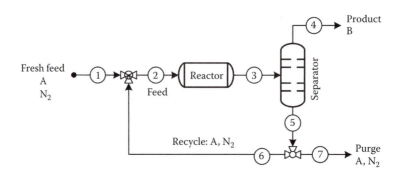

FIGURE 4.5
Process with purge stream.

4.3.4 Makeup

A makeup stream is required to replace losses to leaks, carryover, etc., within the recycle loop. Proper sizing and location of makeup and purge streams can prevent many problems within process plants. A process flow sheet with recycle and makeup stream is shown in Figure 4.6.

Example 4.2 Two Distillation Columns in Series

Two separation columns in sequence are used to separate the components of a feed consisting of 30% benzene (B), 55% toluene (T), and 15% xylene (X). The analysis of the overhead stream from the first column yields 94.4% B, 4.54% T, and 1.06% X. The second column is designed to recover 92% of the toluene in the original feed in its overhead stream at a composition of 0.946. The bottoms are intended to contain 92.6% of xylene in the original feed at a composition of 0.776. Compute the compositions of all streams.

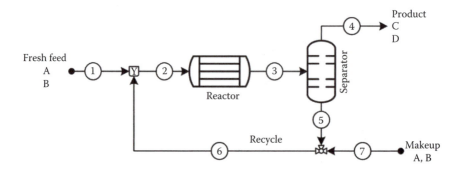

FIGURE 4.6
Process with makeup stream.

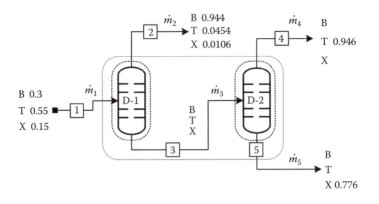

EXAMPLE FIGURE 4.2.1
Process with multiunits.

Solution

Known quantities: Composition of feed stream, top product of the first column, toluene and xylene compositions of the top and bottom streams of the second column are known.

Find: Flow rate and compositions of all streams.

Analysis: The process flowchart is shown in Example Figure 4.2.1.

The degrees of freedom analysis is summarized in the following table:

	System		
	D-1	D-2	Process
Number of unknowns	5	7	6
Number of independent equations	3	3	3
Number of relations	—	—	2
NDF	2	4	1

Since the lowest degree of freedom is one for the overall process, and none of the streams have known mass or molar flow rates, a basis should be considered. Accordingly, a basis of 100 kg/h is assumed, which reduces the degree of freedom of the overall process to zero.

Basis: 100 kg/h of fresh feed.

System: Overall process

Total mass balance:

$$100 = \dot{m}_2 + \dot{m}_4 + \dot{m}_5$$

Component mass balance:

$$B: \quad (0.3)(100) = (0.944)\dot{m}_2 + x_{B,4}\dot{m}_4 + x_{B,5}\dot{m}_5$$

T: $(0.55)(100) = (0.0454)\dot{m}_2 + (0.946)\dot{m}_4 + (1 - 0.776 - x_{B,5})\dot{m}_5$

X: $(0.15)(100) = (0.0106)\dot{m}_2 + (1 - 0.946 - x_{B,4})\dot{m}_4 + (0.776)\dot{m}_5$

Relations: The second column is designed to recover 92% of the toluene in the original feed in its overhead stream:

$$0.92 \times (0.55 \times 100) = 0.946\dot{m}_4 \rightarrow \dot{m}_4 = 53.5 \text{ kg/h}$$

The bottom stream of the second column is intended to recover 92.6% of xylene in the fresh feed:

$$(0.926)(0.15 \times 100) = 0.776\dot{m}_5 \rightarrow \dot{m}_5 = 17.90 \text{ kg/h}$$

Substitute $\dot{m}_4$ and $\dot{m}_5$ in the total mass balance equation $\rightarrow$ $\dot{m}_2 = 28.6$ kg/h.
Substitute $\dot{m}_2$, $\dot{m}_4$, and $\dot{m}_5$ in toluene (T) balance equation $\rightarrow$ $x_{B,5} = 0.051$.
Substitute $\dot{m}_2$, $\dot{m}_4$, and $\dot{m}_5$ in xylene (X) balance equation $\rightarrow x_{B,4} = 0.04$.

Example 4.3 Separation Column

Problem
Fresh feed containing 55 wt% A and 45 wt% B flowing at 100 kg/h enters a separator that removes a portion of pure component A only as a bottom product. The top product stream of the separator contains 10 wt% of component A and the balance is B. A small part of the separator product stream is recycled and joined in the fresh feed stream. The other portion is purged. The separator is designed to remove exactly two-thirds of component A fed to it (not the fresh feed). The recycle loop is used to achieve this goal. Compute all unknown stream flow rates and compositions.

Solution

Known quantities: Feed flow rate and composition and product composition.

Find: All unknown stream flow rates and compositions.

Analysis: The process flow sheet is shown in Example Figure 4.3.1.
The degrees of freedom analysis:

	Systems			
	Mixing	Separator	Splitter	Process
Number of unknowns	3	4	3	2
Number of independent equations	2	2	1	2
Number of relations	—	1	—	—
NDF	1	1	2	0

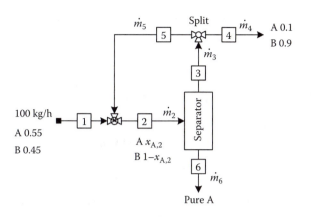

EXAMPLE FIGURE 4.3.1
Separation process.

Basis: 100 kg/h of fresh feed.

System: Overall process

Total mass balance:

$$100 = \dot{m}_6 + \dot{m}_4$$

Component balance (A):

$$0.55(100) = \dot{m}_6 + 0.1\dot{m}_4$$

Substituting m_6 from overall balance equation,

$$0.55(100) = (100 - \dot{m}_4) + 0.1\dot{m}_4$$

Solving for $\dot{m}_4$,

$$\dot{m}_4 = 50 \text{ kg/h}$$
$$\dot{m}_6 = 50 \text{ kg/h}$$

System: Separator.

Total mass balance:

$$\dot{m}_2 = 50 + \dot{m}_3$$

Component balance (A):

$$x_{A,2}\dot{m}_2 = 50 + \dot{m}_3(0.1)$$

Relation: Separator removed 2/3 of A fed to it in the bottom stream:

$$\left(\frac{2}{3}\right)(x_{A,2}\dot{m}_2) = \dot{m}_6 = 50 \text{ kg/h}$$

Rearranging $\rightarrow x_{A,2}\dot{m}_2 = 75 \text{ kg/h}$

From the component balance and the relation,

$$75 = 50 + \dot{m}_3(0.1)$$

$$\dot{m}_3 = 250 \text{ kg/h}$$

Substituting $\dot{m}_3 = 250 \text{ kg/h}$ into overall mass balance equation around the separator,

$$\dot{m}_2 = 50 + \dot{m}_3$$

$$\dot{m}_2 = 50 + 250, \quad \dot{m}_2 = 300 \text{ kg/h}$$

$$x_{A,2} = \frac{75 \text{ kg/h}}{\dot{m}_2} = \frac{75 \text{ kg/h}}{300 \text{ kg/h}} = 0.25$$

System: Splitter
Overall mass balance:

$$\dot{m}_3 = \dot{m}_4 + \dot{m}_5$$

$$250 = 50 + \dot{m}_5$$

$$\dot{m}_5 = 200 \text{ kg/h}$$

Example 4.4 Absorber Column

Problem
The raw feed to a sulfur removal process contains 15 wt% CO_2, 5% H_2S, and 1.5% CO, and the balance is CH_4. The original absorber design placed a maximum flow rate limit of 80 kg/h. The product stream of the whole process contains 1% H_2S, 0.3% CO, and the balance is carbon dioxide and methane. Any feed flow rate in excess of 80 kg/h is bypassed. The absorber absorbs hydrogen sulfide and carbon monoxide only. If the fresh feed to the unit is 100 kg/h, perform a degrees of freedom analysis and find the flow rates of the product streams.

Solution

Known quantities: Fresh feed flow rate and composition, absorber feed rate and composition, product gas composition of carbon monoxide, CO_2, and hydrogen sulfide, H_2S.

Find: Flow rates of the product streams.

Analysis: The process flow sheet is shown in Example Figure 4.4.1.

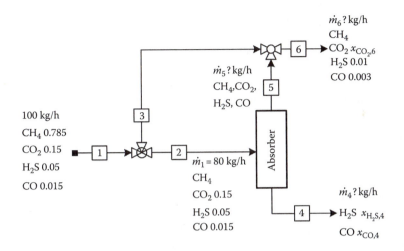

EXAMPLE FIGURE 4.4.1
Absorption process with bypass stream.

Assumptions: Steady state, physical process.

The degrees of freedom analysis is summarized in the following table:

	Systems		
	Absorber	Mixer	Process
Number of unknowns	6	6	4
Number of independent equations	4	4	4
Number of relations	0	0	0
NDF	2	2	0

Basis: 100 kg/h of fresh feed.

System: Overall process

Total material balance:

$$100 = \dot{m}_4 + \dot{m}_6$$

Component balances (CO_2):

$$0.15(100) = (x_{CO_2,6})\dot{m}_6 + 0$$

Component balances (H_2S):

$$0.05(100) = 0.01\dot{m}_6 + (x_{H_2S,4})\dot{m}_4$$

Component balances (CO):

$$0.015(100) = 0.003\dot{m}_6 + x_{CO,4}\dot{m}_4$$

$$x_{CO,4} = 1 - x_{H_2S,4}$$

Substitute $x_{CO,4}$ in the CO component balance equation:

$$0.015(100) = 0.003\dot{m}_6 + (1 - x_{H_2S,4})\dot{m}_4$$

Simplifying and adding to H_2S component balance equation, $x_{CO,4}$ is eliminated:

$$0.015(100) = 0.003\dot{m}_6 + \dot{m}_4 - x_{H_2S,4}\dot{m}_4$$

$$0.05(100) = 0.01\dot{m}_6 + x_{H_2S,4}\dot{m}_4$$

Adding the two equations,

$$1.5 + 5 = 0.013\dot{m}_6 + \dot{m}_4$$

Expressing $\dot{m}_4$ as a function of $\dot{m}_6$,

$$6.5 - 0.013\dot{m}_6 = \dot{m}_4$$

Substituting $\dot{m}_4$ in the total overall material balance,

$$100 = \dot{m}_4 + \dot{m}_6$$

$$100 = (6.5 - 0.013\dot{m}_6) + \dot{m}_6 \rightarrow 93.5 = 0.987\dot{m}_6$$

$$\dot{m}_6 = \frac{93.5}{0.987} = 94.73 \text{ kg/h}$$

Solving for $\dot{m}_4$,

$$\dot{m}_4 = 6.5 - 0.013(94.73) = 5.27 \text{ kg/h}$$

Note that the compositions of all components in streams 1, 2, and 3 (from the splitter) are identical. Therefore, only one material balance can be written for a splitter.

Example 4.5 Separation of Hexane and Pentane Mixture

Problem
Hexane (H) and pentane (P) are separated in a distillation column with a reflux ratio of 0.60 (reflux ratio = Reflux/Distillate = R/D). If the feed contains 50 wt% hexane, the distillate is 5wt% hexane, and the bottom stream is 96wt% hexane, determine the distillate, bottom, recycle, and overhead flow rates on the basis that the mass flow rate of the feed stream is 100 kg/h. Rework the problem given 100 kmol/h of feed rate.

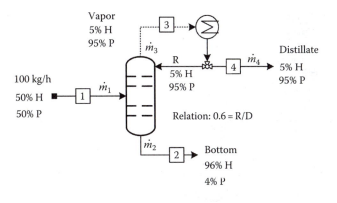

EXAMPLE FIGURE 4.5.1
Binary separation process.

Solution

Known quantities: Feed flow rate and its composition and product composition.

Find: The distillate, bottom, recycle, and overhead flow rates on the basis that mass flow rate of the feed stream is 100 kg/h. Rework the problem based on 100 kmol/h of feed rate.

Analysis: The process flow sheet is shown in Example Figure 4.5.1.

The degrees of freedom analysis is shown in the following table:

	Systems		
	Still	**Splitter**	**Process**
Number of unknowns	3	3	2
Number of independent equations	2	1	2
Number of relations	0	1	0
NDF	1	1	0

Zero degree of freedom indicates that this is where we should start, and the problem is uniquely defined.

Basis: 100 kg/h of feed.

System: Overall process system.

Total mass balance:

$$100 \text{ kg/h} = \dot{m}_4 + \dot{m}_2$$

Component balance (H):

$$50 \text{ kg/h} = 0.05 \times \dot{m}_4 + 0.96 \times \dot{m}_2$$

Rearranging and solving for B by substituting $\dot{m}_4$ from the overall total mass balance

$$50 \text{ kg} = 0.05 \times \left(100 - \dot{m}_2\right) + 0.96 \times \dot{m}_2$$

Simplifying

$$45 \text{ kg} = 0.91 \times \dot{m}_2$$

System: Splitter

Total balance:

$$\dot{m}_3 = R + D$$

Relation:

$$R/D = 0.6$$

Solving these two equations for the two unknowns by substituting R in terms of $\dot{m}_4$ from the relation $R = 0.6\dot{m}_4$,

$$R = 0.6\dot{m}_4 = 0.6(50.5) = 30.3 \text{ kg/h}$$

The overhead from the distillation column, $\dot{m}_3$, is given by

$$\dot{m}_3 = R + \dot{m}_4 = 30.3 + 50.5 = 80.8 \text{ kg/h}$$

Scale up:

Basis: Feed rate 100 kg/h.

Molar flow rate of feed stream ($\dot{n}_1$):

$$\text{Moles of hexane:} \quad \dot{n}_H = \left(\frac{50 \text{ kg H}}{86 \text{ kg H/kmol H}}\right) = 0.581 \text{ kmol/h}$$

$$\text{Moles of pentane:} \quad \dot{n}_P = \frac{50 \text{ kg P}}{72 \text{ kg P/kmol P}} = 0.694 \text{ kmol/h}$$

$$\dot{n}_1 = \dot{n}_H + \dot{n}_P = 1.276 \text{ kmol/h}$$

Overhead molar flow rate, $\dot{n}_3$:

$$(0.05)(80.8)/86 + (0.95)(80.8)/72 = 1.07 \text{ kmol/h}$$

Distillate molar flow rate, $\dot{n}_4$:

$$(0.05)(50.5)/86 + (0.95)(50.5)/72 = 0.70 \text{ kmol/h}$$

Bottom molar flow rate, $\dot{n}_2$:

$$(0.96)(49.5)/86 + (0.04)(49.5)/72 = 0.58 \text{ kmol/h}$$

Scale factor

$$(100 \text{ kmol F/h})/(1.276 \text{ kmol F/h}) = 78.4$$

Results after scaling up

$\dot{n}_1 = (1.276 \text{ kmol/h})(78.4) = 100 \text{ kmol/h}$
$\dot{n}_2 = (0.58 \text{ kmol/h})(78.4) = 45.5 \text{ kmol/h}$
$\dot{n}_3 = (1.07 \text{ kmol/h})(78.4) = 83.9 \text{ kmol/h}$
$\dot{n}_4 = (0.70 \text{ kmol/h})(78.4) = 54.9 \text{ kmol/h}$

The compositions splitter inlet and exist streams do not change. The exit streams mass flow rate can be different; accordingly, there is only one independent material balance equation, which is the total mass balance. This will always be true for a splitter.

Example 4.6 Evaporator–Crystallizer Unit

Problem
Fresh feed stream flowing at 100 kg/h contains 20% by weight KNO_3 (K) in H_2O (W). The fresh feed stream is combined with a recycle stream and is fed to an evaporator. The concentrated liquid solution exited the evaporator contains 50% KNO_3 is fed to a crystallizer. The crystals obtained from the crystallizer are 96% KNO_3 and 4% water. The liquid from the crystallizer constitutes the recycle stream and contains 0.6 kg KNO_3 per 1.0 kg of H_2O. Calculate all stream flow rate values and compositions.

Solution

Known quantities: Feed flow rate and composition, cake composition.

Find: All stream values and compositions.

Analysis: The labeled process flow sheet is shown in Example Figure 4.6.1.

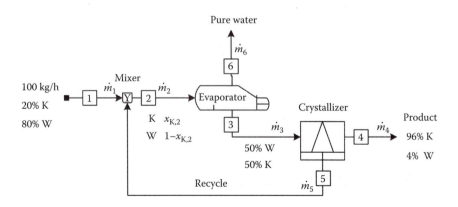

EXAMPLE FIGURE 4.6.1
Slurry filtration process.

The degrees of freedom analysis is shown in the following table:

	Systems			
	Evaporator	Crystallizer	Mixer	Process
Number of unknowns	4	4	3	2
Number of independent equations	2	2	2	2
Number of relations	0	1	0	0
NDF	2	1	1	0

Since the lowest NDF is for the entire process, accordingly, we start with the overall process system.

Basis: 100 kg/h feed.

System: Overall process

From the relation in the recycle stream, the mass fractions of K and W in the recycle stream can be calculated:

Mass fractions of K in recycle stream:

$$x_{K,5} = \left(\frac{0.6 \text{ kg K/h}}{1 \text{ kg W/h} + 0.6 \text{ kg K/h}} \right) = 0.375$$

Mass fraction of water in the recycle stream:

$$x_{W,5} = 1 - x_{K,5} = 1 - 0.375 = 0.625$$

Overall material balance:

$$100 \text{ kg/h} = \dot{m}_6 + \dot{m}_4$$

Component balance (K):

$$20 \frac{\text{kg}}{\text{h}} = (0.96)\dot{m}_4$$

$$\dot{m}_4 = 20.83 \text{ kg/h,}$$

Substitute in the total material balance to get the value of $\dot{m}_6$:

$$\dot{m}_6 = 79.17 \text{ kg/h}$$

Check your answer with W balance:

$$80 \text{ kg/h} = 79.17 \text{ kg/h} + (0.04)(20.83 \text{ kg/h}) = 80 \text{ kg/h}$$

System: Crystallizer.

Total material balance:

$$\dot{m}_3 = 20.83 + \dot{m}_5$$

Component balance (K):

$$0.5\dot{m}_3 = (0.96)(2083) + (0.375)\dot{m}_5$$

Substitute $\dot{m}_3 = 20.83 + \dot{m}_5$ in the component balance of K:

$$0.5(20.83 + \dot{m}_5) = (0.96)(20.83) + (0.375)\dot{m}_5$$

Solving for $\dot{m}_5$,

$$\dot{m}_5 = 76.65 \text{ kg/h} \quad \text{and} \quad \dot{m}_3 = 97.48 \text{ kg/h}$$

Now the NDF for the evaporator and the mixer has been reduced to zero, and either unit could be solved next.

System: Mixer.

Overall material balance:

$$100 \text{ kg/h} + 76.65 \text{ kg/h} = \dot{m}_2$$

$$\dot{m}_2 = 176.68 \text{ kg/h}$$

Component balance (K):

$$20 \text{ kg/h} + (0.375)\left(76.65 \text{ kg/h}\right) = x_{K,2}\left(176.65 \text{ kg/h}\right)$$

$$x_{K,2} = 0.276$$

Example 4.7 Absorber–Stripper Process Unit

Problem

The absorber–stripper system shown in Example Figure 4.7.1 is used to remove carbon dioxide and hydrogen sulfide from a feed consisting of 30% CO_2 and 10% H_2S in nitrogen. In the absorber, a solvent selectively absorbs hydrogen sulfide and carbon dioxide. The absorber overhead contains only 1% CO_2 and no H_2S. N_2 is insoluble in the solvent. The rich solvent stream leaving the absorber is flashed, and the overhead stream consists of 20% solvent, and contains 25% of the CO_2 and 15% of the H_2S from the raw feed to the absorber. The liquid stream leaving the flash unit is split into equal portions, one being returned to the absorber. The other portion, which contains 5% CO_2, is fed to the stripper. The liquid

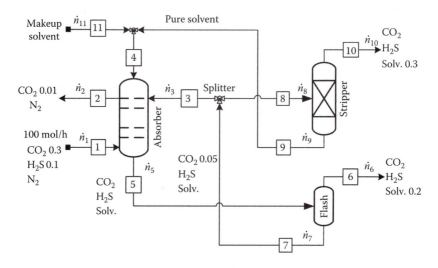

EXAMPLE FIGURE 4.7.1
Schematic of multiunit process.

stream leaving the stripper consists of pure solvent and is returned to the absorber along with *makeup* solvent. The stripper overhead contains 30% solvent. Calculate all flow rates and compositions of unknown streams. The gas feed rate is 100 mol/h.

Solution

Known quantities: Feed flow rate and composition.

Find: Flow rates and compositions of unknown streams.

Analysis: The labeled process flow sheet is shown in Example Figure 4.7.1.

Basis: Assume fresh feed molar flow rate is 100 mol/h (stream 1).

The degrees of freedom analysis is shown in the following table:

	Systems					
	Absorber	**Flash**	**Stripper**	**Splitter**	**Mixer**	**Process**
Number of unknowns	7	7	5	4	3	6
Number of independent equations	4	3	3	1	1	4
Number of relations	0	0	0	1	0	2
NDF	3	4	2	2	2	0

The lowest NDF value is of the overall process.

Basis: 100 mol/h of feed stream.

System: Overall process.

Material balance:

Total overall material balance:

$$100 + \dot{n}_{11} = \dot{n}_2 + \dot{n}_6 + \dot{n}_{10}$$

Component balances:

$$CO_2: \quad 0.3(100) = 0.01\dot{n}_2 + \dot{n}_{10}x_{CO_2,10} + \dot{n}_6 x_{CO_2,6}$$

$$H_2S: \quad 0.1(100) = 0 + n_{10}(1 - 0.3 - x_{CO_2,10}) + (1 - x_{CO_2,6} - 0.2)n_6$$

$$Solvent: \quad n_{11} = 0.2n_6 + 0.3n_{10}$$

Relation 1: The overhead stream contains 25% of the CO_2 in the raw feed $\rightarrow 0.25 \times (0.3 \times 100) = x_{CO_2,6}\dot{n}_6$.

Relation 2: The overhead stream contains 15% of the H_2S in the raw feed $\rightarrow 0.15 \times (0.1 \times 100) = (1 - x_{CO_2,6} - 0.2)\dot{n}_6$.

Add relation 1 and relation 2:

$$9 = 0.8\dot{n}_6, \quad \dot{n}_6 = 11.25 \text{ mol/h}$$

Substituting in relation 1, $x_{CO_2,6} = 0.67$.

From CO_2 component balance,

$$30 - 0.01\dot{n}_2 - 7.5 = \dot{n}_{10}x_{CO_2,10}$$

$$22.5 = \dot{n}_{10}x_{CO_2,10} + 0.01\dot{n}_2$$

From H_2S component balance,

$$8.5 = 0.7n_{10} - n_{10}x_{CO_2,10}$$

Add the last two equations:

$$31 = 0.7\dot{n}_{10} + 0.01\dot{n}_2$$

$$3100 - 70\dot{n}_{10} = \dot{n}_2$$

Substitute in the overall material balance equation:

$$100 + \dot{n}_{11} = \dot{n}_2 + \dot{n}_6 + \dot{n}_{10}$$

$$100 + 0.2\dot{n}_6 + 0.3\dot{n}_{10} = 3100 - 70\dot{n}_{10} + \dot{n}_6 + \dot{n}_{10}$$

Substituting the values,

$$100 + 2.25 + 0.3\dot{n}_{10} = 3100 - 70\dot{n}_{10} + 11.25 + \dot{n}_{10}$$

$$\dot{n}_{10} = 43.42 \text{ mol/h}$$

Final results:

$$\dot{n}_2 = 60.61, \quad \dot{n}_6 = 11.25, \quad \dot{n}_{10} = 43.42, \quad \dot{n}_{11} = 15.28,$$
$$x_{CO_2,10} = 0.504, \quad x_{CO_2,6} = 0.67$$

System: Stripper.
Total material balance:

$$\dot{n}_8 = \dot{n}_9 + 43.42$$

Component balance (CO_2):

$$CO_2: \quad 0.05 \times \dot{n}_8 = 0.504 \times \dot{n}_{10} = 0.504 \times 43.42$$

$$\dot{n}_8 = 437.67 \text{ mol/h}$$

$$\dot{n}_9 = 394.25 \text{ mol/h}$$

Component balance (solvent):

$$x_{s,8} \times \dot{n}_8 = 0.3 \times \dot{n}_{10} + \dot{n}_9$$

$$x_{s,8} \times 437.67 = 0.3 \times 11.25 + 394.25$$

Answer: $\dot{n}_8 = 437.67 \text{ mol/h}, \quad x_{s,8} = 0.91$
System: Splitter.
Total mole balance:

$$\dot{n}_7 = 437.472 + \dot{n}_3$$

Relation: The liquid stream leaving the flash unit is split into two equal portions → $\dot{n}_3 = \dot{n}_8 = 437.67 \text{ mol/h}$.

$$\dot{n}_7 = 437.472 + 437.67$$

This leads to $\dot{n}_7 = 875.14 \text{ mol/h}$.
System: Flash.
Total mole balance: $\dot{n}_5 = \dot{n}_6 + \dot{n}_7$

$$\dot{n}_5 = 11.25 + 875.14$$

$$\dot{n}_5 = 886.39 \text{ mol/h}$$

System: Absorber.

Total mole balance:

$$100 + \dot{n}_3 + \dot{n}_4 = \dot{n}_2 + \dot{n}_5$$

$$100 + 437.67 + \dot{n}_4 = 60.61 + 886.39$$

$$\rightarrow \dot{n}_4 = 409.33 \text{ mol/h.}$$

Example 4.8 Toluene–Xylene–Benzene Mixture

Problems
A fresh feed stream contains liquid mixture containing 35.0 mol% toluene (T), 27.0 mol% xylene (X), and the remainder benzene (B) is fed to a distillation column. The fresh feed molar flow rate is 100 kmol/h. The bottom product contains 97.0 mol% X and no B, and 93.0% of the xylene in the feed is recovered in this stream. The overhead product is fed to a second column. The overhead product from the second column contains 5.0 mol% T and no X, and 96.0% of the benzene fed to the system is recovered in this stream. Draw and completely label a flow sheet of the process. Calculate (i) the composition of the bottoms stream from the second column, and (ii) the percentage of toluene contained in the process feed that emerges in the bottom product from the second column.

Solution

Known quantities: Compositions of feed and bottom of the first column, and composition of the distillate of the second column.

Find: (i) The composition of the bottoms stream from the second column and (ii) the percentage of toluene contained in the process feed that emerges in the bottom product from the second column.

Analysis: The labeled process flow sheet is shown in Example Figure 4.8.1. Since no stream flow rates are specified, assume a basis.

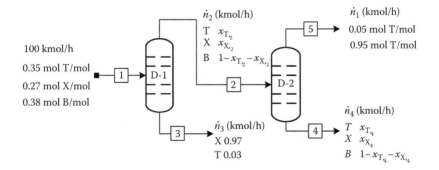

EXAMPLE FIGURE 4.8.1
Multiple component separation process.

Basis: Feed molar flow rate, $\dot{n}_f = 100 \text{ kmol/h}$.

Relations:

(1) 93% of feed X is recovered in the bottom stream of the first column → $0.93(100 \times 0.27) = \dot{n}_3(0.97)$

$$\dot{n}_3 = 25.89 \text{ kmol/h}$$

(2) 96% of feed B is recovered in the overhead stream of the second column → $(0.96)(100 \times 0.38) = \dot{n}_5(0.95)$

$$\dot{n}_5 = 38.40 \text{ kmol/h}$$

Total balances on the overall system:

Total material balance:

$$100 = \dot{n}_3 + \dot{n}_5 + \dot{n}_4$$

Solving for $\dot{n}_4$,

$$100 = \dot{n}_3 + \dot{n}_5 + \dot{n}_4$$

Substituting values of $\dot{n}_1$ and $\dot{n}_3$, then solving for $\dot{n}_4$ → $100 = 25.89 + 38.4 + \dot{n}_4$

$$\dot{n}_4 = 35.71 \text{ kmol/h}$$

Component balance:

$$\text{Benzene:} \quad 38 = 0.95 \, \dot{n}_5 + x_{B_4}\dot{n}_4$$

$$\text{Xylene:} \quad 27 = 0.97 \times \dot{n}_3 + x_{X_4}\dot{n}_4$$

In these two component balance equations, the variables $\dot{n}_3$, $\dot{n}_5$, and $\dot{n}_4$ are known and the remaining two variables are unknown. Solving for $x_{B,4}$,

$$\text{Benzene:} \quad 38 = 0.95 \times 38.40 + x_{B,4}\,35.71$$

$$\text{Xylene:} \quad 27 = 0.97 \times 25.89 + x_{X,4}\,35.71$$

$$x_{B,4} = 0.043 \text{ mol B/mol}$$

$$x_{X,4} = 0.053 \text{ mol X/mol}$$

The bottom stream from the second column contains 4.3 mol% benzene, 5.3 mol% xylene, and 90.4 mol% toluene.

Fraction of toluene recovered in the bottom stream

$$= \frac{(0.904)(35.71)}{35} = 92.3\%$$

Example 4.9 Benzene–Toluene Separation Process

Problem

A feed to a distillation column consisting of 50 wt% benzene and 50 wt% toluene is separated into two product streams by fractional distillation. The fresh feed stream flow rate is 100 mol/min. The vapor leaving the top of the column, which contains 97 wt% benzene, is completely condensed and split into two equal fractions: one is removed as the overhead product stream, and the other (the reflux) is recycled to the top of the column. The overhead product stream contains 89% of the benzene fed to the column. The liquid leaving the bottom of the column is fed to a partial reboiler in which 45% of the liquid is vaporized. The vapor generated in the reboiler (the boil up) is recycled to the column to become the rising vapor stream in the column, and the residual reboiler liquid is removed as the bottom product stream. The relation that governs the compositions of the streams leaving the reboiler is

$$y_B = 2.25x_B$$

where y_B and x_B are the mole fractions of benzene in the vapor and liquid streams, respectively. Draw and completely label a flowchart of the separation process. Calculate the molar amounts of the overhead and bottom products, and the mole fraction of benzene in the bottom product.

Solution

Known quantities: As per the schematic diagram.

Find: The molar amounts of the overhead and bottom products and the mole fraction of benzene in the bottom product.

Analysis: The process flow sheet is depicted in Example Figure 4.9.1.

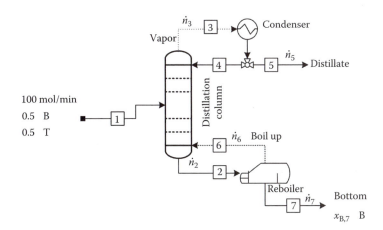

EXAMPLE FIGURE 4.9.1
Schematic of a distillation column with a reboiler and a total condenser.

The degrees of freedom analysis is summarized here:

	Systems			
	Condenser	Reboiler	Column	Overall
Number of unknowns	2	6	6	3
Number of independent equations	1	2	2	2
Number of relations	—	2	1	1
NDF	1	2	3	0

Calculate all flow rates and compositions of unknown streams, starting with the system whose NDF $= 0$.

Basis: 100 mol/min of fresh feed.

System: Overall process.

Total balance:

$$100 \text{ mol} = \dot{n}_5 + \dot{n}_7$$

Component balance (B):

$$0.5(100 \text{ mol}) = 0.97\dot{n}_5 + x_{B,7}\dot{n}_7$$

Auxiliary relation:
The overhead product stream contains 89% of the benzene fed to the column $\rightarrow 0.89(0.5 \times 100) = 0.97\dot{n}_5$
 We have three equations with three unknowns, which can be calculated as follows:
 From the auxiliary relation,

$$\dot{n}_5 = \frac{0.89(0.5 \times 100)}{0.97} = 45.88 \text{ mol/min}$$

Substituting n_5 in the overall material balance and solving for $\dot{n}_7$,

$$100 \text{ mol/min} = 45.88 + \dot{n}_7$$

$$\dot{n}_7 = 54.12 \text{ mol/min}$$

Substituting n_5 and n_7 in the component balance equation and solving for $x_{B,7}$,

$$0.5(100 \text{ mol}) = 0.97 \times 45.88 + x_{B,7}54.12$$

$$x_{B,7} = 0.10$$

System: Condenser.

The total material balance around the condenser is as given from the problem statement; the vapor leaving the top of the column, which contains 97 wt% benzene, is completely condensed and split into two equal fractions, $\dot{n}_4 = \dot{n}_5$. Hence,

$$\dot{n}_3 = \dot{n}_4 + \dot{n}_5 = 2\dot{n}_5$$

Calculating $\dot{n}_3$,

$$\dot{n}_3 = 2\dot{n}_5 = 2(45.88)$$

$$\dot{n}_3 = 91.76 \text{ mol/min}$$

System: Partial reboiler.

Total material balance for the reboiler:

$$\dot{n}_2 = \dot{n}_6 + \dot{n}_7$$

$$\dot{n}_2 = \dot{n}_6 + 54.12$$

Component balance (B):

$$z_{B,2}\dot{n}_2 = y_{B,6}\dot{n}_6 + 0.1(54.12)$$

Relation 1: The liquid leaving the bottom of the column is fed to a partial reboiler in which 45% of it is vaporized:

$$\dot{n}_6 = 0.45 n_2 \quad \text{and} \quad \dot{n}_7 = 0.55\dot{n}_2 \quad \text{where } \dot{n}_7 = 54.12 \text{ mol/min}$$

$$\dot{n}_7 = 0.55 n_2, \quad \dot{n}_2 = \frac{54.12}{0.55}$$

$$\dot{n}_2 = 98.4 \text{ mol/min}$$

$$\dot{n}_6 = 0.45\dot{n}_2 = 0.45 \times 98.4 = 44.28 \text{ mol/min}$$

Relation 2: The relation that governs the compositions of the streams leaving the reboiler is $y_{B,6} = 2.25 x_{B,7}$.

Substitute relation 1 into the overall material balance equation:

$$\dot{n}_7 = 0.45\dot{n}_2$$

$$\dot{n}_2 = 54.12 + \dot{n}_7$$

$$n_2 = 54.12 + 0.45 n_2$$

Rearranging and solving for n_2,

$$\dot{n}_2(1-0.45) = 54.12$$

$$\dot{n}_2 = \frac{54.12}{0.55} = 98.4 \text{ mol/min}$$

Solving for $\dot{n}_6$,

$$\dot{n}_6 = 0.45\dot{n}_2 = 0.45 \times 98.4 = 44.28 \text{ mol/min}$$

Calculate the composition around the reboiler using the equilibrium relation:

$$y_{B,6} = 2.25x_{B,7} = y_{B,6} = 2.25(0.10) = 0.225$$

Substitute known quantities in the benzene component balance around the reboiler:

$$z_{B,2}\dot{n}_2 = y_{B,6}\dot{n}_6 + 0.1(54.12)$$

Substituting the values of $\dot{n}_2$, $\dot{n}_6$, and $y_{B,6}$ obtained from the equilibrium relation,

$$z_{B,2}98.4 = 0.225 \times 44.28 + 0.1(54.12)$$

Calculate the benzene mole fraction of the liquid stream leaving the bottom of the column before entering the partial reboiler:

$$z_{B,2} = 0.156$$

Example 4.10 Absorber and Separation System

Problem

Contaminated air containing 3% acetone and 2% water is fed to an absorber column as shown in Example Figure 4.10.1. The mass flow rate of air is 100 kg/h. Pure water is used as an absorbent to absorb acetone from air. The air leaving the absorber should be free of acetone and contains 0.5% water. The bottom liquid of the absorber is fed to a distillation column to separate acetone from water. The bottom stream of the distillation column was found to contain 4% acetone and the balance is water. The vapor from the overhead of the distillation column was totally condensed and split into two portions, one portion is recycled to the column and the second fraction forms the distillate product stream. The concentration of the condensate is found to be 99% acetone. All percentages are in weight percent. Calculate the flow rate of all unknown streams.

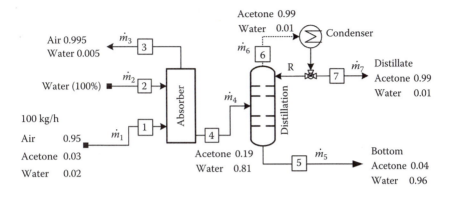

EXAMPLE FIGURE 4.10.1
Absorption–separation process.

Solution

Known quantities: Inlet and exit stream compositions of the absorber and distillation column are known.

Find: The flow rate of all unknown streams.

Analysis: Perform degrees of freedom analysis first, and then start solving the system with zero degree of freedom.

The degrees of freedom analysis is shown in the following table. The process consists of two subsystems and one overall system. The absorber is the system with zero degree of freedom.

	Systems		
	Absorber	Distillation	Process
Number of unknowns	3	3	4
Number of independent equations	3	2	3
Number of relations	—	—	—
NDF	0	1	1

Basis: 100 kg/h of contaminated air.

System: Absorber.

Total material balance:

$$\dot{m}_2 + 100 = \dot{m}_3 + \dot{m}_4$$

Component balance (air):

$$0.95(100) = 0.995\dot{m}_3 \rightarrow \dot{m}_3 = 95.5 \text{ kg/h}$$

Component balance (acetone):

$$0.03(100) = 0.19\dot{m}_4 \rightarrow \dot{m}_4 = 15.8 \text{ kg/h}$$

From the total balance equation around the absorber, $\dot{m}_2$ is calculated by substituting $\dot{m}_3$ and $\dot{m}_4$:

$$\dot{m}_2 + 100 = 95.5 + 15.8 \rightarrow \dot{m}_2 = 11.3 \text{ kg/h}$$

System: Distillation column.
Total material balance:

$$15.8 = \dot{m}_7 + \dot{m}_5$$

Component balance (acetone):

$$0.19(15.8) = 0.99\dot{m}_7 + 0.04\dot{m}_5$$

The earlier equations can be solved as follows:

$$\dot{m}_7 = 15.8 - \dot{m}_5$$

Substitute $\dot{m}_7$ in the acetone component mass balance:

$$0.19(15.8) = 0.99(15.8 - \dot{m}_5) + 0.04\dot{m}_5$$

Simplifying,

$$0.19(15.8) = 0.99 \times 15.8 - 0.99\dot{m}_5 + 0.04\dot{m}_5$$

$$0.95\dot{m}_5 = 12.63,$$

$$\dot{m}_5 = \frac{12.63}{0.95} = 13.3 \text{ kg/h}$$

$$\dot{m}_5 = 13.3 \text{ kg/h}, \quad \dot{m}_7 = 2.5 \text{ kg/h}$$

Homework Problems

4.1   The feed to a separation process contains 50 wt% benzene, 30 wt% toluene, and 20 wt% xylene. The feed enters the process at a flow rate of 100 kg/h. The overhead stream from the first unit contains 95 wt% benzene, 3 wt% toluene, and 2 wt% xylene. The overhead stream from the second unit contains 3 wt% benzene, 95 wt% toluene, and 2 wt% xylene. The flow rate of the overhead from the first unit is 52% of the fresh feed flow rate. The amount of benzene in the overhead from the second unit is 75% of the benzene that enters the second unit.

(a) Determine the flow rates for all of the streams leaving both units. (52, 15, and 33 kg/h)

4.2 It is proposed to separate $CaCO_3$ (solid) from slurry via crystallization. The feed slurry contains equal mass fractions of $CaCO_3$ precipitate in a solution of NaOH and H_2O. The slurry is washed with a dilute solution of 5 wt% NaOH in H_2O and of an equal flow rate as the slurry. The washed slurry, after leaving the unit, contains 2 kg solution per 1 kg of solid ($CaCO_3$). The clear solution (free of solid) withdrawn from the unit can be assumed to have the same concentration in NaOH and water as the solution withdrawn with the solid crystals. On the basis of 100 kg/h of feed slurry, draw and label the process flowchart and calculate the composition of sodium hydroxide in the clear solution. (0.23)

4.3 Fresh apple juice contains 75% water flowing at a mass flow rate of 200 kg/h. The fresh juice is concentrated in an evaporation process. A portion of the fresh feed is bypassed to meet the evaporator maximum capacity. The bypass stream joins the evaporator outlet stream. Ninety percent of the water entering the evaporator is evaporated. The concentrated apple juice contains 25% water. Draw the process flow sheet and calculate the mass flow rates of produced and evaporated streams. (66.7, 133.3 kg/h)

4.4 A solution containing 15.0 wt% NaCl, 5.0 wt% KCl, and 80 wt% water is fed to an evaporator–crystallizer process. The fresh feed stream mass flow rate is 1000 kg/h, the fresh feed is combined with the recycle stream, coming from the crystallizer stream, and enters an evaporator. In the evaporator, pure NaCl is collected at the bottom of the evaporator, and the top stream is pure water vapor. The evaporator outlet stream composition is 17 wt% NaCl and 22 wt% KCl, and 61% water enters the crystallizer unit. Pure KCl is collected as bottom product from the crystallizer. The crystallizer outlet stream is entirely recycled and joins the fresh feed stream. The recycle stream contains 20 wt% NaCl, the balance is water and KCl (i.e., both components are 80 wt%). The bottom products of the evaporator and the crystallizer consist of dry and pure NaCl and KCl, respectively. Draw the process flow sheet and calculate the mass fraction of KCl in the recycle stream. (0.082)

4.5 The raw feed to a sulfur removal system contains 15 mol% CO_2, 7% H_2S, and the balance is CH_4. The fresh feed to the unit is 100 kmol/h. The original absorber design places a maximum flow rate limit of 80 kmol/h, and yields a product stream with only 1% H_2S, and the balance is CO_2 and CH_4. The absorber removes H_2S only. The excess feed flow rate is bypassed and blended with the product stream. Draw and label the process flowchart. Find the CO_2 mole fraction in the absorber exit gas stream. (0.16)

4.6 A separator is designed to remove 75% of the CO_2 in flue gas (assume N_2) that is fed to adsorption unit, not the fresh feed. The CO_2

is removed from the bottom of the separator. The exit composition of the stream leaving the separator contains 10% CO_2, and the balance is nitrogen. A recycle stream is used, where part of the exit of the separator is recycled and unites with the fresh feed. Draw and label a process flowchart. Calculate the flow rate and compositions of all unknown streams.

4.7   A distillation column separates a mixture of 100 kg/h of 50% ethanol and the balance is water. The distillate at the top of the column was found to contain 95% ethanol, and the bottom products contain 96% water. The vapor flow rate from the top of the column is 80 kg/h. The condenser condenses vapor completely. A portion of the condenser product is recycled to the column as reflux. The rest of the condensate is withdrawn as product. Calculate the bottom stream mass flow rate. (49.45 kg/h)

4.8   A feed stream flowing at 1000 kg/h contains 20% KNO_3 and 80% H_2O. The stream enters an evaporator where part of water is evaporated and the stream exits with 50% KNO_3. The evaporator outlet stream enters a crystallizer. The bottom product stream of the crystallizer contains 4% H_2O and the balance is KNO_3. The outlet stream from the crystallizer is recycled and combined with the fresh feed stream. The recycle stream consists of 0.6 kg KNO_3/kg H_2O. Draw and label the process flowchart and calculate the recycle to fresh feed ratio. (0.767)

4.9   Thousand kilograms of soya beans are composed of 18% oil, 35% protein, 10.0% moisture, and the balance are carbohydrates, fiber, and ash. The feed is crushed and pressed, which reduces oil content in beans to 6%, and then extracted with hexane to produce a meal containing 0.5% oil, and finally dried to 8% moisture. Assuming that there is no loss of protein and water with the oil, determine the water loss during drying. (4.25 kg)

4.10   A process consists of two distillation columns. The composition of the feed to the first tower is 30.0 mol% benzene (B), 30.0 mol% toluene (T), and 40.0 mol% xylene (X). The overhead stream from the first tower contains no xylene (X). The benzene content of the overhead stream from the first tower is 55.0 mol%. The bottoms stream from the second tower has a flow rate of 100 mol/h and contains 15.0 mol% toluene and no benzene. While the overhead stream from the second tower contains 50.0 mol% benzene, there are also other components, T or X, or both, in this stream. The flow rate of the overhead stream from the first tower is twice the flow rate of the overhead stream from the second tower. Determine the molar flow rates of the fresh feed stream and the overhead stream leaving the first column. (228, 85.8 mol/h)

4.11   The feed stream to a processing system consists of 40 wt% A, 25 wt% B, and 35 wt% C. This stream enters the first process unit from which two streams emerge. The overhead stream from unit one contains

60 wt% A and no C. The bottom stream from unit one (stream 3) goes to a second unit. Two streams emerge from unit 2. The top stream contains 20 wt% B and no C. The bottom stream from unit 2 contains 30 wt% A, 20 wt% B, and 50 wt% C, and leaves unit 2 at a rate of 140 kg/h. Draw a flow chart for this system. Determine the flow rates for streams 1 through 4. (200, 50, 150, 10 kg/h)

4.12 A fresh feed stream flowing at a rate of 100 kg/min consists of the mass fractions 0.4 A, 0.1 B, and 0.5 C and enters an evaporator where the evaporator product stream leaving from the top of the evaporator contains 0.8 mol fraction A and the balance is B and C. Sixty percent of the evaporator product stream is recycled and combined with the fresh feed stream. The evaporator product stream consists of 0.3 mass fractions A and 0.125 mass fractions B, and the balance is C. Draw and label the process flow sheet and determine the flow rate and mass fraction of A in the evaporator feed streams. (220 kg/h, 0.35)

4.13 A fresh pressed juice contains 5% of total solids and it is desired to raise this percentage to 10% of total solids by evaporation and then to add sugar to give 2% of added sugar in the concentrated juice. Calculate the quantity of water that must be removed and of sugar that must be added with respect to each 100 kg of pressed juice. (50 kg, sugar added 1.02 kg)

References

1. Reklaitis, G.V. (1983) *Introduction to Material and Energy Balances*, John Wiley & Sons, New York.
2. Felder, R.M. and R.W. Rousseau (1978) *Elementary Principles of Chemical Processes*, John Wiley & Sons, New York.
3. Himmelblau, D.M. (1996) *Basic Principles and Calculations in Chemical Engineering*, 6th edn., Prentice-Hall, Upper Saddle River, NJ.
4. Whirwell, J.C. and R.K. Toner (1969) *Conservation of Mass and Energy*, Blaisdell, Waltham, MA.
5. Shaheen, E.I. (1975) *Basic Practice of Chemical Engineering*, Houghton Mifflin, Boston, MA.
6. Hougen, O.A., K.M. Watson, and R.A. Ragatz (1954) *Chemical Process Principles: Material and Energy Balances*, 2nd edn., John Wiley & Sons, New York.

5

Material Balances on Reactive Systems

Chapter 4 dealt with material balances on single- and multiunit systems with no chemical reactions. Normally, reactor is at the heart of chemical processes. In this chapter, first the concepts of stoichiometry as it relates to the material balance are discussed. Then, the application of the reactor in reactive systems is formulated. Next, the different types of balances are described, namely, the element material balance method and the molecular species method. As a special application, material balance on combustion systems is thoroughly discussed.

Learning Objectives

1. Write a balanced chemical reaction and use stoichiometry to determine the corresponding amounts of participants in a reaction (Section 5.1).

2. Understand the formulations of the material balance (Section 5.2).

3. Write balance equations based on the extent of reaction (Section 5.3).

4. Write balance equations involving atomic species (Section. 5.4).

5. Write balance equations using molecular species (Section 5.5).

6. Use of the extent of reaction for a system of chemical reactions (Section 5.6).

7. Use of component balance for a system of chemical reactions (Section 5.7).

8. Apply the degrees of freedom analysis for a reactive system (Section 5.8).

9. Define the features of combustion processes and properly apply material balances on them (Section 5.9).

5.1 Stoichiometry Basics

In the following, reaction stoichiometry is addressed in light of a few concepts that include the stoichiometric equation, stoichiometric coefficients, and stoichiometric ratios [1].

5.1.1 Stoichiometric Equation

It is an equation that relates the relative number of molecules or moles of participants (reactants and products) in a chemical reaction. To be valid, the equation must be balanced. For example, the following stoichiometric equation is not balanced:

$$C_2H_5OH + O_2 \rightarrow CO_2 + H_2O$$

The following equation is balanced because the number of atoms is the same on both sides of the equation (C, H, and O):

$$C_2H_5OH + 3O_2 \rightarrow 2CO_2 + 3H_2O$$

5.1.2 Stoichiometric Coefficients

These are the values preceding each molecular species, i, in a balanced stoichiometric equation. Values are defined as positive for products and negative for reactants. For the reaction,

$$2SO_2 + O_2 \rightarrow 2SO_3$$

$$v_{SO_2} = -2, \quad v_{O_2} = -1, \quad v_{SO_3} = 2$$

5.1.3 Stoichiometric Ratio

It is the ratio of stoichiometric coefficients in a balanced stoichiometric equation. Consider the oxidation of sulfur dioxide:

$$2SO_2 + O_2 \rightarrow 2SO_3$$

The following stoichiometric ratio is employed in solving material balance problems that involve this chemical reaction:

$$\frac{2 \text{ mol } SO_3 \text{ generated}}{1 \text{ mol } O_2 \text{ consumed}}$$

Two reactants, A and B, are in stoichiometric proportion, if the ratio (moles of A present)/(moles of B present) equals their stoichiometric ratio determined from the stoichiometric equation.

5.1.4 Limiting Reactant

A reactant is limiting if it is present in less than its stoichiometric proportion relatively to every other reactant. It is the reactant that would be the first to be consumed completely, if the reaction were complete. In order to find the limiting reactant, you balance the stoichiometric equation and then take the ratio of the reactant amount (mole, flow rate) in the feed to reactant stoichiometric coefficient, that is,

$$\frac{\dot{n}_i^o}{v_i} = \frac{\text{(molar flow rate of component } i \text{ in the feed)}}{\text{(stoichiometric cofficient of component } i)}$$

The ratio with the lowest value corresponds to the limiting reactant.

5.1.5 Excess Reactants

All reactants, other than the limiting species, are termed excess reactants. An excess reactant is not fully used up when the reaction is complete. The fractional excess is the ratio of the amount by which the feed exceeds stoichiometric requirements divided by the stoichiometric requirement. The fractional excess of the reactant is the ratio of the excess to the stoichiometric requirement:

$$\text{Fraction excess of A} = \frac{(n_A)_{\text{feed}} - (n_A)_{\text{stoich}}}{(n_A)_{\text{stoich}}} \text{ or } \left(= \frac{(\dot{n}_A)_{\text{feed}} - (\dot{n}_A)_{\text{stoich}}}{(\dot{n}_A)_{\text{stoich}}} \right)$$

where
   $(n_A)_{\text{feed}}$ is the amount (mole, flow rate) of an excess reactant, A, present in the feed to a reactor
   $(n_A)_{\text{stoich}}$ is the stoichiometric requirement of A, or the amount needed to react completely with the limiting reactant

Percentage excess of A is 100 times the fractional excess.

Example 5.1 Balance Chemical Reaction

Problem

1. How many moles of atomic hydrogen and oxygen would be released in 1 mol of H_2O if the latter were broken up into its constituent parts?
2. A water drop is 0.05 g. How many moles are there in the drop? What would be the mass of air in the same number of moles (Mw of air is 29)?
3. How many kilograms of H_2 that can be obtained by the electrolysis of 1 kg of water?

4. Balance the equation for glucose oxidation (i.e., determine α, β, and γ):

$$C_6H_{12}O_6 + \alpha O_2 \rightarrow \beta CO_2 + \gamma H_2O$$

Solution

1. The number of moles of atomic hydrogen and oxygen that would be released if 1 mol of water were broken up into its constituent parts can be calculated as follows. Consider 1 mol of water broken up into its constituent parts.

$$H_2O \rightarrow 2H + O$$

There are two hydrogen atoms and one oxygen atom in a water molecule. Thus, 1 mol of water will release 2 mol of atomic hydrogen and 1 mol of atomic oxygen.

2. The number of moles in an average rain drop contains 0.05 g water, which can be calculated considering the Mw of water (H_2O), that is, $(2 \times 1) + (1 \times 16) = 18$ g/mol. The number of moles is, therefore,

$$(n) = m/Mw = 0.05 \text{ g}/(18 \text{ g/mol}) = 2.78 \times 10^{-3} \text{ mol}$$

The mass of air in the same number of moles as that of water is calculated as follows:

$$2.78 \times 10^{-3} \text{ mol} \left| \frac{29 \text{ g}}{\text{mol}} \right. = 0.08 \text{ g of air}$$

3. Electrolysis is the use of electrical energy to turn water into H_2 and O_2. The relevant reaction is

$$H_2O \rightarrow H_2 + \frac{1}{2}O_2$$

Thus, 1 mol of H_2O yields 1 mol of H_2. The number of moles in 1 kg of H_2O is $n = m/Mw = 1$ kg/(18 kg/kmol). Thus, (1/18) kmol of H_2O yields (1/18) kmol of H_2. Mass of (1/18) kmol $H_2 =$ number of moles of H_2 multiplied by the Mw of hydrogen.

$$\text{Mass of } (1/18) \text{ kmol } H_2 = \frac{1}{18} \text{ kmol} \left| \frac{2 \text{ kg}}{1 \text{ kmol}} \right. = \frac{1}{9} \text{ kg}$$

4. β and γ are determined by balancing the number of each atom on both sides of the equation. In other words, the number of atoms on the left side of the equation must equal the number of atoms on the right side:

$$C_6H_{12}O_6 + \alpha O_2 \rightarrow \beta CO_2 + \gamma H_2O$$

$$C: 6 = \beta$$

$$H: 12 = 2\gamma \Rightarrow \gamma = 6$$

$$O: 6 + 2\alpha = 2\beta + \gamma, \quad \alpha = 6$$

Thus, the balanced equation is

$$C_6H_{12}O_6 + 6O_2 \rightarrow 6CO_2 + 6H_2O$$

Example 5.2 Limiting Reactant

Problem

For the following cases, determine which reactant is limiting and which is in excess as well as the percent excess for that component.

1. 2 mol of nitrogen (N_2) reacts with 4 mol of hydrogen (H_2) to form ammonia (NH_3) via the reaction:

$$N_2 + 3H_2 \rightarrow 2NH_3$$

2. 100 kg ethanol (C_2H_5OH) reacts with 100 kg of acetic acid (CH_3COOH) to form ethyl acetate:

$$C_2H_5OH + CH_3COOH \rightarrow CH_3COOHC_2H_5 + H_2O$$

3. 64 g of methanol (CH_3OH) reacts with 0.5 mol of oxygen (O_2) to form formaldehyde:

$$CH_3OH + \frac{1}{2}O_2 \rightarrow HCHO + H_2O$$

Solution

Known quantities: Reactant amount is known.

Find: Limiting reactant.

Analysis: Divide feed component flow rate to its stoichiometric coefficient, and the lower value is the limiting reactant.

1. The feed rate to stoichiometric ratio of both reactants is as follows:

$$N_2 + 3H_2 \rightarrow 2NH_3$$

$$\frac{2.0}{1} \quad \frac{4.0}{3}$$

This means that hydrogen (H_2) is the limiting reactant, since the ratio of hydrogen to its stoichiometric coefficient is lower than that of nitrogen. Accordingly, nitrogen is the component in excess.

The percent excess of nitrogen (N_2) can be calculated from

$$\% \text{ excess of } N_2 = \frac{(N_2)_{feed} - (N_2)_{stoich}}{(N_2)_{stoich}} \times 100\%$$

$$(N_2)_{stoich} = 4 \text{ mol } H_2 \left| \frac{1 \text{ mol } N_2}{3 \text{ mol } H_2} \right| = \frac{4}{3} \text{ mol } N_2$$

$$\% \text{ excess of } N_2 = \frac{\left(2 - \dfrac{4}{3}\right) \text{mol } N_2}{\dfrac{4}{3} \text{mol } N_2} \times 100\% = 50\%$$

2. Molecular weight of ethanol is 46 kg/kmol, acetic acid 60 kg/kmol, water 18 kg/kmol, and ethyl acetate 88 kg/kmol. First, you need to convert mass to mole. Thus,

$$C_2H_5OH: \quad 100 \text{ kg} \left(\frac{1 \text{ kmol}}{46 \text{ kg}}\right) = 2.17 \text{ kmol } C_2H_5OH$$

$$CH_3COOH: \quad 100 \text{ kg} \left(\frac{1 \text{ kmol}}{60 \text{ kg}}\right) = 1.67 \text{ kmol } CH_3COOH$$

The ratio of the feed rate to stoichiometric coefficient for each reactant is as follows:

$$C_2H_5OH + CH_3COOH \rightarrow CH_3COOHC_2H_5 + H_2O$$

$$\frac{2.17}{1} \qquad\qquad \frac{1.67}{1}$$

This means that acetic acid (CH_3COOH) is the limiting reactant. The component in excess is ethanol (C_2H_5OH). The percent excess of ethanol (C_2H_5OH) is

$$\% \text{ excess} = \frac{2.17 - 1.67}{1.67} \times 100\% = 30.0\%$$

3. Molecular weight (g/mol) data: $O_2 = 32$, $CH_3OH = 32$, $H_2O = 18$, $HCHO = 30$
First, convert masses to mole. Thus,

$$CH_3OH: \quad 64 \text{ g} \left(\frac{1 \text{ mol}}{32 \text{ g}}\right) = 2 \text{ mol } CH_3OH$$

The feed to stoichiometric ratio for each reactant is

$$CH_3OH + \frac{1}{2}O_2 \rightarrow HCHO + H_2O$$

$$\frac{2}{1} \qquad\qquad \frac{0.5}{1/2}$$

Oxygen (O_2) is the limiting reactant. The percent by which methanol is in excess is as follows:

The stoichiometric amount of methanol

$$(n_{CH_3OH})_{stoich} = (0.5 \text{ mol } O_2)_{feed} \left(\frac{1 \text{ mol } CH_3OH \text{ consumed}}{1/2 \text{ mol } O_2 \text{ consumed}} \right) = 1 \text{ mol } CH_3OH$$

The percent excess of methanol

$$\% \text{ excess of } CH_3OH = \frac{(2 \text{ mol})_{feed} - (1 \text{ mol})_{stoich}}{(1 \text{ mol})_{stoich}} \times 100\% = 100\%$$

5.1.6 Fractional Conversion

Chemical reactions do not occur instantaneously but rather proceed quite slowly. Therefore, it is not practical to design a reactor for the complete conversion of the limiting reactant. Instead, the reactant is separated from the reactor outlet stream and recycled back to the reactor inlet. The fractional conversion of a reactant A is the ratio of the amount reacted to the amount fed to the reactor:

$$f_A = \frac{(n_A)_{reacted}}{(n_A)_{in}} = \frac{(n_A)_{in} - (n_A)_{out}}{(n_A)_{in}}$$

The percentage conversion of component A is

$$\frac{(n_A)_{reacted}}{(n_A)_{in}} \times 100\% = \frac{(n_A)_{in} - (n_A)_{out}}{(n_A)_{in}} \times 100\%$$

If no component is specified, then the fractional conversion (f) is based on limiting reactant:

$$f = \frac{(n_{Lr})_{reacted}}{(n_{Lr})_{fed}} = \frac{(n_{Lr})_{in} - (n_{Lr})_{out}}{(n_{Lr})_{in}}$$

where n_{Lr} is the number of moles of limiting reactant.

5.2 General Material Balance

Mass balances are either integral mass balances or differential mass balances. An integral mass balance is a black box approach and focuses on the overall behavior of a system whereas a differential mass balances focuses on mechanisms within the system [2].

5.2.1 Differential Balance

A differential balance is a material balance at a given instant in time and deals with rates, that is, amount/unit time. Typically, a differential material balance may be written more precisely in a mathematical form as

$$\frac{dm}{dt} = \dot{m}_{in} - \dot{m}_{out} + \dot{G} - \dot{C}$$

where
　dm/dt denotes the rate of change of the material
　$\dot{G}$ and $\dot{C}$ denote the rate of generation and consumption, respectively

There are a couple of special cases:
　For the steady-state case,

$$0 = in - out + generation - consumption$$

$$0 = \dot{m}_{in} - \dot{m}_{out} + \dot{G} - \dot{C}$$

For the case without chemical reaction,

$$in = out; \quad \dot{m}_{in} = \dot{m}_{out}$$

5.2.2 Integral Balance

An integral balance deals with the entire time of the process at once and uses amounts rather than rates at steady state; none of the process variables change with time. For transient process that begins with time $= t_0$ and ends at a later time t_f, the general integral material balance equation is

$$m_{t_f} - m_{t_0} = \int_{t_0}^{t_f} \dot{m}_{in} \, dt - \int_{t_0}^{t_f} \dot{m}_{out} \, dt + \int_{t_0}^{t_f} \Re \, dT$$

where
　m_{t_f} is the mass of system content at final time
　m_{t_0} is the mass of system content at initial time
　$\Re$ is the generation/consumption term

This is significant for component with reactive systems only. The term is positive for produced material and negative if the material is consumed. The stoichiometric equation of the reaction imposes constraints on the

TABLE 5.1

Validity of "Input = Output" for a Steady-State Process

Type of Balance	Without Chemical Reaction	With Chemical Reaction
Total mass	Yes	Yes
Total moles	Yes	No
Mass of a chemical compound	Yes	No
Moles of a chemical compound	Yes	No
Moles of an atomic species	Yes	Yes

relative amounts of reactants and products in the input and output streams. The simple relation "input equals output" holds for steady-state processes under the circumstances explained in Table 5.1.

5.2.3 Formulation Approaches of Mass Balance

There are three approaches in the formulation of material balances with reaction, that is, the extent of reaction method, the atomic balance method, and the molecular species method. For multiple reactions, sometimes, it is more convenient to use the atomic balance approach. Generally, atomic species balances lead to the most straightforward solution procedure, especially when more than one reaction is involved. Molecular species balances require more complex calculations than either of the other two approaches and should be used only for simple systems involving one reaction. Each approach provides the same results, but one method may be more convenient than the other for a given problem [1,2].

5.3 Extent of Reaction Method for a Single Reaction

The extent of reaction (ξ or $\dot{\xi}$) is the amount (in moles or molar flow rate) of a species converted in a reaction divided by the species stoichiometric coefficient. Therefore, the extent of reaction is a quantity that characterizes the reaction since it is based on the stoichiometric equation. As such, the extent can be very useful in simplifying material balance calculations. The extent of reaction ξ (or $\dot{\xi}$) has the same units as n (or $\dot{n}$) divided by the moles (stoichiometric) reacting.

For a continuous process and single reaction at steady state:

$$\dot{n}_i = \dot{n}_i^\circ + v_i \dot{\xi}$$

where $\dot{n}_i^\circ$ and $\dot{n}_i$ are the molar flow rates of species i in the feed and outlet streams, respectively. For a batch process,

$$n_i = n_i^\circ + v_i \xi$$

where n_i° and n_i are the initial and final molar amounts of species i, respectively.

Example 5.3 Production of Ethylene Oxide

Problem

Ethylene oxide is produced by the reaction of ethylene with oxygen as per the following reaction:

$$2C_2H_4 + O_2 \leftrightarrow 2C_2H_4O$$

The feed to the reactor contains 5 mol ethylene, 3 mol oxygen, and 2 mol ethylene oxide. Draw and label the process flow sheet. Write the material balance equations as a function of the extent of reaction.

Solution

Known quantities: Inlet component molar amounts.

Find: Write material balance equations as a function of the extent of reaction.

Analysis: The process flow sheet is shown in Example Figure 5.3.1. Use the extent of reaction method.

From the definition of the extent of reaction for a single reaction,

$$\dot{n}_i = \dot{n}_i^\circ + v_i \dot{\xi}$$

Compound mole balance:

$$C_2H_4 : \quad n_{C_2H_4} = n_{C_2H_4}^\circ - 2\xi = 5 - 2\xi$$

$$O_2 : \quad n_{O_2} = n_{O_2}^\circ - \xi = 3 - \xi$$

$$C_2H_4O : \quad n_{C_2H_4O} = n_{C_2H_4O}^\circ + 2\xi = 2 + 2\xi$$

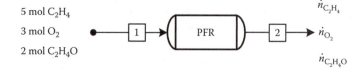

5 mol C_2H_4
3 mol O_2
2 mol C_2H_4O

EXAMPLE FIGURE 5.3.1
Production of ethylene oxide.

Total material balance is the sum of the component balance equations:

$$n = n^\circ - \xi = 10 - \xi$$

where

$$n^\circ = n^\circ_{C_2H_4} + n^\circ_{O_2} + n^\circ_{C_2H_4O}$$

5.4 Element or Atomic Balance Method

Element balances have no generation or consumption terms (atoms are not generated or consumed), and the mass balance is simplified to input equals output for continuous, steady-state processes. The element balance is based on the number of moles of that element regardless of the number of moles of the compound. The number of moles of each compound must be multiplied by the number of atoms of the element in the compound in order to obtain the number of moles of the element.

For instance, in the ethane dehydrogenation process (Figure 5.1), there are 2 mol of carbon atom for every mole of ethane.

The atomic balance for each element, C and H, is expressed as

$$C: \quad 2n^\circ_{C_2H_6} = 2n_{C_2H_4} + 2n_{C_2H_6}$$

$$H: \quad 6n^\circ_{C_2H_6} = 4n_{C_2H_4} + 6n_{C_2H_6} + 2n_{H_2}$$

5.5 Molecular or Component Balance Approach

When applying molecular or component balances, consumption and generation terms need to be considered according to the problem at hand. Therefore, the general mass balance for steady-state flow processes becomes

$$m_{acc} = 0 = \sum m_{in} - \sum m_{out} + m_{gen} - m_{cons}$$

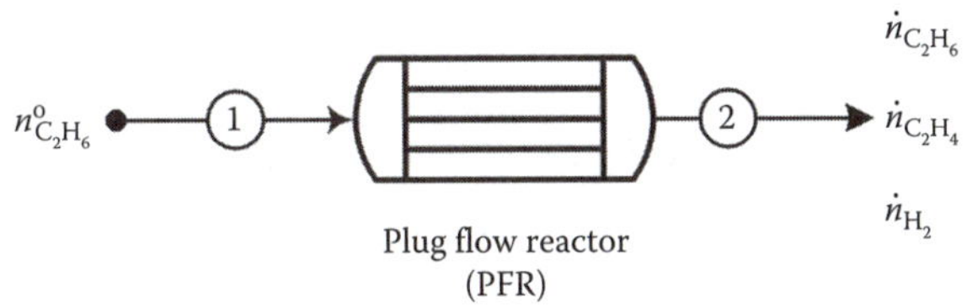

FIGURE 5.1
Schematic of plug flow.

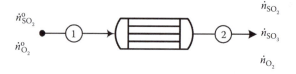

FIGURE 5.2
Schematic of combustion reactor.

In the oxidation of sulfur dioxide process, $SO_2 + \dfrac{1}{2}O_2 \rightarrow SO_3$, suppose that 5 mol/h of O_2 is consumed in the process shown in Figure 5.2.

From the stoichiometric equation and using molecular species balance, the number of moles of SO_3 generated is

$$(5 \; \cancel{\text{mol of } O_2 \text{ consumed}}/h) \dfrac{1 \, \text{mol/h } SO_3 \text{ generated}}{\dfrac{1}{2} \; \cancel{\text{mol } O_2 \text{ consumed}}/h}$$

$$= 10 \, \text{mol/h } SO_3 \text{ produced}$$

Using the molecular species balance, the number of moles of SO_3 leaving the reactor is

$$n_{SO_3} = 0 + 10 - 0 = 10 \, \text{mol}$$

Example 5.4 Extent of Reaction, Atomic Balance, and Molecular Species Balance Methods

Problem
Ammonia is burned to form nitric oxide and water:

$$4NH_3 + 5O_2 \rightarrow 4NO + 6H_2O$$

The fractional conversion of oxygen is 0.5. The inlet molar flow rate is 5 mol/h of NH_3 and 5 mol/h of oxygen. Calculate the exit component molar flow rates using the three methods:

a. Extent of reaction method
b. Atomic balance approach
c. Molecular species balance approach

Solution

Known quantities: Inlet molar flow rate of ammonia and oxygen.

Find: Exit component molar flow rate.

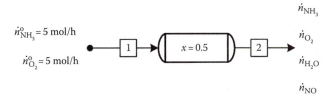

EXAMPLE FIGURE 5.4.1
Schematic of ammonia combustion reactor.

Analysis: The process flow sheet is shown in Example Figure 5.4.1.

Basis: All results are based on 1 h of operation.

a. *Extent of reaction method*

The material balance can be written using the extent of reaction method as follows:

$$n_i = n_i^o + v\xi$$

where $v_{NH_3} = -4, v_{O_2} = -5, v_{NO_2} = 4$, and $v_{H_2O} = 6$

Material balance of each component is then

$$NH_3: \quad n_{NH_3} = n_{NH_3}^o - 4\xi$$

$$O_2: \quad n_{O_2} = n_{O_2}^o - 5\xi$$

$$NO: \quad n_{NO} = n_{NO}^o + 4\xi$$

$$H_2O: \quad n_{H_2O} = n_{H_2O}^o + 6\xi$$

The total number of moles at the outlet of the reactor:

$$n = n^o + (-4 - 5 + 4 + 6)\xi = n^o + \xi$$

where

$$n^o = n_{NH_3}^o + n_{O_2}^o + n_{NO}^o + n_{H_2O}^o$$

The total material balance equation is

$$n = n^o + \xi$$

Inlet molar feed rates:

$$n_{NH_3}^o = 5 \, mol/h, \quad n_{O_2}^o = 5 \, mol/h, \quad n_{NO}^o = 0, \quad n_{H_2O}^o = 0$$

The reactor single pass conversion based on oxygen component is given by

$$\text{Conversion} = f = \frac{n_{O_2}^0 - n_{O_2}}{n_{O_2}^0}, \text{ substituting known quantities.}$$

$$0.5 = \frac{5 - n_{O_2}}{5}, \text{ the exit number of moles of oxygen is } n_{O_2} = 2.5 \text{ mol.}$$

Substituting n_{O_2} in oxygen component mole balance equation and solve for ξ

$$n_{O_2} = n_{O_2}^0 - 5\xi$$

$$2.5 = 5 - 5\xi$$

The extent of reaction is $\xi = 0.5$

Substituting the value of $\xi = 0.5$ and the initial molar flow rates of each component into components mole balance equations, the final results are then

$$n_{O_2} = 2.5 \text{ mol/h}, \quad n_{NH_3} = 3 \text{ mol/h},$$

$$n_{H_2O} = 3 \text{ mol/h}, \quad n_{NO} = 2 \text{ mol/h}$$

b. *Atomic balance approach*

Atomic balance on atoms involved in the reaction (N, O, H)—this is based on reactor inlet and outlet streams and not on the stoichiometry of the reaction equation:

$$\text{N}: \quad 5 = n_{NH_3} + n_{NO}$$

$$\text{O}: \quad 2(5) = 2(n_{O_2}) + n_{H_2O} + n_{NO}$$

$$\text{H}: \quad 3(5) = 3n_{NH_3} + 2n_{H_2O}$$

The single pass conversion, f_{O_2}

$$f_{O_2} = \frac{n_{O_2}^0 - n_{O_2}}{n_{O_2}^0} = 0.5 = \frac{5 - n_{O_2}}{5}, \quad n_{O_2} = 2.5 \text{ mol}$$

Substituting n_{O_2} in the O atomic balance and rearranging equations

$$5 = n_{NH_3} + n_{NO} \tag{1}$$

$$5 = n_{H_2O} + n_{NO} \tag{2}$$

$$15 = 3n_{NH_3} + 2n_{H_2O} \tag{3}$$

Subtracting Equation 2 from Equation 1 leads to

$$0 = n_{NH_3} - n_{H_2O}, \quad \text{hence, } n_{NH_3} = n_{H_2O}$$

Substitution of $n_{NH_3} = n_{H_2O}$ in Equation 3

$$15 = 3n_{NH_3} + 2n_{NH_3}$$

$$5n_{NH_3} = 15, \quad n_{NH_3} = 3 \text{ mol}$$

Since $n_{NH_3} = n_{H_2O}$, accordingly, $n_{H_2O} = 3$ mol
    Substitute n_{H_2O} in Equation 2 to get the value of n_{NO}

$$5 = n_{H_2O} + n_{NO}$$

$$5 = 3 + n_{NO}, \quad n_{NO} = 2 \text{ mol}$$

The final results are

$$n_{O_2} = 2.5 \text{ mol/h}, \quad n_{NH_3} = 3 \text{ mol/h}, \quad n_{NH_3} = 3 \text{ mol/h},$$

$$n_{H_2O} = 3 \text{ mol/h}, \quad n_{NO} = 2 \text{ mol/h}$$

c. *Molecular species approach*
    Molecular species balances can be done as follows:

The limiting reactant is oxygen:

$$\text{Conversion} = f = 0.5 = \frac{\text{Moles reacted}}{\text{Moles in the feed}} = \frac{\text{Mole reacted}}{5}$$

Moles of O_2 reacted $= 0.5 \times 5 = 2.5$ mol
    Moles of O_2 exiting the reactor $= 5 - 2.5 = 2.5$ mol

Moles of NH_3 consumed $= 2.5$ mol O_2 consumed

$$\times \frac{4 \text{ mol } NH_3 \text{ consumed}}{5 \text{ mol of } O_2 \text{ consumed}} = 2 \text{ mol}$$

Moles of NH_3 leaving the reactor $=$ in $-$ consumed $= 5 - 2 = 3$ mol

Moles H_2O generated $= 2.5$ mol O_2 consumed

$$\times \frac{6 \text{ mol } H_2O \text{ generated}}{5 \text{ mol of } O_2 \text{ consumed}} = 3 \text{ mol}$$

Moles of NO generated $= 2.5$ mol O_2 consumed

$$\times \frac{4 \text{ mol NO generated}}{5 \text{ mol of } O_2 \text{ consumed}} = 2 \text{ mol}$$

The final values of exit stream component molar flow rates are

$$n_{O_2} = 2.5 \text{ mol/h}, \quad n_{NH_3} = 3 \text{ mol/h},$$

$$n_{H_2O} = 3 \text{ mol/h}, \quad n_{NO} = 2 \text{ mol/h}$$

The extent of reaction method and molecular species balance require the specification of the stoichiometric equation. By contrast, the stoichiometric equation is not needed in atomic balance. All of the three methods lead to the same results.

Example 5.5 Extent of Reaction, Fractional Conversion, and Yield

Problem

Ethylene oxide (C_2H_4O) is produced by the reaction of ethylene (C_2H_4) with oxygen:

$$2C_2H_4 + O_2 \rightarrow 2C_2H_4O$$

The feed to the reactor contains 5 mol/h of ethylene, 3 mol/h of oxygen, and 2 mol/h of ethylene oxide. Set up the compound balances in general, and then use these to solve the cases suggested as follows:

a. Suppose the amount of C_2H_4 coming out of the reactor is 5 mol. How much reaction has occurred?
b. How much reaction has occurred if $\dot{n}_{C_2H_4} = 6 \text{ mol/h}$?
c. How much reaction has occurred if $\dot{n}_{C_2H_4} = 0 \text{ mol/h}$?
d. Which component is the limiting reagent?
e. What is the theoretical yield?
f. What is the fractional conversion if $\dot{n}_{C_2H_4} = 2 \text{ mol/h}$?

Solution

Known quantities: Feed component molar flow rates are known.

Find: The number of moles for each component leaving the reactor.

Analysis: The flow sheet of the process is shown in Example Figure 5.5.1.

Material balances:
   Based on following reaction:

$$2C_2H_4 + O_2 \rightarrow 2C_2H_4O$$

$\dot{n}^0_{C_2H_4} = 5 \text{ mol/h}$ $\dot{n}_{C_2H_4} = ?$

$\dot{n}^0_{O_2} = 3 \text{ mol/h}$ •——[1]→[Reactor]→[2]→ $\dot{n}_{O_2} = ?$

$\dot{n}^0_{C_2H_4O} = 2 \text{ mol/h}$ $\dot{n}_{C_2H_4O} = ?$

EXAMPLE FIGURE 5.5.1
Schematic of ethylene combustion reaction.

The extent of reaction method employed to find moles of exit components:

$$CH_4: \quad \dot{n}_{C_2H_4} = \dot{n}^o_{C_2H_4} - 2\dot{\xi}$$

$$O_2: \quad \dot{n}_{O_2} = \dot{n}_{O_2} - \dot{\xi}$$

$$C_2H_4O: \quad \dot{n}_{C_2H_4O} = \dot{n}^o_{C_2H_4O} + 2\dot{\xi}$$

Every term in these equations is expressed in terms of a single unknown, the extent of reaction ($\dot{\xi}$). If we know the moles in and out for any one component, we can solve for the extent of reaction and complete all of the material balances. Let us look at each of the specific cases using these balances:

a. From the C_2H_4 component balance, given that the exit molar flow rate of C_2H_4 is 5 mol/h, 5 mol/h = 5 mol/h – 2ξ, so ξ = 0 mol. Thus, no reaction has occurred.

b. From the balance for C_2H_4, given that the exit molar flow rate of C_2H_4 is 6 mol/h, 6 mol/h = 5 mol/h – 2ξ, or ξ = –0.5 mol. Thus, the reaction proceeds to the left by 0.5 mol/h.

c. From the balance for C_2H_4, given that the exit molar flow rate of C_2H_4 is 0 mol/h, 0 mol = 5 mol/h – 2ξ, or ξ = 2.5 mol/h. Thus, the reaction proceeds to the right by 2.5 mol/h.

d. To find the limiting reagent, we compare the ratio of moles of the reactants in the feed to their stoichiometric coefficients. In this case, $\dot{n}^o_{C_2H_4}$ = 5 mol/h and $\dot{n}^o_{O_2}$ = 3 mol/h, so

$$\frac{n^o_{C_2H_4}}{v_{C_2H_4}} = \frac{5}{2} < \frac{n^o_{O_2}}{v_{O_2}} = \frac{3}{1}$$

There is too little C_2H_4 relatively to O_2; therefore, C_2H_4 is the limiting reactant.

e. The theoretical yield is the maximum amount of product that could be produced if the reaction proceeded until the entire limiting reagent had been used up.

Because C_2H_4 is the limiting reagent, we use its material balance and plug in 0 mol/h for the outlet quantity. This refers to the preceding case(c) whereby ξ = 2.5 mol. We then use this value in the material balance for the product to find the theoretical maximum yield, the molar flow rate of C_2H_4O produced due to reaction.

$$\dot{n}_{C_2H_4O} = 2\xi = 2(2.5)$$

$$\dot{n}_{C_2H_4O} = 5 \text{ mol/h}$$

f. The fractional conversion when $\dot{n}_{C_2H_4} = 2$ mol/h is calculated as follows:

Again we use the limiting reagent to determine the fractional conversion. The fractional conversion is defined by

$$f = \frac{(n_{Lro} - n_{Lr})}{n_{Lro}} = \frac{\text{moles limiting used}}{\text{moles limiting in the feed}} = \frac{(5-2)}{5} = \frac{3}{5} = 0.6$$

The fractional conversion will often be given as a subsidiary relation in material balances involving reactions. If the fractional conversion is known, then one can use it to determine the amount of the limiting reagent leaving the reactor:

$$\dot{n}_{Lr} = (1-f) \times n_{Lro}$$

For example, if $f=0.6$, then $\dot{n}_{Lr} = (1-0.6)(5\text{ mol}) = 2$ mol/h, as is the preceding case.

Example 5.6 Acrylonitrile Production

Problem
The feed contains 10.0 mol% propylene (C_3H_6), 12.0 mol% ammonia (NH_3), and 78.0 mol% air. A fractional conversion of 30.0% of the limiting reactant is achieved. Taking 100 mol of feed as a basis, determine which reactant is limiting, the percentage by which each of the other reactants is in excess, and the molar amounts of all product gas. Use all methods of solution. Acrylonitrile (C_3H_3N) is produced by the reaction of propylene, ammonia, and oxygen:

$$C_3H_6 + NH_3 + \frac{3}{2}O_2 \rightarrow C_3H_3N + 3H_2O$$

Solution

Known quantities: Feed molar flow rate and inlet stream component molar percent.

Find: The number of moles for each component leaving the reactor.

Analysis: The process flow sheet is shown in Example Figure 5.6.1. Use the element balance, component balance, and extent of reaction methods.

Basis: 1 h of operation.

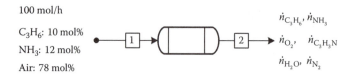

100 mol/h

C$_3$H$_6$: 10 mol%
NH$_3$: 12 mol%
Air: 78 mol%

$\dot{n}_{C_3H_6}, \dot{n}_{NH_3}$
$\dot{n}_{O_2}, \dot{n}_{C_3H_3N}$
$\dot{n}_{H_2O}, \dot{n}_{N_2}$

EXAMPLE FIGURE 5.6.1
Schematic of acrylonitrile production reactor.

System: Reaction vessel.

The limiting reactant is found by determining the lowest n_i/v_i ratio:

$$\frac{10 \text{ mol}}{1} : \frac{12 \text{ mol}}{1} : \frac{0.78(100)0.21}{3/2}$$

$$10: 12: 10.9$$

The limiting reactant is propylene. Consequently, the percent excess of ammonia is

$$\%NH_3 \text{ in excess} = \frac{n_{\text{feed}} - n_{\text{stoich}}}{n_{\text{stoich}}} = \frac{12-10}{10} \times 100\% = 20\%$$

The percent excess of oxygen:

$$\%O_2 \text{ in excess} = \frac{n_{\text{feed}} - n_{\text{stoich}}}{n_{\text{stoich}}} = \frac{16.38-15}{15} \times 100\% = 9.2\%$$

Extent of reaction method

$$C_3H_6 + NH_3 + \frac{3}{2}O_2 \rightarrow C_3H_3N + 3H_2O \quad \xi$$

Using the extent of reaction method for single reaction, $n_i = n_i^o + v_i \xi$

$$C_3H_6: \quad n_{C_3H_6} = 10 - \xi$$

$$NH_3: \quad n_{NH_3} = 12 - \xi$$

$$O_2: \quad n_{O_2} = (0.21 \times 78) - \frac{3}{2}\xi$$

$$C_3H_3N: \quad n_{C_3H_3N} = 0.0 + \xi$$

$$H_2O: \quad n_{H_2O} = 0.0 + 3\xi$$

Conversion of the limiting reactant (C_3H_6):

$$0.3 = \frac{10 - n_{C_3H_6}}{10}, \quad n_{C_3H_6} = 7 \text{ mol}$$

Substituting $n_{C_3H_6}$ in C_3H_6 component balance equation

$$7 = 10 - \xi, \quad \xi = 3$$

Substitute $\xi = 3$ in the rest of component mole balance equations to get the number of moles of all components leaving the reactor, the values supposed to be as follows:

$$n_{C_3H_6} = 7 \text{ mol}, \quad n_{NH_3} = 9 \text{ mol}, \quad n_{O_2} = 11.88 \text{ mol}, \quad n_{C_3H_3N} = 3 \text{ mol}$$

Atomic balance approach

In the atomic balance method, there is no need for the stoichiometric equation. The atomic balance degrees of freedom analysis is summarized as follows:

DFA	System: Reactor
Number of unknowns	6 ($n_{C_3H_6}$, n_{NH_3}, n_{O_2}, $n_{C_3H_3N}$, n_{H_2O}, n_I)
Number of independent equations	4 (C, H, O, N)
Number of relations	2 (conversion, n_I is inert)
Number degree of freedom $= 6 - 4 - 2$	0

Setting up the atomic balance equations:
Input = output

$$C: \quad 3 \times 10 = 3n_{C_3H_6} + 3n_{C_3H_3N}$$

$$H: \quad 6(10) + 3(12) = 6n_{C_3H_6} + 3n_{C_3H_3N} + 2n_{H_2O} + 3n_{NH_3}$$

$$O: \quad 2 \times 0.21(78) = 2n_{O_2} + n_{H_2O}$$

$$N: \quad 2 \times 0.79(78) + 12 = n_{NH_3} + n_{C_3H_3N} + 2n_I$$

The single pass conversion:

$$f = 0.3 = \frac{10 - n_{C_3H_6}}{10}, \quad n_{C_3H_6} = 7 \text{ mol}$$

Substitute $n_{C_3H_6}$ in C atomic balance:

$$30 = 3 \times 7 + 3n_{C_3H_3N}, \quad n_{C_3H_3N} = 3 \text{ mol}$$

Substitute $n_{C_3H_3N} = 3$ mol and $n_{N_2} = 0.79 \times 78 = 61.62$ mol:

$$2 \times 0.79(78) + 12 = n_{NH_3} + 3 + 2 \times 61.62$$

Solving for n_{NH_3}

$$2 \times 0.79(78) + 12 = n_{NH_3} + 3 + 2 \times 61.62$$

$$n_{NH_3} = 9 \text{ mol}$$

Solving the rest of equations gives the following results (in mol):

$$n_{C_3H_6} = 7, \quad n_{NH_3} = 9, \quad n_{O_2} = 11.88, \quad n_{C_3H_3N} = 3, \quad n_{H_2O} = 9, \quad n_{N_2} = 61.62$$

Molecular species balance approach

In this method, we again have to use the stoichiometric equation.

Accumulation = (input − output) + (generation − consumption)

$$\text{Conversion}: \quad 0.3 = \frac{10 - n_{C_3H_6}}{10}, \quad n_{C_3H_6} = 7 \text{ mol}$$

Accordingly, the amount of C_3H_6 consumed $= 10 - 7 = 3$ mol

Molecular balance (NH_3):

$$0.0 = input - output + generation - consumption$$

$$0.0 = 12 \text{ mol} - n_{NH_3} + 0.0 - \left(\frac{1 \text{ mol } NH_3 \text{ consumed}}{1 \text{ mol } C_3H_6 \text{ consumed}} \right) \times 3 \text{ mol } C_3H_6 \text{ consumed}$$

$$n_{NH_3} = 9 \text{ mol}$$

Molecular balance (O_2):

$$0.0 = in - out + generation - consumption$$

$$0.0 = 0.21(78) \text{ mol in} - n_{O_2} + 0.0 - \left(\frac{3/2 \text{ mol } O_2 \text{ consumed}}{1 \text{ mol } C_3H_6 \text{ consumed}} \right)$$

$$\times 3 \text{ mol } C_3H_6 \text{ consumed}$$

$$n_{O_2} = 11.88 \text{ mol}$$

Molecular balance (C_3H_3N):

$$0.0 = in - out + generation - consumption$$

$$0.0 = 0.0 - n_{C_3H_3N} + \left(\frac{1 \text{ mol } C_3H_3N \text{ generated}}{1 \text{ mol } C_3H_6 \text{ consumed}} \right) \times 3 \text{ mol } C_3H_6 \text{ consumed} - 0.0$$

$$n_{C_3H_3N} = 3 \text{ mol}$$

Molecular balance (H_2O):

$$0.0 = in - out + generation - consumption$$

$$0.0 = 0.0 - n_{H_2O} + \left(\frac{3 \text{ mol } H_2O \text{ generated}}{1 \text{ mol } C_3H_6 \text{ consumed}} \right) \times 3 \text{ mol } C_3H_6 \text{ consumed} - 0.0$$

$$n_{H_2O} = 9 \text{ mol}$$

5.6 Extent of Reaction and Multiple Reactions

Generally, the synthesis of chemical products does not involve a single reaction but rather multiple reactions. For instance, the goal would be to maximize the production of the desirable product and minimize the production of unwanted by-products. For example, ethylene is produced by the dehydrogenation of ethane:

$$C_2H_6 \rightarrow C_2H_4 + H_2$$

$$C_2H_6 + H_2 \rightarrow 2CH_4$$

$$C_2H_4 + C_2H_6 \rightarrow C_3H_6 + CH_4$$

Yield and selectivity are used to describe the degree to which a desired reaction predominates over competing side reactions in a multireaction system. Yield has various definitions:

$$\text{Yield} = \frac{\text{moles of desired product formed}}{\text{moles formed if there were no side reactions and limiting reactant reacted completely}}$$

$$\text{Yield} = \frac{\text{moles of desired product formed}}{\text{moles of reactant fed}}$$

$$\text{Yield} = \frac{\text{moles of desired product formed}}{\text{moles of reactant consumed}}$$

The selectivity of a component is the number of moles of desired component to the number of moles of undesired component.

$$\text{Selectivity} = \frac{\text{moles of desired product formed}}{\text{moles of undesired product formed}}$$

The concept of extent of reaction can also be applied to multiple reactions, with each reaction having its own extent. If a set of reactions takes place in a batch or continuous, steady-state reactor, we can write the following equation for species i:

$$n_i = n_i^o + \sum_j v_{ij}\xi_j$$

where
  v_{ij} is the stoichiometric coefficient of substance i in reaction j
  ξ_j is the extent of reaction for reaction j

For a single reaction, the preceding equation reduces to the equation reported previously in Section 5.3.

Example 5.7 Ethane Combustion

Problem

The following reactions take place in a reactor, where CO is the undesired product:

$$C_2H_6 + 3.5O_2 \rightarrow 2CO_2 + 3H_2O$$

$$C_2H_6 + 2.5O_2 \rightarrow 2CO + 3H_2O$$

The feed to reactor consists of 100 mol C_2H_6 and 500 mol O_2. The product stream was analyzed and found to contain 20 mol C_2H_6, 120 mol CO_2, 40 mol CO, 240 mol O_2, and 240 mol H_2O. The reactor conversion is 80%. Calculate the yield and selectivity.

Solution

Known quantities: Feed and product molar flow rates.

Find: Yield and selectivity.

Analysis: Use the three definitions for yield and selectivity:

$$\text{Yield} = \frac{\text{moles of desired product formed}}{\text{moles formed if there were no side reactions and limiting reactant reacted completely}}$$

$$= \frac{120 \text{ mol } CO_2}{200 \text{ mol } CO_2 \text{ should be formed}} = 0.6$$

$$\text{Yield} = \frac{\text{moles of desired product formed}}{\text{moles of reactant fed}} = \frac{120 \text{ mol } CO_2}{100 \text{ mol } C_2H_6} = 1.2$$

$$\text{Yield} = \frac{\text{moles of desired product formed}}{\text{moles of reactant consumed}} = \frac{120 \text{ mol } CO_2}{100 - 20} = 1.5$$

Selectivity is the number of moles of desired product (carbon dioxide) to the number of moles of undesired product formed:

$$\text{Selectivity} = \frac{\text{moles of desired product formed}}{\text{moles of undesired product formed}} = \frac{120 \text{ mol } CO_2}{40 \text{ mol } CO} = 3$$

Example 5.8 Oxidation Reaction

Problem

Ethylene is oxidized to ethylene oxide (desired) and carbon dioxide (undesired). Express the moles (or molar flow rates) of each of the five

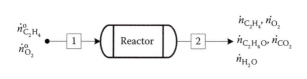

EXAMPLE FIGURE 5.8.1
Schematic of ethylene oxidation reaction.

species in the product stream in terms of the extent of reaction. The following reactions are taking place:

$$C_2H_4 + \frac{1}{2}O_2 \rightarrow C_2H_4O$$

$$C_2H_4 + 3O_2 \rightarrow 2CO_2 + 2H_2O$$

Solution

Known quantities: No quantities are given.

Find: Express the moles of components leaving the reactor in terms of extent of reactions.

Analysis: The process flow sheet is shown in Example Figure 5.8.1. Assign an extent of reaction for each reaction; ξ_1 for the first reaction and ξ_2 for the second reaction.

The first reaction: $C_2H_4 + \frac{1}{2}O_2 \rightarrow C_2H_4O \quad \xi_1$

The second reaction: $C_2H_4 + 3O_2 \rightarrow 2CO_2 + 2H_2O \quad \xi_2$

Mole balance using the extent of reaction approach:

$$C_2H_4: \quad n_{C_2H_4} = n^o_{C_2H_4} - \xi_1 - \xi_2$$

$$O_2: \quad n_{O_2} = n^o_{O_2} - \frac{1}{2}\xi_1 - 3\xi_2$$

$$C_2H_4O: \quad n_{C_2H_4O} = 0 + \xi_1$$

$$CO_2: \quad n_{CO_2} = 0 + 2\xi_2$$

$$H_2O: \quad n_{H_2O} = 0 + 2\xi_2$$

Example 5.9 Production of Ethylene

Problem

The feed stream contains 85.0 mol% ethane (C_2H_6), and the balance is inert (I). The fractional conversion of ethane is 0.5, and the fractional yield of ethylene (C_2H_4) is 0.40. Calculate the molar composition of the product gas and the selectivity of ethylene for methane production.

The following two multiple reactions take place in a continuous reactor at steady state:

$$C_2H_6 \rightarrow C_2H_4 + H_2$$

$$C_2H_6 + H_2 \rightarrow 2CH_4$$

Solution

Known quantities: Feed stream molar composition is known.

Find: The molar composition of the product gas and the selectivity of ethylene for methane production.

Analysis: Assign an extent of reaction for each reaction, namely, ξ_1 for the first reaction and ξ_2 for the second reaction. For simplicity, you may assign symbols for the components involved in the reaction.

Basis: 100 mol of feed.

The primary reaction: $C_2H_6 \rightarrow C_2H_4 + H_2 \quad \xi_1$
The secondary reaction: $C_2H_6 + H_2 \rightarrow 2CH_4 \quad \xi_2$
The process flow sheet is shown in Example Figure 5.9.1.
Component mole balance equations using the extent of reaction method approach:

$$C_2H_6: \quad n_{C_2H_6} = 85 - \xi_1 - \xi_2 \tag{5.1}$$

$$C_2H_4: \quad n_{C_2H_4} = 0.0 + \xi_1 \tag{5.2}$$

$$H_2: \quad n_{H_2} = 0.0 + \xi_1 - \xi_2 \tag{5.3}$$

$$CH_4: \quad n_{CH_4} = 0.0 + 2\xi_2 \tag{5.4}$$

The fractional conversion of ethane, C_2H_6, is

$$f_{C_2H_6} = 0.5 = \frac{85 - n_{C_2H_6}}{85}$$

Moles of unreacted ethane in the exit stream

$$n_{C_2H_6} = 42.25 \text{ mol}$$

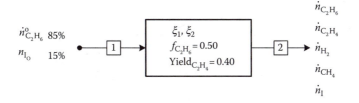

EXAMPLE FIGURE 5.9.1
Schematic of ethylene production.

The yield of ethylene (C_2H_4) equals the number of moles of the desired component produced to the number of moles produced of the same component, if there were no side reactions and the reactant is completely used up:

$$0.40 = \frac{n_{C_2H_4}}{85}$$

The number of moles of ethylene in the exit stream is $n_{C_2H_4} = 34$ mol
Substitute $n_{C_2H_4}$ in Equation 5.2 to determine ξ_1:

$$34 = 0.0 + \xi_1$$

The extent of reaction for the first reaction is $\xi_1 = 34$ mol
Substitute ξ_1 and $n_{C_2H_6}$ in Equation 5.1 to determine ξ_2:

$$42.5 = 85 - 34 - \xi_2$$

The extent of reaction for the second reaction is $\xi_2 = 8.5$ mol
Substituting ξ_1 and ξ_2 in Equation 5.3 gives

$$n_{H_2} = 0.0 + \xi_1 - \xi_2$$

$$n_{H_2} = 0.0 + 34 - 8.5$$

The number of moles of hydrogen in the exit stream is $n_{H_2} = 25.5$ mol
Substituting ξ_2 in Equation 5.4 to find out n_{CH_4}:

$$n_{CH_4} = 0.0 + 2 \times 8.5$$

The number of moles of methane in the exit stream is $n_{CH_4} = 17$ mol

5.7 Molecular Species Approach for Multiple Reactions

When using the molecular species approach for multiple reactions, we have to choose a single chemical species in each equation that appears in that reaction only. We can then use the number of moles of that species to keep track of how much of that reaction occurs. Consider the following multiple reactions:

$$C_6H_6 + Cl_2 \rightarrow C_6H_5Cl + HCl$$

$$C_6H_5Cl + Cl_2 \rightarrow C_6H_4Cl_2 + HCl$$

$$C_6H_4Cl_2 + Cl_2 \rightarrow C_6H_3Cl_3 + HCl$$

The schematic diagram for the multiple reactions is shown in Figure 5.3.

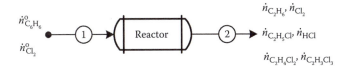

FIGURE 5.3
Benzene clarification reactions.

We have to choose a single unique chemical species in each reaction that does not appear in other reactions. There is a unique chemical compound to the first and the last reactions only. However, reaction 2 has no species that are unique to it. The amount of reaction that occurs in the first reaction can be determined by computing how much benzene (C_6H_6) reacts. All other species consumed or produced via the first reaction can be expressed in terms of this quantity and the stoichiometric coefficients.

$$\mathcal{R}_{C_6H_6} = \dot{n}^o_{C_6H_6} - \dot{n}_{C_6H_6}$$

The amount of reaction that occurs in the third reaction can be determined by computing how much trichlorobenzene ($C_6H_3Cl_3$) is generated. All other species consumed or produced via the third reaction can be expressed in terms of this quantity and the stoichiometric coefficients.

$$\mathcal{R}_{C_6H_3Cl_3} = \dot{n}_{C_6H_3Cl_3}$$

The amount of the second reaction that occurs is the total amount of HCl minus that is formed by the first and last reactions. Thus the amount of the second reaction that occurs is

$$\mathcal{R}_{HCl} = \left(\dot{n}_{HCl} - \dot{n}^o_{HCl} \right) - \left(\dot{n}^o_{C_6H_6} - \dot{n}_{C_6H_6} \right) - \dot{n}_{C_6H_3Cl_3}$$

Material balances on the remaining chemical species:
Material balance on Cl:

$$\dot{n}_{Cl_2} = \dot{n}^o_{Cl_2} - \left(\dot{n}^o_{C_6H_6} - \dot{n}_{C_6H_6} \right) - \left(\dot{n}_{HCl} + \dot{n}_{C_6H_6} - \dot{n}_{C_6H_3Cl_3} - \dot{n}^o_{C_6H_6} \right) - \left(\dot{n}_{C_6H_3Cl_3} \right)$$

Material balance on C_6H_5Cl:

$$\dot{n}_{C_6H_5Cl} = \dot{n}^o_{C_6H_5Cl} + \left(\dot{n}^o_{C_6H_6} - \dot{n}_{C_6H_6} \right) - \left(\dot{n}_{HCl} + \dot{n}_{C_6H_6} - \dot{n}_{C_6H_3Cl_3} - \dot{n}^o_{C_6H_6} \right)$$

Material balance on $C_6H_4Cl_2$:

$$\dot{n}_{C_6H_4Cl_2} = \dot{n}^o_{C_6H_4Cl_2} + \left(\dot{n}_{HCl} + \dot{n}_{C_6H_6} - \dot{n}_{C_6H_3Cl_3} - \dot{n}^o_{C_6H_6} \right) - \left(\dot{n}_{C_6H_3Cl_3} \right)$$

Example 5.10 Multiple Reactions

Problem

10 mol/h of benzene (C_6H_6) and 20 mol/h of chlorine (Cl_2) are fed to a reactor. The exit stream was analyzed and found to contain 1.0 mol/h of Cl_2, 2.0 mol/h of C_6H_5Cl, and 4.0 mol/h of $C_6H_4Cl_2$. Solve the problem using the extent of reaction method, atomic species balance, and molecular species approach. Calculate the percent conversion of benzene. The chlorination of benzene occurs via the following reactions:

$$C_6H_6 + Cl_2 \rightarrow C_6H_5Cl + HCl$$

$$C_6H_5Cl + Cl_2 \rightarrow C_6H_4Cl_2 + HCl$$

$$C_6H_4Cl_2 + Cl_2 \rightarrow C_6H_3Cl_3 + HCl$$

Solution

Known quantities: Exit stream contains 1.0 mol of chlorine (Cl_2), 2.0 mol/h of monochlorobenzene (C_6H_5Cl), and 4.0 mol/h of dichlorobenzene ($C_6H_4Cl_2$).

Find: The percent conversion of benzene.

Analysis: The labeled process flow sheet is shown in Example Figure 5.10.1. Use the three balance methods.

Basis: 1 h of operation.

Extent of reaction approach

Let ξ_1, ξ_2, and ξ_3 be the extent of the first, second, and third reactions, respectively.

$$C_6H_6 + Cl_2 \rightarrow C_6H_5Cl + HCl \quad \xi_1$$

$$C_6H_5Cl + Cl_2 \rightarrow C_6H_4Cl_2 + HCl \quad \xi_2$$

$$C_6H_4Cl_2 + Cl_2 \rightarrow C_6H_3Cl_3 + HCl \quad \xi_3$$

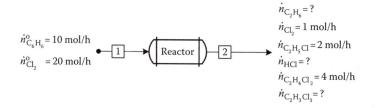

EXAMPLE FIGURE 5.10.1
Schematic chlorination of benzene reactor.

The number of moles of each species in the reactor exist stream may be written as

$$C_6H_6: \quad n_{C_6H_6} = n_{C_6H_6}^o - \xi_1 \tag{1}$$

$$Cl_2: \quad n_{Cl_2} = n_{Cl_2}^o - \xi_1 - \xi_2 - \xi_3 \tag{2}$$

$$C_6H_5Cl: \quad n_{C_6H_5Cl} = n_{C_6H_5Cl}^o + \xi_1 - \xi_2 \tag{3}$$

$$HCl: \quad n_{HCl} = n_{HCl}^o + \xi_1 + \xi_2 + \xi_3 \tag{4}$$

$$C_6H_4Cl_2: \quad n_{C_6H_4Cl_2}^o = n_{C_6H_4Cl_2}^o + \xi_2 - \xi_3 \tag{5}$$

$$C_6H_3Cl_3: \quad n_{C_2H_3Cl_3} = n_{C_2H_3Cl_3}^o + \xi_3 \tag{6}$$

Substituting in the known values in Equations 1 through 6

$$n_{C_6H_6} = 10 - \xi_1$$

$$1 = 20 - \xi_1 - \xi_2 - \xi_3$$

$$2 = 0 + \xi_1 - \xi_2$$

$$n_{HCl} = 0 + \xi_1 + \xi_2 + \xi_3$$

$$4 = 0 + \xi_2 - \xi_3$$

$$n_{C_2H_3Cl_3} = 0 + \xi_3$$

Solving the third equation for ξ_1 and the fifth for ξ_3 as follows:

$$\xi_1 = 2 + \xi_2$$

$$\xi_3 = \xi_2 - 4$$

Substitute ξ_1 and ξ_3 into the second equation to get

$$0 = 19 - (2 + \xi_2) - \xi_2 - (\xi_2 - 4)$$

Solving this equation gives $\xi_2 = 7$ mol
    Substituting ξ_2 into these preceding equations gives $\xi_1 = 9$ mol and $\xi_3 = 3$ mol. Substituting these values into the species balances gives

$$n_{C_6H_6} = 1 \text{ mol}, \quad n_{HCl} = 19 \text{ mol}, \quad n_{C_6H_3Cl_3} = 3 \text{ mol}$$

The conversion of benzene is

$$f_{C_6H_6} = \frac{10 \text{ mol} - 1 \text{ mol}}{10 \text{ mol}} = 0.90$$

Selectivity is defined as

$$\text{Selectivity} = \frac{\text{moles of desired product formed}}{\text{moles of undersired product formed}}$$

The desired product is C_6H_5Cl, and the undesired product is $C_6H_3Cl_3$. Accordingly,

$$\text{Selectivity} = \frac{2}{3} = 0.667$$

Atomic species approach

An alternative solution is the atomic species approach. We can now write the atomic species balances for C, H, and Cl.

$$\text{C:} \quad 6(10) = 6\dot{n}_{C_6H_6} + 6(2) + 6(4) + 6\dot{n}_{C_6H_3Cl_3}$$

$$\text{H:} \quad 6(10) = 6\dot{n}_{C_6H_6} + 5(2) + 4(4) + 3\dot{n}_{C_6H_3Cl_3} + \dot{n}_{HCl}$$

$$\text{Cl:} \quad 2(20) = 2(1) + 1(2) + 2(4) + 3\dot{n}_{C_6H_3Cl_3} + \dot{n}_{HCl}$$

Simplifying these three equations gives

$$6\dot{n}_{C_6H_6} + 6\dot{n}_{C_6H_3Cl_3} = 24$$

$$6\dot{n}_{C_6H_6} + 3\dot{n}_{C_6H_3Cl_3} + \dot{n}_{HCl} = 34$$

$$3\dot{n}_{C_6H_3Cl_3} + \dot{n}_{HCl} = 28$$

There are now three equations in three unknowns. Subtract the last from the second to get

$$6\dot{n}_{C_6H_6} = 6, \quad \text{hence } \dot{n}_{C_6H_6} = 1 \text{ mol}$$

Substitute this value into the first equation to get $\dot{n}_{C_6H_3Cl_3} = 3$ mol, and further substitute this value into the last equation to get $\dot{n}_{HCl} = 19$ mol. Since these are the same numbers as those obtained via the extent of reaction method, the conversion and selectivity values will be 90% and 0.667, respectively.

Molecular species approach

The third alternative solution is the molecular species approach. The three chemical reactions that occur are as follows:

$$C_6H_6 + Cl_2 \rightarrow C_6H_5Cl + HCl$$

$$C_6H_5Cl + Cl_2 \rightarrow C_6H_4Cl_2 + HCl$$

$$C_6H_4Cl_2 + Cl_2 \rightarrow C_6H_3Cl_3 + HCl$$

In using the molecular species approach, we pick a single chemical species in each reaction that is unique to that reaction. We can single out such a species in the first and last reactions. However, the second reaction has no species that are unique to it.

The amount of reaction that occurs in the first reaction can be determined by computing how much benzene reacts. The amount of benzene consumed ($\Re_{C_6H_6}$) from reaction 1 is given by

$$\Re_{C_6H_6} = \left(\dot{n}^o_{C_6H_6} - \dot{n}_{C_6H_6} \right) = 10 - \dot{n}_{C_6H_6}.$$

The amount of $C_6H_3Cl_3$ generated ($\Re_{C_6H_3Cl_3}$) from the last reaction is given by

$$\Re_{C_6H_3Cl_3} = \left(\dot{n}_{C_6H_3Cl_3} - \dot{n}^o_{C_6H_3Cl_3} \right) = \dot{n}_{C_6H_3Cl_3} - 0$$

The amount of HCl generated ($\Re_{HCl}$) in the second reaction is given by

$$\Re_{HCl} = \left(\dot{n}_{HCl} - \dot{n}^o_{HCl} \right) - \left(10 - \dot{n}_{C_6H_6} \right) - \dot{n}_{C_6H_3Cl_3} = \dot{n}_{HCl} - 0 - 10 + \dot{n}_{C_6H_6} - \dot{n}_{C_6H_3Cl_3}$$

We can now write the balances on the remaining chemical species.

Chlorine (Cl_2) is fed to the reacting system, reacts in reactions 1, 2, and 3, and also leaves the system. Thus, a balance on chlorine gives

$$\dot{n}_{Cl_2} = \dot{n}^o_{Cl_2} - \left(10 - \dot{n}_{C_6H_6} \right) - \left(\dot{n}_{HCl} + \dot{n}_{C_6H_6} - \dot{n}_{C_6H_3Cl_3} - 10 \right) - \left(\dot{n}_{C_6H_3Cl_3} \right)$$

Material balance on monochlorobenzene (C_6H_5Cl):

$$\dot{n}_{C_6H_5Cl} = \dot{n}^o_{C_6H_5Cl} + \left(10 - \dot{n}_{C_6H_6} \right) - \left(\dot{n}_{HCl} + \dot{n}_{C_6H_6} - \dot{n}_{C_6H_3Cl_3} - 10 \right)$$

Material balance on dichlorobenzene ($C_6H_4Cl_2$):

$$\dot{n}_{C_6H_4Cl_2} = \dot{n}^o_{C_6H_4Cl_2} + \left(\dot{n}_{HCl} + \dot{n}_{C_6H_6} - \dot{n}_{C_6H_3Cl_3} - 10 \right) - \left(\dot{n}_{C_6H_3Cl_3} \right)$$

Substituting known values and rearranging these three equations gives

$$1 = 20 - \dot{n}_{HCl}$$

$$2 = 20 - 2\dot{n}_{C_6H_6} - \dot{n}_{HCl} + \dot{n}_{C_6H_3Cl_3}$$

$$4 = \dot{n}_{HCl} + \dot{n}_{C_6H_6} - 2\dot{n}_{C_6H_3Cl_3} - 10$$

The first equation gives $\dot{n}_{HCl} = 19$ mol. Substituting this value into the last two equations gives

$$-2\dot{n}_{C_6H_6} + \dot{n}_{C_6H_3Cl_3} = 1$$

$$\dot{n}_{C_6H_6} - 2\dot{n}_{C_6H_3Cl_3} = -5$$

Solving these two equations simultaneously gives $\dot{n}_{C_6H_6} = 1$ mol/min and $\dot{n}_{C_6H_3Cl_3} = 3$ mol/min. These again are the same values as those obtained from the prior two methods. Therefore, the conversion and selectivity values will be 90% and 0.667, respectively.

5.8 Degrees of Freedom Analysis for Reactive Processes

The degrees of freedom analysis for *molecular species* balances and material balance based on the extent of reaction is almost the same. That is,

$$
NDF = \left\{ \begin{array}{c} \text{number of} \\ \text{unknowns} \end{array} \right\} + \left\{ \begin{array}{c} \text{number of} \\ \text{independent} \\ \text{chemical} \\ \text{reactions} \end{array} \right\} - \left\{ \begin{array}{c} \text{number of} \\ \text{independent molecular} \\ \text{species balances} \end{array} \right\}
$$

$$
- \left\{ \begin{array}{c} \text{number of} \\ \text{other equations} \\ \text{relating} \\ \text{variables} \end{array} \right\}
$$

In *atomic balance* case, the degrees of freedom analysis is expressed as follows:

$$
NDF = \left\{ \begin{array}{c} \text{number of} \\ \text{unknowns} \end{array} \right\} - \left\{ \begin{array}{c} \text{number of} \\ \text{independent} \\ \text{atomic} \\ \text{species} \\ \text{balances} \end{array} \right\} - \left\{ \begin{array}{c} \text{number of} \\ \text{molecular} \\ \text{balances on} \\ \text{independent} \\ \text{nonreactive} \\ \text{species} \end{array} \right\} - \left\{ \begin{array}{c} \text{number} \\ \text{of other} \\ \text{relations} \\ \text{relating} \\ \text{variables} \end{array} \right\}
$$

A set of chemical reactions are independent, if the stoichiometric equation of any one of them cannot be obtained by a linear combination (via addition, subtraction, or multiplication) of the stoichiometric equations of others. Consider the following equations:

$$A \rightarrow 2B \tag{1}$$

$$B \rightarrow C \tag{2}$$

$$A \rightarrow 2C \tag{3}$$

These three reactions are not all independent, since $(3) = (1) + 2 \times (2)$. Only two of the three equations are independent. Generally, there are n balances for n independent species. But if two species are in the same ratio to each other wherever they appear in the process, balances on these species will not be independent equations. For example, air consists of 21 mol% O_2

and 79 mol% N_2. In ammonia synthesis, reactants (N_2:H_2) in certain cases enter the reactor on a stoichiometric basis (1:3).

5.8.1 Chemical Equilibrium

More often than not, reactions do not proceed instantly. In fact, predicting the speed at which a reaction occurs is very important. Furthermore, reactions do not necessarily happen independently. Very often, the reverse "half" reaction of the reaction we are interested in also takes place. Chemical equilibrium is reached when the rates of the forward and reverse reactions are equal to each other (i.e., compositions no longer change with time). While we will not calculate reaction rates, we need to know what affects them because this will affect the equilibrium. Things that we must consider that affect reaction rates and, hence, equilibrium are temperature and concentration.

Example 5.11 Equilibrium Reaction

Problem

Consider the reaction of methane with oxygen:

$$2CH_4 + O_2 \rightleftharpoons 2CH_3OH$$

At equilibrium, the compositions of the components satisfy the relation

$$K(T) = \frac{y_{CH_3OH}^2}{y_{CH_4}^2 y_{O_2}}$$

If you are given the feed compositions ($n_{CH_4}^0$, $n_{O_2}^0$, and $n_{CH_3OH}^0$) and the equilibrium constant $K(T)$, how do you determine the equilibrium compositions?

Solution

Use the equilibrium expression to determine the equilibrium compositions and extent of reaction.

$$\text{The equilibrium constant } K(T) = \frac{y_{CH_3OH}^2}{y_{CH_4}^2 y_{O_2}}$$

where

$$y_{CH_4} = \frac{n_{CH_4}}{n}$$

$$y_{CH_3OH} = \frac{n_{CH_3OH}}{n}$$

$$y_{O_2} = \frac{n_{O_2}}{n}$$

The component and total molar balances are

$$n_{CH_3OH} = n^o_{CH_3OH} + 2\xi$$

$$n_{CH_4} = n^o_{CH_4} - 2\xi$$

$$n_{O_2} = n^o_{O_2} - \xi$$

$$n = n^o - \xi$$

Substitution of known quantities into the equilibrium constant expression gives

$$K(T) = \frac{\left(\dfrac{n^o_{CH_3OH} + 2\xi}{n^o - \xi}\right)^2}{\left(\dfrac{n^o_{CH_4} - 2\xi}{n^o - \xi}\right)^2 \left(\dfrac{n^o_{O_2} - \xi}{n^o - \xi}\right)}$$

Solving for the extent of reaction from this equation is a matter of simple algebra.

Example 5.12 Methane Oxidation Process

Problem
The feed to a plug flow reactor contains equimolar amounts of methane and oxygen. Assume a basis of 100 mol feed/s. The fractional conversion of methane is 0.9, and the fraction yield of formaldehyde is 0.855. Calculate the molar composition of the reactor output stream and the selectivity of formaldehyde production relative to carbon dioxide production. Methane (CH_4) and oxygen react in the presence of a catalyst to form formaldehyde (HCHO). In a parallel reaction, methane is oxidized to carbon dioxide and water:

$$CH_4 + O_2 \rightarrow HCHO + H_2O$$
$$CH_4 + 2O_2 \rightarrow CO_2 + 2H_2O$$

Solution

Known quantities: Total inlet molar flow rate, conversion, and yield.

Find: Calculate the molar composition of the reactor output stream and the selectivity of formaldehyde production relative to carbon dioxide production.

Analysis: Use the extent of reaction to calculate the molar composition of reactor output stream.

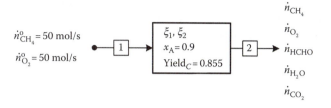

EXAMPLE FIGURE 5.12.1
Methane combustion reaction.

Basis: 100 mol/s of feed to reactor.

ξ_1 is assigned to the first reaction: $CH_4 + O_2 \rightarrow HCHO + H_2O \quad \xi_1$
ξ_2 is assigned to the second reaction: $CH_4 + 2O_2 \rightarrow CO_2 + 2H_2O \quad \xi_2$

The process flow sheet of reactor with multiple reactions is shown in Example Figure 5.12.1.

Degrees of freedom analysis:

Number of unknowns: $7 = 5$ $(n_{CH_4}, n_{O_2}, n_{HCHO}, n_{H_2O}, n_{CO_2}) + 2$ (ξ_1, ξ_2)

Number of independent equations: 5 (number of components)
   Number of auxiliary relations: 2

$$NDF = 7 - 5 - 2 = 0$$

The components material balance equations are

$$CH_4: \quad n_{CH_4} = 50 - \xi_1 - \xi_2$$

$$O_2: \quad n_{O_2} = 50 - \xi_1 - 2\xi_2$$

$$HCHO: \quad n_{HCHO} = 0.0 + \xi_1$$

$$H_2O: \quad n_{H_2O} = 0.0 + \xi_1 + 2\xi_2$$

$$CO_2: \quad n_{CO_2} = \xi_2$$

The fractional conversion of methane is given by

$$0.9 = \frac{50 - n_{CH_4}}{50}$$

Unreacted moles of methane in the outlet stream is

$$n_{CH_4} = 5 \text{ mol}$$

Yield of formaldehyde is 0.855 (moles of formaldehyde produced if there were no side reaction and limiting reactant (CH_4) reacts completely; that is, 50 mol is consumed based on the first reaction).

$$Yield_{HCHO} = 0.855 = \frac{n_{HCHO}}{50}$$

Moles of formaldehyde in the reactor outlet stream is $n_{HCHO} = 42.75$ mol

$$n_{HCHO} = \xi_1, \text{ and, hence, } \xi_1 = 42.75 \text{ mol } \xi_1 = 42.75 \text{ mol}$$

Substitute ξ_1 and n_{CH_4} in the first equation to get

$$5 = 50 - 42.75 - \xi_2$$

The extent of reaction for the second reaction is $\xi_2 = 2.25$ mol.

Knowing values of ξ_1 and ξ_2, substitute those values when needed in component mole balance equation gives you the molar flow rate of exit stream component.

5.9 Combustion Reactions

Combustion is the rapid reaction of a fuel with oxygen to produce energy. Combustion is a very important industrial chemical reaction. Fuels include coal (C, H, S, and others), fuel oil (high Mw hydrocarbons and some S), gaseous fuel (natural gas—mostly methane), or liquefied petroleum gas (propane and/or butane). Maximum energy is produced when fuel is completely burned (oxidized). The product gas is called stack gas or flue gas. Complete combustion results in all C oxidized to CO_2, all H oxidized to H_2O, and all S oxidized to SO_2. In incomplete combustion, C is oxidized to CO and CO_2 [3]. Complete combustion of butane:

$$C_4H_{10} + \frac{13}{2}O_2 \rightarrow 4CO_2 + 5H_2O$$

Side reaction; incomplete combustion of butane:

$$C_4H_{10} + \frac{9}{2}O_2 \rightarrow 4CO + 5H_2O$$

5.9.1 Theoretical and Excess Air

For obvious economic reasons, air (79% N_2, 21% O_2) is the source of oxygen in combustion reactions. Combustion reactions are always conducted with excess air, thus ensuring good conversion of the expensive fuel.

Theoretical oxygen is the moles or molar flow rate of O_2 required for complete combustion of all the fuel. Theoretical air is the quantity of air that contains the theoretical oxygen [4].

$$\text{Theoretical air} = \frac{1}{0.21} \times \text{theoretical } O_2$$

Excess air is the amount of air fed to the reactor that exceeds the theoretical air.

$$\text{Percent excess air} = \frac{(\text{moles air})_{\text{fed}} - (\text{moles air})_{\text{theoretical}}}{(\text{moles air})_{\text{theoretical}}} \times 100\%$$

Example 5.13 Coal Combustion Process

Problem
Consider the combustion process of coal as shown in Example Figure 5.13.1. Calculate the flow rate of all streams and their compositions. Assuming all the coal is consumed, calculate the percent excess air and the ratio of water vapor and dry gas. Note that the feed composition is given in mole fraction. The following reactions are taking place:

$$C + O_2 \rightarrow CO_2$$

$$S + O_2 \rightarrow SO_2$$

$$2H + \frac{1}{2}O_2 \rightarrow H_2O$$

Solution

Known quantities: Inlet and exit stream mole fractions are known.

Find: The percent excess air and the ratio of water vapor and dry gas.

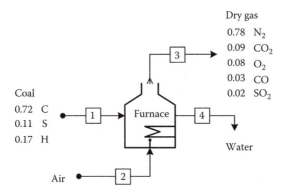

EXAMPLE FIGURE 5.13.1
Schematic of coal combustion process.

Analysis: The combustion process flow sheet is shown in Example Figure 5.13.1.

Basis: 100 mol/min of dry gas.

Molar flow rate of nitrogen in the dry gas:

$$N_2 : \ (0.78)\left(100 \ \frac{mol}{min} \right) = 78 \ mol \ N_2/min$$

Since nitrogen is an inert gas, its amount is the same as that in the inlet air. Accordingly, the inlet air molar flow rate is

$$\text{Total inlet air} = \frac{78 \ mol \ N_2/min}{0.79} = 98.7 \ mol/h$$

C atomic balance:

C atoms in feed = C atom in product

$$\text{C atoms in the feed} = (100 \ mol/min)(0.09 + 0.03)$$

$$= 12.0 \ mol \ C/min \ \text{in the feed stream}$$

Amount of feed stream per 100 mol dry gas:

$$0.72 \times n_1 = 12.0 \ mol$$

Rearranging

$$n_1 = \frac{12.0 \ mol}{0.72} = 16.7$$

Moles of atomic S in the feed

$$(0.11)(16.7) = 1.84 \ mol/min$$

Moles of atomic H in the feed

$$(0.17)(16.7) = 2.84 \ mol/min$$

The number of moles of oxygen required for complete combustion of C, S, and H is:

$$C: \ 12 \ mol \ C \times \frac{1 \ mol \ O_2}{1 \ mol \ C} = 12 \ mol \ O_2$$

$$S: \ 1.8 \ mol \ S \times \frac{1 \ mol \ O_2}{1 \ mol \ S} = 1.8 \ mol \ O_2$$

$$H: \ 2.8 \ mol \ H \times \frac{0.5 \ mol \ O_2}{2 \ mol \ H} = 0.7 \ mol \ O_2$$

Total inlet theoretical O_2 in the fed air = 12 + 1.8 + 0.7 = 14.5 mol/h:

$$\left(\frac{78 \ mol \ N_2}{min} \right)\left(\frac{0.21 \ mol \ O_2}{0.79 \ mol \ N_2} \right) = 20.73 \ mol \ O_2 \ \text{in feed/min}$$

Total inlet air:

$$\left(\frac{78 \text{ mol N}_2}{\text{min}} \right)\left(\frac{1 \text{ mol Air}}{0.79 \text{ mol N}_2} \right) = 98.7 \text{ mol/min}$$

Excess air (equal to excess O_2)
  Total theoretical air $= 14.5/0.21 = 69 \text{ mol/min}$

$$\text{The percent excess air} = \frac{98.7 - 69}{69} \times 100\% = 43\% \text{ excess air}$$

Assume all H_2O in the product gas is from the combustion process. Form 2.8 mol H reacted, 1.4 mol of H_2O is produced. Ratio of water vapor to dry gas:

$$\frac{1.4 \text{ mol H}_2\text{O}}{100 \text{ mol dry gas}} = 0.014 \text{ mol H}_2\text{O/mol dry gas}$$

Example 5.14 Fermentation Bioprocess

Problem

The production of 1.8 kg/h acetic acid using fermentation of ethanol is described by the following reaction:

$$C_2H_5OH + O_2 \rightarrow CH_3COOH + H_2O$$

The maximum acetic acid concentration tolerated by the cells is 10%. Air is pumped into the fermentor at a rate of 200 mol/h. Calculate the minimum amount of ethanol required, and the minimum amount of water that must be used to dilute ethanol to avoid acid inhibition [5].

Solution

Known quantities: Air inlet molar flow rate, exit acetic acid flow rate.

Find: Minimum amount of ethanol required and composition of the fermentor off-gas.

Analysis: The process flowchart is shown in Example Figure 5.14.1. As the following reaction is taking place

$$C_2H_5OH + O_2 \rightarrow CH_3COOH + H_2O$$

Moles of acetic acid produced:

$$1.8 \text{ kg/h} \left| \frac{1}{60 \text{ kg/kmol}} \right. = 0.03 \text{ kmol/h}$$

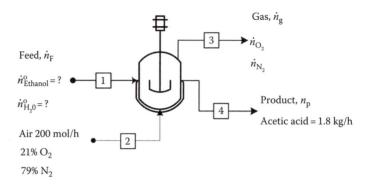

EXAMPLE FIGURE 5.14.1
Fermentation of ethanol.

Component material balance using the extent of reaction method:

$$n_{CH_3COOH} = 0.0 + \xi$$

$$n_{H_2O} = n^o_{H_2O} + \xi$$

$$n_{C_2H_5OH} = n^o_{C_2H_5OH} - \xi$$

$$n_{O_2} = n^o_{O_2} - \xi$$

From the first equation and substituting the molar flow rate of produced acetic acid, we get

$$0.03 = 0 + \xi$$

$$\xi = 0.03 \text{ kmol}$$

For the complete conversion of ethanol, all ethanol is reacted and no ethanol is present in the exit stream:

$$n_{C_2H_5OH} = n^o_{C_2H_5OH} - \xi$$

Substituting the extent of reaction and setting the amount of exit ethanol to zero give

$$0 = n^o_{C_2H_5OH} - 0.03$$

$$n^o_{C_2H_5OH} = 0.03 \text{ kmol/h}$$

This is the minimum amount of ethanol required.

Water required:

Since the maximum acetic acid concentration tolerated by cells is 10%, the minimum amount of water required to avoid acid inhibition (w) is given by

$$0.10 = \frac{1.8 \text{ kg/h Acetic acid}}{(w \text{ kg of water} + 1.8 \text{ kg/h of acetic acid})}$$

The required mass flow rate of water is 16.2 kg/h.

$$\text{Moles of water out } (n_{H_2O}) = \frac{\text{mass}}{\text{Mw}} = \frac{16.2 \text{ kg/h}}{18 \text{ kg/kmol}} = 0.9 \text{ kmol}$$

Using the mole balance equation for water: $n_{H_2O} = n^o_{H_2O} + \xi$

Substitute the values of exit water molar flow rate and extent of reaction

$$0.9 \text{ kmol/h} = n^o_{H_2O} + 0.03 \text{ kmol/h}$$

Moles of inlet water required to avoid acetic acid inhibition is $n^o_{H_2O} = 0.87 \text{ kmol/h}$

Example 5.15 Methane Combustion

Problem

Air and an unknown amount of methane (CH_4) gas are fed to a combustion process. All the water produced during the process was condensed and removed from the bottom of the combustion chamber. The exhaust gases, free of water, leave the top of the combustion process. An analysis of the exhaust gases gives 5 mol% O_2, 9 mol% CO2, 0.5 mol% CO, and the remainder N2. Based on 100 mol/min of exhaust gases (dry basis), determine the % excess oxygen fed to this process. The following combustion reaction takes place:

$$CH_4 + 2O_2 \rightarrow CO_2 + 2H_2O$$

Solution

Known quantities: Exhaust gas composition and its molar flow rate (dry basis).

Find: The percentage excess oxygen fed to this process.

Analysis: The system looks like Example Figure 5.15.1. Use atomic balance.

Basis: 100 mol/min of exhaust dry gas stream (stream 3).

From the fact that the mole fractions in any stream must sum up to 1.0, we know that

$$y_{N_2,3} = 1 - 0.05 - 0.09 - 0.005 = 0.855.$$

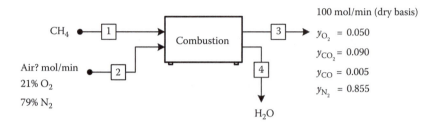

EXAMPLE FIGURE 5.15.1
Methane combustion process.

Since N_2 is inert, the number of moles of N_2 in stream 2 must be 85.5 mol/min. Thus, the number of moles of O_2 in stream 2 can be calculated:

$$\dot{n}_{N2,2} = 85.5 = 0.79 \times \dot{n}_2$$

Solving for $\dot{n}_2$

$$\dot{n}_2 = \frac{85.5}{0.79} = 108.23 \, \text{mol/min}$$

The inlet oxygen molar flow rate is

$$\dot{n}_{O_2,2} = 0.21 \times \dot{n}_2$$

where $\dot{n}_2$ is the inlet molar flow rate of air.

$$\dot{n}_{O_2,2} = 0.21 \times 108.23 = 22.73 \, \text{mol/min}$$

A carbon balance gives

$$(1)\dot{n}_{CH_4,1} = (1)9 + (1)0.5 = 9.5 \, \text{mol/min}$$

For the complete combustion of CH_4, the stoichiometry is

$$CH_4 + 2O_2 \rightarrow CO_2 + 2H_2O$$

That is, 2 mol of O_2 is needed for each mole of CH_4. Accordingly, the number of moles of O_2 needed for complete combustion is

$$\text{Theoretical oxygen} = 2(9.5) = 19.0 \, \text{mol/min}$$

Since there is 22.73 mol fed to the furnace and the needed oxygen is 19 mol/min, the number of moles of excess oxygen is

$$\text{Moles of excess } O_2 = 22.73 - 19.0 = 3.73 \, \text{mol/min}$$

Thus, the percent excess oxygen is

$$\% \, \text{excess} = \frac{3.73}{19.0} 100\% = 19.62\%$$

Homework Problems

5.1 Methanol is dehydrogenated in a catalytic reactor to form formaldehyde in the reaction:

$$CH_3OH \rightarrow HCHO + H_2$$

Formaldehyde in the reactor product is separated from the unreacted methanol and hydrogen. The production rate of formaldehyde is 30 kmol. If the single pass conversion is 60%, calculate the required feed rate of methanol to the process in kmol/h. (50 kmol/h)

5.2 Formaldehyde is produced by the dehydrogenation of ethanol in a catalytic reactor. The following reaction takes place in the reactor:

$$CH_3OH \rightarrow HCHO + H_2$$

Methanol feed rate to the reactor is 50 kmol/h. If the single pass conversion is 60%, calculate the production rate of formaldehyde. (30 kmol/h)

5.3 The feed to a continuous ammonia formation reactor is 100 mol/s nitrogen, 300 mol/s hydrogen, and 1 mol/s argon. The ammonia production reaction is

$$N_2 + 3H_2 \rightarrow 2NH_3$$

The percent conversion of hydrogen in the reactor is 60.0%. Find the molar flow rate of ammonia produced. (120 mol/s)

5.4 The molar feed rate to a continuous ammonia synthesis reactor is 100 mol/min nitrogen, 150 mol/min hydrogen. The ammonia formation reaction is

$$N_2 + 3H_2 \rightarrow 2NH_3$$

The percent conversion of the limiting reactant in the reactor is 60.0%. What are the limiting reactant and its molar flow rate in the reactor effluent stream? (60 mol/min)

5.5 Ethylene monomer is the main source for the production of well-known polyethylene polymers. Ethylene is produced by the catalytic cracking of ethane according to the following reaction:

$$C_2H_6 \rightarrow C_2H_4 + H_2$$

The inlet molar flow rate to a catalytic reactor is 100 mol/min of C_2H_6. The amount of hydrogen in the product stream was found to be 40 mol/min. Perform a degrees of freedom analysis and calculate the molar flow rates of unreacted ethane in the reactor exit stream. Use the molecular species balance. (60 mol/min)

5.6   Butane (C_4H_{10}) is fed to a fired heater at a molar flow rate of 100 mol/h and 50% excess air. Fifty percent of the butane is consumed, and the product gas contains 9 mol CO_2 per mol CO. Calculate the molar flow rate of water in the stack gas. (250 mol/h)

5.7   Butane is completely burned in a process fed with 25.0% excess air. Based on 1 mol/h of butane, determine the composition in mole fractions of oxygen in product stream. (0.04)

5.8   Methane at a molar flow rate of 55 mol/min is burned with dry air. A partial analysis of the exit gas reveals that the mole fraction of nitrogen is 0.85 on a dry basis, and the molar flow rate of the dry exit gas is 600 mol/min. The fractional conversion of methane is 90%, and the molar ratio of CO_2 to CO is 10. Based on the given information, determine the percent excess air fed to the combustion chamber and the mole fraction of water in the exit gas stream. (23%)

5.9   A gaseous mixture containing 80 mol% methane and the balance methyl ethyl ketone (C_4H_8O) is burned with 50% excess air. During the reaction, 95% of the methane and 75% of the methyl ethyl ketone were burned; however, some of the methane that burns forms CO. The exhaust gas contains 0.5 mol% CO on a dry basis. On the basis of 1 mol of hydrocarbon gas, calculate the moles of oxygen in the exhaust gas. (1.75 mol)

5.10   A solid ore containing 82 wt% FeS_2 and 18 wt% inert is fed to a furnace. Dry air is fed in 40% excess of the amount theoretically required to oxidize all of the sulfur in the ore to SO_3. A pyrite conversion of 85% is obtained, with 40% of FeS_2 converted to form sulfur dioxide and the rest forming sulfur trioxide. Two streams leave the roaster: a gas stream containing SO_3, SO_2, O_2, and N_2, and a solid stream containing unconverted pyrite, ferric oxide (Fe_2O_3), and inert. Using 100 kg/min of ore as a basis, calculate the rate of Fe_2O_3 production (kg/min) in the outlet solid stream, and the total molar flow and composition of the outlet gas stream. The following reactions take place in the furnace: (46.5 kg)

$$FeS_2 \ (s) + \frac{15}{2} O_2 \ (g) \rightarrow Fe_2O_3 (s) + 4 \, SO_3 \ (g)$$

$$2FeS_2 \ (s) + \frac{11}{2} O_2 \ (g) \rightarrow Fe_2O_3 \ (s) + 4 \, SO_2 \ (g)$$

5.11   100 kmol/h of a stream containing 20 mol% CO, 20 mol% H_2O, and 60 mol% H_2 is combined with additional steam and fed to a water–gas shift reactor to produce more hydrogen according to the reaction:

$$CO + H_2O \longrightarrow CO_2 + H_2$$

The equilibrium constant of the water–gas shift reaction is 200, and the mole fraction of hydrogen at the outlet is 0.75. Write and balance the chemical equation. How much additional steam is fed to the water–gas shift reactor? (5.1 kmol/h)

5.12 Sulfuric acid (H_2SO_4) is formed by first burning sulfur in air to produce sulfur dioxide (SO_2). Sulfur dioxide is then catalytically reacted with more air to form sulfur trioxide. Finally, the sulfur trioxide is combined with water to form sulfuric acid. The flow diagram for this process is shown in Problem Figure 5.12.1. The components in each stream are as shown in the flow diagram. The stream of produced concentrated sulfuric acid is 98% sulfuric acid and 2 wt% water. This process is to produce 200,000 kg/day of the concentrated acid. How much sulfur in kg/day must enter the process in order to meet this production demand? Use the component balance approach. (64,000 kg/day)

The overall reaction stoichiometry is given by

$$S + \frac{3}{2}O_2 + H_2O \rightarrow H_2SO_4$$

5.13 Ethane is burned with air in a continuous steady-state combustion reactor to yield a mixture of carbon monoxide, carbon dioxide, and water. The feed to the reactor contains 10 mol/min C_2H_6 and 200 mol/min air. The percentage conversion of ethane is 80%, and the gas leaving the reactor contains 8 mol CO_2 per mol of CO. Determine the molar flow rate of the ethane gas in the product stream. (2 mol/min)

5.14 The feed to a batch reactor consists of 10 mol of benzene (C_6H_6) and 20 mol of chlorine (Cl_2). The exit stream is found to contain 1.0 mol of chlorine (Cl_2), 2.0 mol of monochlorobenzene (C_6H_5Cl), and 4.0 mol of dichlorobenzene ($C_6H_4Cl_2$). The chlorination of benzene occurs via the sequence

$$C_6H_6 + Cl_2 \rightarrow C_6H_5Cl + HCl$$

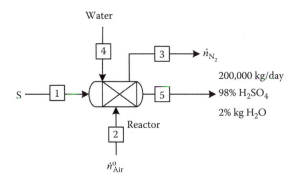

PROBLEM FIGURE 5.12.1
Production of sulfuric acid.

$$C_6H_5Cl + Cl_2 \rightarrow C_6H_4Cl_2 + HCl$$

$$C_6H_4Cl_2 + Cl_2 \rightarrow C_6H_3Cl_3 + HCl$$

Use the molecular species approach to find the conversion of benzene. (0.9)

5.15 A reactor is designed to take in a feed of 200 mol/min of propane and 50% excess steam, and convert them to CO and H_2 with a 65% conversion rate. Determine the mole flow rate of propane in the reactor product stream. (70 mol/min)

$$C_3H_8 + 3H_2O \rightleftharpoons 3CO + 7H_2$$

5.16 The following reversible reactions occur in a reactor:

$$A + 2B \leftrightarrow C + D$$

$$2D + B \leftrightarrow E$$

$$E + B \leftrightarrow A$$

Suppose 10 mol of A, 50 mol of B, and 5 mol of D are fed to the reactor and the product stream is sampled and found to contain 30 mol% C, 20 mol% A, and 5 mol% E. What is the percent conversion of A and B? (60%, 75.3%)

5.17 Nitrogen gas and hydrogen gas are fed to a reactor in stoichiometric quantities that react to form ammonia. The conversion of nitrogen to ammonia is 25%. How much nitrogen is required to make 245 kmol/h of ammonia? (490 kmol/h)

References

1. Reklaitis, G.V. (1983) *Introduction to Material and Energy Balances*, John Wiley & Sons, New York.
2. Felder, R.M. and R.W. Rousseau (1978) *Elementary Principles of Chemical Processes*, John Wiley & Sons, New York.
3. Himmelblau, D.M. (1974) *Basic Principles and Calculations in Chemical Engineering*, 3rd edn., Prentice-Hall, NJ.
4. Whirwell, J.C. and R.K. Toner (1969) *Conservation of Mass and Energy*, Blaisdell, Waltham, MA.
5. Atkinson, B. and F. Mavituna (1991) *Biochemical Engineering and Biotechnology Handbook*, 2nd edn., Macmillan, Basingstoke, U.K.

6

Multiple-Unit Systems Involving Reaction, Recycle, and Purge

This chapter discusses the application of material balances to systems consisting of two or more process units and involving chemical reactions, recycle, and purge. The chapter begins with the description of the features of the flow sheets of reactive processes. Next, it is extended to the degrees of freedom analysis to such processes based on their flow sheet structure. Finally, the methodology of solving steady-state multiple-unit problems with reactions is reviewed.

Learning Objectives

1. Construct the flow sheet of a multiple-unit process involving reaction, recycle, and purge (Section 6.1).

2. Carry out a degrees of freedom analysis for a reactive multiple-unit process (Section 6.2).

3. Understand how to solve material balances of reactive multiple-unit systems (Section 6.3).

6.1 Multiple-Unit Process Flowcharts

As mentioned in an earlier chapter, a process flow sheet is a graphical representation of a process. The use of flow sheets for multiple-unit processes with reactions is even more important as it tremendously helps in the description of the complex process and formulation of material (and energy) balances. A couple of new issues arise when considering recycle and purge in reactive multiple-unit processes, namely, overall conversion and single-pass conversion [1].

6.1.1 Flow Sheet for Reaction with Recycle

Figure 6.1 shows a process flow sheet where unreacted compounds are recycled for further processing.

6.1.2 Reaction with Product Splitter and Recycle

A recycle stream is introduced to recover and reuse unreacted reactants. Two definitions of reactant conversion are used in the analysis of chemical reactors with product separation and recycle of unconsumed reactants. The overall conversion, which seems to ignore the fact that recycle is occurring and is calculated purely on the basis of the difference between the overall process inputs (feed streams) and outputs (product streams) as shown in Figure 6.2.

$$\text{Overall conversion} = \frac{\begin{array}{c}\text{reactant input to process}\\ -\text{ reactant output from process}\end{array}}{\text{reactant input to process}}$$

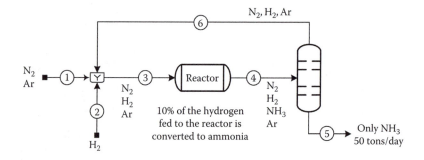

FIGURE 6.1
Schematic of process flow sheet for reaction with recycle.

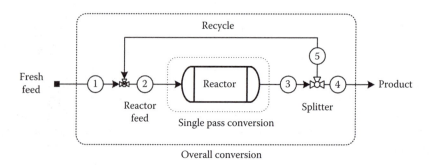

FIGURE 6.2
Schematic of reactor with recycle.

The single-pass conversion is based on the difference between the reactor input and output, inside the recycle loop.

$$\text{Single-pass conversion} = \frac{\begin{array}{c}\text{reactant input to reactor} \\ -\text{ reactant output from reactor}\end{array}}{\text{reactant input to reactor}}$$

6.1.3 Reaction with Recycle and Purge

A problem that can occur in the process that involves recycle is that the material that enters the process in the feed stream, or is generated in the reactor, may remain entirely in the recycle stream rather than being carried in the product stream. To prevent this buildup of material, a portion of the recycle stream is withdrawn as a purge stream. In the process flowchart, a purge point is a simple splitter. A purge stream bleeds material out of a recycle stream so that accumulation is avoided (Figure 6.3). As is usual with a splitter, the recycle stream before and after the purge point has the same composition, and only one independent material balance can be written. The recycle stream allows the operation of the reactor at low single-pass conversion and has high overall conversion for the system.

An industrial example involving recycle and purge is the fluidized bed reactor used for polyethylene production (UNIPOL process). Figure 6.4 shows the process flow sheet. Nitrogen (N_2) in the feed of the polyethylene reactor is inert and is used as a diluent carrier gas for ethylene. If there were no purge, the unreacted nitrogen will be accumulated within the reactor and will affect reactor operating conditions and hence reactor performance.

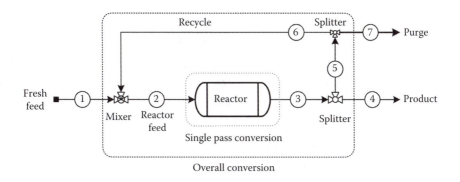

FIGURE 6.3
Schematic of reactive process with recycle and purge.

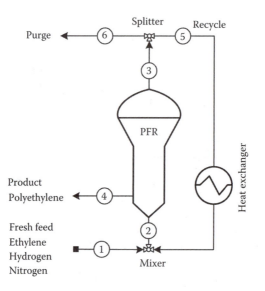

FIGURE 6.4
Process flow sheet of polyethylene fluidized bed reactor.

6.2 Degrees of Freedom Analysis for Reactive Multiple-Unit Processes

Problems in which there is insufficient information are underspecified. Problems with too much information are overspecified. A problem with just the right amount of information is uniquely specified. You have learned, by now, that the degrees of freedom analysis is used to determine whether a process is solvable [2,3]. The degrees of freedom analysis for multiple-unit systems can be easily extended to a reactive multiple-unit process by accounting for the chemical reactions. The basic steps are as follows:

1. Find the number of unknowns in the problem.
2. Add the number of reactions in the system.
3. Subtract the number of independent material balances.
4. Subtract the number of other relationships given in the problem statement.

Example 6.1 Methanol Production

Problem
Methanol is produced by the reaction of carbon dioxide and hydrogen:

$$CO_2 + 3H_2 \rightarrow CH_3OH + H_2O$$

The fresh feed to the process contains hydrogen, carbon dioxide, and 0.40 mol% inert (I). The reactor effluent passes to a separator that removes essentially all of the methanol and water formed, but none of the reactants or inert. The latter substances are recycled to the reactor. To avoid buildup of the inert in the system, a purge stream is withdrawn from the recycle. The feed to the reactor (not the fresh feed to the process) contains 28.0 mol% CO_2, 70.0 mol% H_2, and 2.00 mol% inert. The single-pass conversion of hydrogen is 60.0%. Calculate the molar flow rates and molar compositions of the fresh feed, the total feed to the reactor, the recycle stream, and the purge stream for a production rate of 155 mol CH_3OH/h.

Solution

Known quantities: The composition of the reactor feed, product molar flow rate of methanol. Inert mol% in the fresh feed.

Find: Molar flow rates and molar compositions of the fresh feed, the total feed to the reactor, the recycle stream, and the purge stream.

Analysis: The methanol production process–labeled flow sheet is shown in Example Figure 6.1.1.
  First, performing the degrees of freedom analysis will narrow down the process unit or system to start with.

	Systems			
	Mixing	Reactor	Condenser	Overall Process
Number of unknowns	7	6+1	9	7+1
Number of independent equations	3	5	5	5
Number of auxiliary relations	0	1	0	0
NDF	4	1	4	3

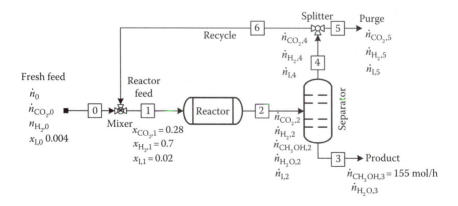

EXAMPLE FIGURE 6.1.1
Methanol production process flow sheet.

None of the process systems carries a number of degrees of freedom of zero. The lowest value of one corresponds to the reactor. This means that one of the unknowns should be specified, or additional information is needed for the problem to be potentially solved.

Method 1:

Basis: 155 mol/h of methanol produced.

System: Reactor.

The entire methanol leaving the reactor is completely separated and exits with the separator in the bottom product stream. In other words, the number of moles of methanol out of the reactor is 155 mol/h, thus reducing NDF to zero.

System: Reactor.

Material balance using the extent of reaction approach:

$$\dot{n}_{CO_2,2} = 0.28\dot{n}_1 - \xi$$

$$\dot{n}_{H_2,2} = 0.7\dot{n}_1 - 3\xi$$

$$\dot{n}_{CH_3OH,2} = 0.0 + \xi$$

$$\dot{n}_{H_2O,2} = 0.0 + \xi$$

$$\dot{n}_{I,2} = \dot{n}_{I,1} = 0.02\dot{n}_1$$

Substitute the number of moles of methanol produced in the methanol mole balance equation

$$\dot{n}_{CH_3OH,2} = 0.0 + \xi$$

$$155 = 0.0 + \xi$$

$$\xi = 155 \text{ mol/h}$$

The conversion can be calculated as

$$0.6 = \frac{\dot{n}_{H_2,1} - \dot{n}_{H_2,2}}{\dot{n}_{H_2,1}}$$

Hydrogen component mole balance

$$\dot{n}_{H_2,2} = \dot{n}_{H_2,1} - 3\xi$$

Rearranging

$$3\xi = \dot{n}_{H_2,1} - \dot{n}_{H_2,2}$$

Substitute the extent of reaction in the conversion relation

$$0.6 = \frac{3\xi}{\dot{n}_{H_2,1}} = \frac{3 \times 155}{\dot{n}_{H_2,1}} = \frac{465}{\dot{n}_{H_2,1}}$$

Rearranging

$$\dot{n}_{H_2,1} = \frac{465}{0.6} = 775 \text{ mol/h}$$

The percentage of hydrogen in the reactor feed (not the fresh feed) is 0.7:

$$\dot{n}_{H_2,1} = 0.7\dot{n}_1 = 775 \text{ mol/h} \Rightarrow \dot{n}_1 = \frac{775 \text{ mol/h}}{0.7} = 1107 \text{ mol/h}$$

Substitute the extent of reaction in the component mole balance equations to get the components molar flow rate (mol/h):

$$n_{CO_2,2} = 155, \quad \dot{n}_{H_2,2} = 310, \quad \dot{n}_{H_2O,2} = 155, \quad \dot{n}_{CH_3OH,2} = 155, \quad n_{I,2} = n_{I,1} = 22$$

System: Separator.

The entire methanol ends up in the separator bottom stream, and the rest of compounds end up in the top stream. Accordingly, the moles of compounds in stream 4 are

$$\dot{n}_{CO_2,4} = 155 \text{ mol/h}, \quad \dot{n}_{H_2,4} = 310 \text{ mol/h}, \quad \dot{n}_{I,4} = 22 \text{ mol/h}$$

System: Splitter.

Inlet and exit stream compositions of the splitter are equal. The composition of stream 4 leaving the top of the separator is

$$y_{CO_2,4} = \frac{155}{155 + 310 + 22} = 0.318$$

$$y_{H_2,4} = \frac{310}{155 + 310 + 22} = 0.636$$

$$y_{I,4} = \frac{22}{155 + 310 + 22} = 0.046$$

System: Mixer.

Balance around the mixing point:

$$\text{Total balance:} \quad \dot{n}_0 + \dot{n}_6 = 1107$$

$$\text{Inert balance:} \quad 0.004 \times \dot{n}_0 + 0.045 \times \dot{n}_6 = 0.02 \times 1107$$

Substitute the total balance in the inert component balance:

$$0.004 \times \dot{n}_0 + 0.045 \times (1107 - \dot{n}_0) = 0.02 \times 1107$$

Rearranging

$$(0.045 - 0.02)1107 = (0.045 - 0.004)\dot{n}_0 \Rightarrow \dot{n}_0 = 685 \text{ mol/h}$$

From the mixer overall material balance,

$$685 + \dot{n}_6 = 1107 \Rightarrow \dot{n}_6 = 422 \text{ mol/h}$$

Method 2:

Basis: 100 mol/h of feed stream (not fresh feed) to reactor.

System: Reactor.

The possible balance can be done around the reactor while taking 100 mol of feed to the reactor (not the fresh feed) as the basis. Taking this basis will reduce the reactor NDF to 0, and, hence, the problem is solvable.

$$CO_2 + 3H_2 \rightarrow CH_3OH + H_2O$$

Material balance using the extent of reaction approach:

$$\dot{n}_{CO_2,2} = 0.28(100) - \xi$$

$$\dot{n}_{H_2,2} = 0.7(100) - 3\xi$$

$$\dot{n}_{CH_3OH,2} = 0.0 + \xi$$

$$\dot{n}_{H_2O,2} = 0.0 + \xi$$

$$\dot{n}_{I,2} = \dot{n}_{I,1} = 0.02(100)$$

The conversion can be calculated as

$$0.6 = \frac{0.7(100) - \dot{n}_{H_2,2}}{0.7(100)}$$

The set of algebraic material balance equations are solved simultaneously, and the following answers are obtained:

$$\dot{n}_{CO_2,2} = 14 \text{ mol/h}, \quad \dot{n}_{H_2,2} = 28 \text{ mol/h}, \quad \dot{n}_{CH_3OH,2} = 14 \text{ mol/h}$$

$$\dot{n}_{H_2O,2} = 14 \text{ mol/h}, \quad \dot{n}_{I,2} = 2 \text{ mol/h}, \quad \xi = 14 \text{ mol/h}$$

Balance around the condenser. Note that the production rate of methanol will not be 155 mol/h because we have taken the unknown stream as the basis. The production rate for the new assumed basis will be different from 155 mol/h. Scaling up will be considered later.

$$\dot{n}_{CO_2,2} = \dot{n}_{CO_2,4} = 14 \text{ mol/h}$$

$$\dot{n}_{H_2,2} = \dot{n}_{H_2,4} = 28 \text{ mol/h}$$

$$\dot{n}_{I,2} = \dot{n}_{I,4} = 2 \text{ mol/h}$$

$$\dot{n}_{CH_3OH,2} = \dot{n}_{CH_3OH,3} = 14 \text{ mol/h}$$

$$\dot{n}_{H_2O,2} = \dot{n}_{H_2O,3} = 14 \text{ mol/h}$$

The recycle stream and the purge stream have the same composition; hence, the composition of the recycle and purge streams are as follows:

$$x_{CO_2,4} = x_{CO_2,5} = x_{CO_2,6} = \frac{14}{14+28+2} = 0.318$$

$$x_{H_2,4} = x_{H_2,5} = x_{H_2,6} = \frac{28}{14+28+2} = 0.636$$

$$x_{I,4} = x_{I,5} = x_{I,6} = \frac{2}{14+28+2} = 0.046$$

Balance around the mixing point.

$$\text{Total balance}: \quad \dot{n}_0 + \dot{n}_6 = 100$$

$$\text{Inert balance}: \quad 0.004 \times \dot{n}_0 + 0.046 \times \dot{n}_6 = 0.02 \times 100$$

Substitute the total balance in the inert component balance.

Inert balance:

$$0.004 \times \dot{n}_0 + 0.046 \times (100 - \dot{n}_0) = 0.02 \times 100$$

Rearranging

$$(0.046 - 0.02)100 = (0.046 - 0.004)\dot{n}_0 \Rightarrow \dot{n}_0 = 61.9 \text{ mol}$$

Total material balance around the mixing point of recycle and fresh feed:

$$61.9 + \dot{n}_6 = 100 \quad \dot{n}_6 = 38.1 \text{ mol}$$

Scaling:

Amount of fresh feed required to produce 155 mol of methanol is equal to

$$155\,\text{mol/h CH}_3\text{OH produced} \times \frac{61.9\,\text{mol/h of fresh feed}}{14\,\text{mol/h of CH}_3\text{OH produced}} = 685\,\text{mol/h}$$

A fresh feed of 685 mol/h is required to produce 155 mol/h.

6.3 Reaction and Multiple-Unit Steady-State Processes

The following points must be considered when solving problems where multiple-unit steady-state processes involve chemical reactions:

1. Some subsystems will involve reactions, and some others will not. For subsystems with reactions (generally the reactor and the overall system),

   a. Use individual component flows around the reactor and not the compositions.

   b. Include stoichiometry and generation/consumption on molar balances.

   c. In subsystems without reaction (such as mixers, splitters, and separators): input = output (moles are conserved, no generation/ consumption terms). More flexibility as to stream specification (component flows or composition).

2. Reactions combined with product separation and recycle:

   a. Recycle stream is introduced to recover and reuse unreacted reactants.

   b. Recycle allows the operation of reactor at low single-pass conversion and has a high overall conversion for the system.

3. Subsystems involving reactions with recycle and purge:

   a. Purge is introduced to prevent the buildup of inert (or incompletely separated products) in the system.

   b. Purge point is a simple splitter, that is, only one independent material balance exists.

Example 6.2 Ammonia Reactors with Recycle Stream

Problem

Nitrogen and hydrogen gases are fed to a reactor in stoichiometric quantities and react to form ammonia. The conversion of nitrogen to

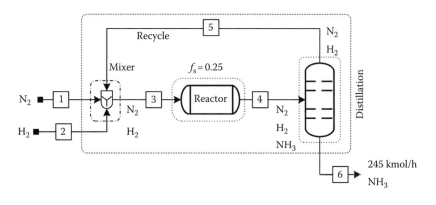

EXAMPLE FIGURE 6.2.1
Process flow sheet for ammonia production.

ammonia is 25%. Since the single pass is low, we decide to separate the product stream and recycle the unreacted gases back to the reactor. How much nitrogen is required to make 245 kmol/h of ammonia? What is the flow rate of the recycle stream?

Solution

Known quantities: Ammonia production rate, feed stream composition.

Find: Nitrogen required for producing 245 kmol/h of ammonia.

Analysis: The process flowchart is shown in Example Figure 6.2.1. Let us use 245 kmol/h of ammonia as the basis. The degrees of freedom analysis shown in the following table reveals that starting with the overall process is the best choice. The degrees of freedom analysis for the ammonia synthesis with recycle stream is shown in the following table:

	Systems		
	Reactor	Distillation	Overall Process
Number of unknowns	$5 + 1(\xi)$	5	$2 + 1(\xi)$
Number of independent equations	3	3	3
Number of auxiliary relations	1	—	0
NDF	2	2	0

The production of ammonia proceeds according to the following reaction:

$$N_2 + 3H_2 \rightarrow 2NH_3$$

Since the control volume over the overall process contains a chemical reaction, the number of moles may not be conserved and atoms may switch from one molecule to another.

Mass in = mass out, and atoms in = atoms out.

System: Overall process.

Overall atomic balance for nitrogen:

$$2.0\dot{n}_1 = 245 \, \frac{\text{kmol N}}{\text{h}},$$

The molar feed rate of nitrogen gas is

$$\dot{n}_1 = \frac{245}{2} = 122.5 \, \text{kmol/h}$$

Overall atomic balance for hydrogen:

$$2.0\dot{n}_2 = 3 \times 245 \, \frac{\text{kmol H}}{\text{h}}$$

The molar flow rate of feed hydrogen:

$$\dot{n}_2 = 367.5 \, \text{kmol/h}$$

System: Reactor.

From the problem statement, we know that 25% of nitrogen entering the reactor has been converted to ammonia. From the definition of conversion,

$$\text{Conversion} = f_A = \frac{\text{mol N}_2 \text{ reacted}}{\text{mol N}_2 \text{ fed}} = \frac{n_{N_2,3} - n_{N_2,4}}{n_{N_2,3}}$$

Ammonia component mole balance around the reactor:

$$\dot{n}_{NH_3,4} = 0 + 2\xi$$

The entire ammonia generated in the reactor and produced in stream 4 exit the bottom stream of separator. Accordingly,

$$\dot{n}_{NH_3,4} = \dot{n}_{NH_3,6}$$

Substitute the values of ammonia produced in the ammonia component balance around the reactor:

$$245 \, \text{kmol/h} = 0 + 2\xi, \quad \xi = \frac{245}{2} = 122.5 \, \text{kmol/h}$$

The extent of reaction:

$$\xi = 122.5 \, \text{kmol/h}$$

Conversion:

$$0.25 = \frac{n_{N_2,3} - n_{N_2,4}}{n_{N_2,3}} = \frac{\xi}{n_{N_2,3}} = \frac{122.5}{n_{N_2,3}}$$

The molar feed rate of nitrogen to the reactor:

$$n_{N_2,3} = 490 \text{ kmol/h}$$

We also know that hydrogen is fed in stoichiometric proportion, so the amount of hydrogen is three times greater than the amount of nitrogen in every stream:

$$n_{H_2,3} = 3n_{N_2,3} = 3 \times 490 = 1470 \text{ kmol/h}$$

Mole balance for nitrogen around the reactor:

$$n_{N_2,4} = n_{N_2,3} - \xi = 490 - 122.5 = 367.5 \text{ kmol/h}$$

The amount of hydrogen is three times greater than the amount of nitrogen in every stream:

$$n_{H_2,4} = 3n_{N_2,4} = 3 \times 367.5 = 1102.5 \text{ kmol/h}$$

The entire nonreacted nitrogen and unreacted hydrogen are recycled. Therefore, the molar flow rate of the recycle stream ($\dot{n}_5$) is

$$\dot{n}_5 = n_{H_2,5} + n_{N_2,5} = 1102.5 + 367.5 = 1470 \text{ kmol/h}$$

Example 6.3 Ammonia Reaction with Inert in the Fresh Feed

Problem
A gas stream contains nitrogen, hydrogen, and 0.2 mol% argon (Ar) as an impurity. Nitrogen and hydrogen are in stoichiometric proportion. If we do only recycling, then Ar concentration will build up and eventually shut down the reaction. To avoid this, we purge a portion of the recycle stream so that the level of Ar in the recycle stream is maintained at 7.0%. The reactor single pass fractional conversion is 0.25. How much nitrogen is required to produce 245 kmol/h of ammonia? What is the flow rate of the recycle stream? What is the flow rate of the purge stream?

Solution

Known quantities: Production rate of ammonia, conversion, Ar composition in feed and purge streams.

Find: The flow rate of purge and recycle streams.

Analysis: This problem is a little more complicated than the preceding problems above and is solved in detail in the following. To start, draw

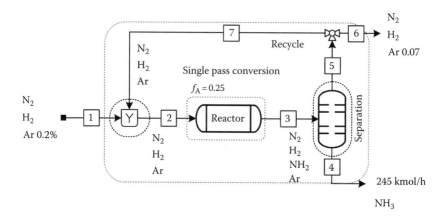

EXAMPLE FIGURE 6.3.1
Process flow sheet for ammonia production with recycle and purge.

the process flow sheet and choose a basis. Perform balances around the overall system, separator, reactor, and splitter. The process flow sheet is shown in Example Figure 6.3.1.

Perform a degrees of freedom analysis and choose the system with zero degree of freedom. The degrees of freedom analysis for ammonia synthesis with purge stream is analyzed in the following table:

	System			
	Reactor	**Separator**	**Splitter**	**Overall Process**
Number of unknowns	$7+1(\xi)$	7	4	$4+1(\xi)$
Number of independent equations	4	4	1	4
Number of auxiliary relations	1	—	—	1
NDF	3	3	2	0

We know some information about some of the streams from common sense and simple calculations. Since hydrogen and nitrogen are fed in stoichiometric proportion, consequently for every mole of nitrogen in the system, 3 mol of hydrogen is present:

Recycle stream:

$$y_{N_2,7} + y_{H_2,7} + 0.07 = 1.0$$

No ammonia is present in the recycle stream, because all of the ammonia is separated in distillation column and condensed:

$$y_{NH_3,7} = 0$$

$$y_{Ar,7} = 0.07 \text{ in stream R}$$

$$3y_{N_2,7} = y_{H_2,7}$$

Solving:

$$y_{N_2,7} + 3y_{N_2,7} + 0.07 = 1.0$$

$$y_{N_2,7} = 0.2325$$

$$y_{H_2,7} = 3y_{N_2,7} = 0.6975$$

Since stream 5 splits into recycle and purge stream, the compositions of streams 6 and 7 are the same as stream 5. Mass and atoms are always conserved, but because the control volume contains a chemical reaction, the atoms may move between molecules and moles may not be conserved. So we most certainly cannot write an overall mole balance. We can however write individual or overall atom balances. We have three types of atoms: N, H, and Ar, so we can write three independent equations.

Overall atomic balances for the system.

Argon balance: $0.002\dot{n}_1 = 0.07n_6 \rightarrow \dot{n}_1 = 35\dot{n}_6$
N atomic balance: $2(0.2495\dot{n}_1) = 245 \text{ kmol/h} + 2 \times (0.2325\dot{n}_6)$

Substitute $\dot{n}_1 \rightarrow 2(0.2495 \times 35\dot{n}_6) = 245 \text{ kmol/h} + 2 \times (0.2325\dot{n}_6)$

Solving for stream 6 (Purge stream) n_6

$$n_6 = 14 \text{ kmol/h}$$

$$\dot{n}_1 = 35 \times 14 = 490 \text{ kmol/h}$$

Component balance around the reactor

Nitrogen:

$$\dot{n}_{N_2,3} = \dot{n}_{N_2,2} - \xi$$

Ammonia:

$$\dot{n}_{NH_3,3} = \dot{n}_{NH_3,2} + 2\xi$$

$$245 \text{ kmol/h} = 0 + 2\xi$$

The extent of reaction:

$$\xi = 122.5 \text{ kmol/h}$$

Conversion of nitrogen:

$$f_A = 0.25 = \frac{\dot{n}_{N_2,2} - \dot{n}_{N_2,3}}{\dot{n}_{N_2,2}} = \frac{\xi}{\dot{n}_{N_2,2}} = \frac{122.5}{\dot{n}_{N_2,2}}$$

$$\dot{n}_{N_2,2} = \frac{122.5}{0.25} = 490 \text{ kmol/h}$$

$$\dot{n}_{H_2,2} = 3\dot{n}_{N_2,2} = 3 \times 490 = 1470 \text{ kmol/h}$$

Moles of nitrogen in stream 3:

$$\dot{n}_{N_2,3} = \dot{n}_{N_2,2} - \xi = 490 - 122.5 = 367.5 \text{ kmol/h}$$

Since hydrogen and nitrogen are fed in stoichiometric quantities,

$$\dot{n}_{H_2,3} = 3\dot{n}_{N_2,3} = 3 \times 367.5 = 1102.5 \text{ kmol/h}$$

$$\dot{n}_{H_2,3} + \dot{n}_{N_2,3} = \dot{n}_{H_2,5} + \dot{n}_{N_2,5} = 1102.5 + 367.5 = 1470 \text{ kmol/h}$$

Since the compositions of streams 5, 6, and 7 are equal, the total moles of nitrogen and hydrogen in stream 5 is 93 mol and the balance is argon:

$$1470 \text{ kmol/h} = 0.93\dot{n}_5$$

The molar flow rate of stream 5:

$$\dot{n}_5 = 1470/0.93 = 1580.65 \text{ kmol/h}$$

The molar flow rate of the recycle stream 7 can be obtained by performing an overall material balance around the splitter:

$$\dot{n}_5 = \dot{n}_6 + \dot{n}_7$$

Substituting the values of $\dot{n}_6$ and $\dot{n}_5$

$$1580.65 = 14 + \dot{n}_7$$

The recycle stream, $\dot{n}_7 = 1580.65 - 14 = 1566.65 \text{ kmol/h}$.

Example 6.4 Ammonia Synthesis with Two Fresh Feed Streams

Problem
Nitrogen stream contains 0.2% argon (Ar) as an impurity is combined with stoichiometric amounts of nitrogen and hydrogen. If we do only recycling, then, since argon is an inert gas, its concentration will build up and eventually shut down the reaction. To avoid this, we purge a portion of the recycle stream so that the level of argon in the recycle stream is maintained at 7.0%. Determine (a) the purge molar flow rate, (b) the amount of nitrogen in the fresh feed stream that is required to make 245 kmol/h of ammonia, and (c) the flow rate of the recycle stream.

Solution

Known quantities: Molar flow rate of produced ammonia, argon mole fraction in feed and purge streams, and reactor single-pass conversion.

Find: The molar flow rate of the purge stream and recycle stream.

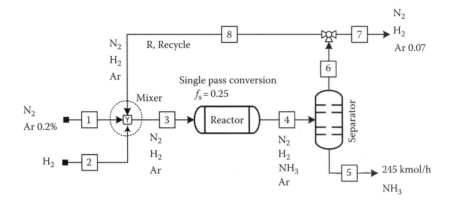

EXAMPLE FIGURE 6.4.1
Process flow sheet for ammonia production with inert in fresh feed.

Analysis: Draw and label the process flow sheet and choose a basis. Perform balances around the overall system, separator, reactor, and splitter. The labeled process flow sheet is shown in Example Figure 6.4.1.

Perform a degrees of freedom analysis and choose the system with zero degree of freedom. Degrees of freedom analysis for ammonia synthesis with purge stream is exposed in the following table:

	System within Control Volume			
	Reactor	**Separator**	**Splitter**	**Overall Process**
Number of unknowns	$7 + 1(\xi)$	7	3	$4 + 1(\xi)$
Number of independent equations	4	4	1	4
Number of auxiliary relations	1	—	—	1
NDF	3	3	2	0

Basis: 245 kmol/h of ammonia.

Since hydrogen and nitrogen are fed in stoichiometric proportion, consequently for every mole of nitrogen in any stream of the system, 3 mol of hydrogen is present per mole of nitrogen. Accordingly, the composition of the recycle stream (stream 7)

$$y_{N_2,7} + y_{H_2,7} + y_{Ar,7} = 1.0$$

No ammonia is present in the recycle stream, because all of the ammonia is condensed and separated:

$$y_{NH_3,7} = 0$$

$$y_{Ar,7} = 0.07 \text{ in stream R}$$

$3y_{N_2,7} = y_{H_2,7}$, because both nitrogen and hydrogen are fed in stoichio-metric proportion

$$y_{N_2,7} + 3y_{N_2,7} + 0.07 = 1.0, \quad y_{N_2,7} = 0.2325$$

$$y_{H_2,7} = 3y_{N_2,7} = 0.6975$$

Since stream 6 splits into purge (stream 7) and recycle (stream 8), the compositions of streams 7 and 8 are the same as stream 6.

System: Entire process.

(a) Start with overall atomic balances for argon:

$$0.002\dot{n}_1 = 0.07\dot{n}_7$$

$$\dot{n}_1 = 35\dot{n}_7$$

Nitrogen mole fraction in the fresh feed stream is $1 - 0.002 = 0.998$.

N atomic balance:

$$2(0.998\dot{n}_1) = 245 \text{ kmol/h} + 2 \times (0.2325\dot{n}_7)$$

Substitute $\dot{n}_1 = 35\dot{n}_7$:

$$2(0.998 \times 35\dot{n}_7) = 245 \text{ kmol/h} + 2 \times (0.2325\dot{n}_7)$$

Solving for stream 7 (purge stream) $\dot{n}_7$

$$\dot{n}_7 = 3.55 \text{ kmol/h}$$

$$\dot{n}_1 = 35 \times 3.55 = 124 \text{ kmol/h}$$

(b) The number of moles of nitrogen required to produce 245 kmol/h of ammonia is

$$\dot{n}_{N_2,1} = 0.998(124 \text{ kmol/h}) = 123.75 \text{ kmol/h}$$

(c) Component balance around the reactor:

$$\dot{n}_{N_2,4} = \dot{n}_{N_2,3} - \xi$$
$$\dot{n}_{NH_3,4} = \dot{n}_{NH_3,3} + 2\xi$$
$$245 \text{ kmol/h} = 0 + 2\xi$$

The extent of reaction:

$$\xi = 122.5 \text{ kmol/h}$$

Conversion of nitrogen:

$$f_A = 0.25 = \frac{\dot{n}_{N_2,3} - \dot{n}_{N_2,4}}{\dot{n}_{N_2,3}} = \frac{\xi}{\dot{n}_{N_2,3}} = \frac{122.5}{\dot{n}_{N_2,3}}$$

$$\dot{n}_{N_2,3} = \frac{122.5}{0.25} = 490 \text{ kmol/h}$$

The molar flow rate of hydrogen is three times that of nitrogen, due to the stoichiometric proportion of feed rates of N_2 and H_2:

$$\dot{n}_{H_2,3} = 3\dot{n}_{N_2,3} = 3 \times 490 = 1470 \text{ kmol/h}$$

Moles of nitrogen in stream 4:

$$\dot{n}_{N_2,4} = \dot{n}_{N_2,3} - \xi = 490 - 122.5 = 367.5 \text{ kmol/h}$$

Since hydrogen and nitrogen are fed in stoichiometric quantities,

$$\dot{n}_{H_2,4} = 3\dot{n}_{N_2,4} = 3 \times 367.5 = 1102.5 \text{ kmol/h}$$

$$\dot{n}_{H_2,4} + \dot{n}_{N_2,4} = \dot{n}_{H_2,6} + \dot{n}_{N_2,6} = 1102.5 + 367.5 = 1470 \text{ kmol/h}$$

The total number of moles of nitrogen and hydrogen in stream 4 is the same as that of stream 6 because only ammonia is separated and removed as liquid from the bottom of the separator.

The compositions of streams 6, 7, and 8 are identical. The total mole fraction of nitrogen and hydrogen in stream 6 is 0.93, and the balance is argon (0.07). The following equation is used to find the total molar flow rate of stream number 6:

$$1470 \text{ kmol/h} = 0.93\dot{n}_6$$

The molar flow rate of stream 6 ($\dot{n}_6$) is

$$\dot{n}_6 = 1470/0.93 = 1580.65 \text{ kmol/h}$$

The molar flow rate of the recycle stream 8 can be obtained by performing an overall material balance around the splitter:

$$\dot{n}_6 = \dot{n}_7 + \dot{n}_8$$

Substituting the values of $\dot{n}_7$ and $\dot{n}_6$

$$1580.65 = 3.55 + \dot{n}_8$$

The recycle stream molar flow rate, $\dot{n}_8 = 1577.1 \text{ kmol/h}$

Example 6.5 Ammonia Synthesis with Ammonia in the Recycle Stream

Problem

A stream of 99.8% nitrogen and 0.2% argon is combined with stoichiometric hydrogen and fed to a reactor to produce 100 kmol/h of ammonia. Ten percent of the hydrogen fed to the reactor is converted to ammonia. The products are then separated in a condenser. The liquid ammonia is withdrawn from the bottom of the condenser as a product. The uncondensed hydrogen, nitrogen, and ammonia are recycled and combined with the nitrogen and hydrogen stream entering the reactor. A portion of the recycle stream is purged to keep the mole fraction of argon in the recycle stream at 0.07. The ammonia produced is not condensed completely, and part of it is recycled. The ammonia mole fraction in the purge stream was found to be 0.263. Determine the recycle stream molar flow rate.

Solution

Known quantities: Product stream flow rate and argon mole fraction in purge stream.

Find: All stream flow rates and composition.

Analysis: The process flow sheet is shown in Example Figure 6.5.1.

Basis: 100 kmol/h of ammonia produced.

The following reaction takes place:

$$N_2 + 3H_2 \rightarrow 2NH_3$$

Composition of fresh feed (0.998 N_2 and 0.002 Ar)

Purge stream composition (stream 7):

$$0.07 + x_{N_2,7} + 3x_{N_2,7} + 0.263 = 1.0$$

$$x_{N_2,7} = 0.167$$

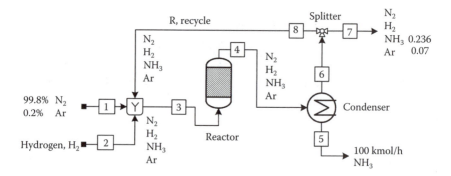

EXAMPLE FIGURE 6.5.1
Process flow sheet for ammonia production.

Stream 6 is split into stream 7 and 8 → the compositions of the three streams are identical:

$$x_{N_2,6} = x_{N_2,7} = x_{N_2,8} = 0.167$$

$$x_{H_2,6} = x_{H_2,7} = x_{H_2,8} = 0.527$$

$$x_{Ar,6} = x_{Ar,7} = x_{Ar,8} = 0.070$$

$$x_{NH_3,6} = x_{NH_3,7} = x_{NH_3,8} = 0.236$$

System: Overall balance.
Argon atomic balance:

$$y_{Ar,1}\dot{n}_1 = y_{Ar,7}\dot{n}_7$$

$$0.002\dot{n}_1 = 0.07\dot{n}_7,$$

$$\dot{n}_1 = \frac{0.07}{0.002}\dot{n}_7 = 35\dot{n}_7 \Rightarrow \dot{n}_1 = 35\dot{n}_7$$

The overall nitrogen atomic balance:

$$2 \times 0.998 \times \dot{n}_1 = 100 + \left(2 \times 0.167\dot{n}_7\right) + 0.263\dot{n}_7$$

Substitute $\dot{n}_1 = 35\dot{n}_7$ in this equation and solve for $\dot{n}_7$:

$$0.998 \times 2 \times \left(35\dot{n}_7\right) = 100 + 0.597\dot{n}_7$$

Solving for $\dot{n}_7$ (purge stream)

$$\dot{n}_7 = 1.44 \text{ kmol/h} \Rightarrow \dot{n}_1 = 35\dot{n}_7 \Rightarrow \dot{n}_1 = 50.4 \text{ kmol/h}$$

Flow rate of nitrogen in the fresh nitrogen stream:

$$\dot{n}_{N_2,1} = 0.998 \times 50.5 = 50.4 \text{ kmol/h}$$

Hydrogen is fed in stoichiometric ratio with nitrogen; accordingly, the molar flow rate of the hydrogen fresh feed stream is

$$\dot{n}_{H_2,2} = 3 \times \dot{n}_{N_2,1} = 3 \times 50.4 = 151 \text{ kmol/h}$$

System: Reactor.

$$\text{Fractional conversion of hydrogen} = 0.10 = \frac{\dot{n}_{H_2,3} - \dot{n}_{H_2,4}}{\dot{n}_{H_2,3}}$$

Hydrogen flow rate in stream 4 in order to determine $\dot{n}_{H_2,4}$ as a function of $\dot{n}_8$:

$$\dot{n}_{H_2,4} = 0.5 \times \left(\dot{n}_8 + 1.44 \, \text{kmol/h} \right)$$

Note that 0.5 is the mole fraction of hydrogen in stream 6, and 1.44 kmol/h is the molar flow rate of purge stream.

Hydrogen mole balance around the mixing point to find $\dot{n}_{H_2,3}$ as a function of $\dot{n}_8$:

$$\dot{n}_{H_2,3} = 151 \, \text{kmol/h} + 0.5 \dot{n}_8$$

Substitute in conversion relation to find an a relation with one unknown ($\dot{n}_8$):

$$0.10 = \frac{151 \, \text{kmol/h} + 0.5 \dot{n}_8 - 0.5 \times \left(\dot{n}_8 + 1.44 \, \text{kmol/h} \right)}{151 \, \text{kmol/h} + 0.5 \dot{n}_8} = \frac{151 - 0.5 \times 1.44}{151 \, \text{kmol/h} + 0.5 \dot{n}_8}$$

$$0.05 \dot{n}_8 = 135.18, \quad \dot{n}_8 = \frac{135.18}{0.05} = 2703.6 \, \text{kmol/h}$$

The recycle stream molar flow rate, $\dot{n}_8 = 2703.6 \, \text{kmol/h}$

Example 6.6 Methanol Production

Problem
Methanol is produced by the reaction of CO with H_2. A portion of the methanol leaving the reactor is condensed, and the unconsumed CO and H_2 and the uncondensed methanol are recycled back to the reactor. The reactor effluent flows at a rate of 275 mol/min and contains the following components: 0.227 H_2, 0.330 CO, and 0.443 methanol (all in mole fractions). The mole fraction of methanol in the recycle stream is 0.004. Calculate the molar flow rates of CO and H_2 in the fresh feed, and the production rate of methanol.

Solution

Known quantities: Reactor exit molar flow rate and compositions.

Find: All streams flow rates and compositions.

Analysis: Note that the mass fractions of reactor exit are given. We must convert these to mole fractions to solve the problem (see Example Figure 6.6.1).

Basis: 1 min of operation.

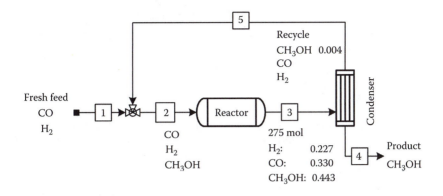

EXAMPLE FIGURE 6.6.1
Methanol production process.

The degrees of freedom analysis is shown in the following table.

	System			
	Mixing	**Reactor**	**Condenser**	**Overall**
Number of unknowns	7	3+1	3	3+1
Number of independent equations	3	3	3	3
Auxiliary relations	0	0	0	0
NDF	4	1	0	1

The condenser got the zero degree of freedom; accordingly, it is considered the first system to start with.

System: Condenser.

Total material balance:

$$275 \text{ mol/min} = \dot{n}_4 + \dot{n}_5$$

Component balance (methanol):

$$0.443(275) = 121.83 = \dot{n}_4 + 0.004\dot{n}_5$$

Substitute $\dot{n}_4$ in the methanol component balance:

$$121.83 = 275 - \dot{n}_5 + 0.004\dot{n}_5$$

The last two equations can be solved simultaneously to obtain

$$\dot{n}_4 = 121.21 \text{ mol/min (production rate of methanol)}$$

$$\dot{n}_5 = 153.79 \text{ mol/min}$$

System: Overall process.

Element balances are generally recommended when doing balancing around the block. Component balances around the block can be applied, but you must include the reactor and the corresponding consumption/ generation terms.

C atomic balance:

$$\dot{n}_{CO,1} = \dot{n}_4$$

$$\dot{n}_{CO,1} = 121.21 \text{ mol/min}$$

H atomic balance:

$$2\dot{n}_{H_2,1} = 4\dot{n}_4,$$

$$2n_{H_2,1} = 4 \times 121.21 = 484.84 \Rightarrow n_{H_2,1} = \frac{484.84}{2} = 242.42 \text{ mol/min} \Rightarrow$$

$$n_{H_2,1} = \frac{484.84}{2} = 242.42 \text{ mol/min}$$

The fresh feed $= n_{CO,1} + n_{H_2,1} = 121.21 + 242.42 = 363.63 \text{ mol/min}$.

Example 6.7 Methanol Production

Problem

In the process of methanol production, a gas stream flowing at a molar flow rate of 100 mol/min containing 32 mol% CO, 64 mol% H_2, and 4 mol% of N_2 is fed to the reactor. The product from the reactor is condensed where liquid methanol is separated. Unreacted gases are recycled in a ratio of 5:1 mol of recycle to moles of fresh feed. Parts of the recycled gases are purged. Calculate all unknown stream flow rates and compositions. The reaction equation is

$$CO + 2H_2 \rightarrow CH_3OH$$

Solution

Known quantities: Feed flow rate and compositions. Recycle ration is 5:1.

Find: Unknown stream flow rate and compositions.

Analysis: The process flow sheet is shown in Example Figure 6.7.1.

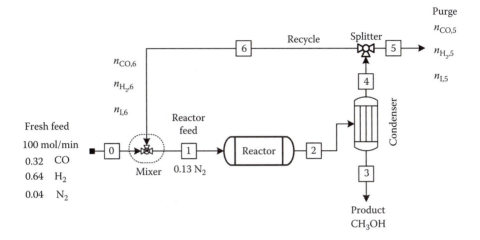

EXAMPLE FIGURE 6.7.1
Methanol production process with inert in fresh feed.

Basis: 100 mol/min fresh feed.

System: Mixing point.

Relation:

5 mol of recycle/1 mol of fresh feed

Recycle stream flow rate = 500 mol/min

The reactor feed, not the fresh feed, consists of the recycle stream plus the fresh feed stream. The reactor feed stream ($\dot{n}_1$)

$$\dot{n}_1 = \dot{n}_0 + \dot{n}_6$$

$$\dot{n}_1 = 500 + 100 = 600 \text{ mol/min}$$

N_2 balance at feed mixing point:

$$(100)(0.04) + 500(y_{N2,6}) = 600(0.13)$$

Nitrogen mole fraction in reactor feed,

$$y_{N2,6} = 0.148$$

Note that for the case of splitter, the mole fractions for all streams are the same:

$$y_{N2,4} = y_{N2,5} = y_{N2,6}$$

System: Overall process.

Overall component balances (nitrogen):

$$N_2: \quad (0.04)100 = (0.148)n_5 \rightarrow n_5 = 27.03 \text{ mol/min}$$

Overall atomic balance:

$$C: \quad 0.32 \times 100 = x_{CO_2,5}(27.03) + \dot{n}_3$$

$$H: \quad 2(0.64 \times 100) = 4(n_3) + 2(27.03)(1 - 0.148 - x_{CO_2,5})$$

Simplifying gives

$$32 = x_{CO_2,5}(27.03) + \dot{n}_3$$

$$128 = 4n_3 + 54.06(0.852 - x_{CO_2,5})$$

Substitute $\dot{n}_3 = 32 - x_{CO_2,5}(27.03)$ in the H atomic balance equation:

$$128 = 4\left(32 - x_{CO_2,5}27.03\right) + 54.06(0.852 - x_{CO_2,5})$$

$$128 = 128 - 108.12x_{CO_2,5} + 46.06 - x_{CO_2,5}54.06$$

Simplifying and rearranging give

$$162.18x_{CO_2,5} = 46.06$$

$$x_{CO_2,5} = 0.284 \quad \text{and} \quad n_3 = 24.3 \text{ mol CH}_3\text{OH}$$

Homework Problems

6.1 Propylene oxide (C_3H_8O) is produced by reacting propylene (C_3H_6) with oxygen and hydrogen. A side reaction results in reacting propylene with hydrogen and produces propane. A mixture of 35 mol/h air, 4 mol/h propylene, and 12.6 mol/h hydrogen (H_2) along with 5000 mol/h carbon dioxide (CO_2 is used as inert solvent) is fed to a continuous reactor. The percent single-pass conversion of limiting reactant is 10%. The fractional yield of C_3H_6 to C_3H_8O is 0.80 (i.e., fraction of moles of C_3H_6 reacted to produce C_3H_8O). The reactor effluents are separated where gases leave the top of the column. Propylene, propane, and propylene oxide (C_3H_6, C_3H_8, and C_3H_8O) are condensed and removed from the bottom of the column as liquids. Calculate the mole flow rate of each component in the effluent streams. ($\dot{n}_{H_2} = 12.2$, $\dot{n}_{O_2} = 7.19$ $\dot{n}_{N_2} = 27.65$, $\dot{n}_{CO_2} = 5000$)

The following reactions are taking place:

$$C_3H_6 + H_2 + \frac{1}{2}O_2 \rightarrow C_3H_8O$$

$$C_3H_6 + H_2 \rightarrow C_3H_8$$

6.2 A stream containing 180 kmol/h of propylene (C_3H_6) is mixed with a stream holding 210 kmol/h of CO and 210 kmol/h H_2. The two streams are mixed with a recycle stream coming from the overhead of a separator. The recycle stream contains pure propylene. A single-pass conversion of 30% propylene is attained. The desired product butanal (C_4H_8O) is removed in a bottom stream; unreacted CO and H_2 are removed from the separator column. Unreacted C_3H_6 is recovered and recycled. Calculate the production rate of C_4H_8O and the flow rate of the recycle stream. (180, 420 kmol/h)
The following reaction is taking place:

$$C_3H_6 + CO + H_2 \rightarrow C_4H_8O$$

6.3 Feed mixture containing 5% propane and 95% propylene is used to produce butanal (C_4H_8O) according to the following reaction:

$$C_3H_6 + CO + H_2 \rightarrow C_4H_8O$$

An overall conversion of 0.90 can be achieved. Propane is an inert, and therefore, a purge stream is installed to avoid the accumulation of propane in the process. The single-pass conversion of propylene in the reactor is 0.3. The butanal production rate is maintained at a rate of 180 kmol/h. Unreacted hydrogen and carbon dioxide are separated and sent to surge tanks. Calculate the flow rate of the fresh propane/propylene stream to the process (kmol/h), the mol% of inert in the purge stream, and the total flow rate (kmol/h) of the feed to the reactor. (210.5 kmol/h, 34.5 mol% inert, 1240.5 kmol/h)

6.4 Benzene is produced using hydrodealkylation process. During the hydrodealkylation of toluene ($C_6H_5CH_3$), both benzene (C_6H_6) and methane (CH_4) are produced.

$$C_6H_5CH_3 + H_2 \rightarrow C_6H_6 + CH_4$$

100 kmol/h of pure liquid toluene is fed along with 200 kmol/h of pure hydrogen gas. These streams are combined with recycle streams and fed to a reactor. In the reactor, 50% of the toluene fed to the reactor

is consumed. The effluent from the reactor enters a vapor–liquid sepa-rator. All of the hydrogen and methane gas leave the separator in the vapor stream; all of the liquids toluene and benzene leave the separator in the liquid stream. The liquid stream then enters a second separator in which all of the benzene leaves in one stream and all of the toluene leaves in the other stream. The benzene stream leaves the process as a product stream. The toluene stream is completely recycled by mixing it with the toluene in fresh fed to the process. The vapor stream from the separator is split into two equal streams. One of these streams is recycled and combined with the hydrogen fresh feed stream, and the other stream is sent to another process unit for further processing. Determine the following:

(a) The molar flow rate of benzene produced. (100 kmol/h)
(b) The composition of first separator vapor stream. (50 mol% CH_4 and 50 mol% H_2)
(c) The toluene is actually fed to the reactor. (200 kmol/h)
(d) The component molar flow rate of the vapor stream fed to the reactor. ($H_2 = 300$ kmol/h, $CH_4 = 100$ kmol/h)

6.5 Feed stream flowing at 10 mol/h consists of 90% Zr, and 10% U is reacted with HCl to produce metal chlorides. The HCl stream contains some water in addition to HCl; the stream is a result of merging recy-cle stream and makes up pure HCl stream via the following reactions:

$$U + 3HCl \rightarrow UCl_3 + \frac{3}{2}H_2$$
$$Zr + 4HCl \rightarrow ZrCl_4 + 2H_2$$

U and Zr are completely converted to chlorides. The total HCl fed to the first reactor (not the fresh feed) is twice the amount required by the reaction stoichiometry. The UCl_3 produced is a solid and, hence, is readily removed from the remaining gaseous reaction products. The vapors from this first reactor are passed to a second reactor where the $ZrCl_4$ vapor is reacted with steam to produce solid ZrO_2 according to the following reaction:

$$ZrCl_4 + 2H_2O \rightarrow ZrO_2 + 4HCl$$

The reaction goes to completion, and the ZrO_2 solid is separated from the gaseous reactor products. The gases leaving the second reactor are sent to an absorber, which uses a solvent consisting of 90% H_2O and 10% HCl solution (coming from stripper) to absorb HCl. The liquid leaving bottom of absorber containing 50% HCl and 50% H_2O and an overhead gas containing 90% H_2 and 10% HCl. The absorber liquid

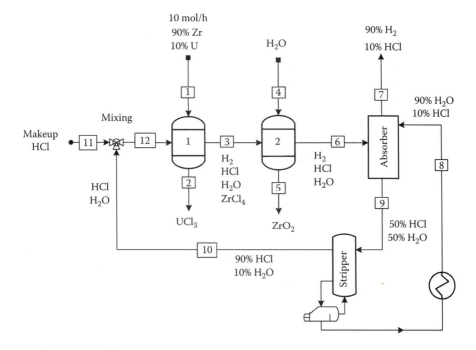

PROBLEM FIGURE 6.5.1
Production of metal chloride.

is sent to a stripper, which boils off most of the HCl to produce 90% HCl and 10% H_2O recycle vapor. The remaining stripper bottoms are cooled and recycled to the absorber for use as a solvent. Calculate the absorber inlet and exit solvent molar flow rates (Problem Figure 6.5.1). ($n_8 = 80.92$ mol/h, $n_9 = 161.852$ mol/h)

6.6 Methanol (CH_3OH) is produced via the partial oxidization of methane (CH_4). The fresh feed contains methane and oxygen that have been mixed in stoichiometric proportions. This feed is mixed with recycle gases. The single-pass conversion of methane is 10.0%. The stream leaving the reactor is passed into a separation process where the entire methanol, and none of the other materials, is removed. All of the remaining materials are recycled to be mixed with the fresh feed. Determine the feed rate of the stream entering the reactor required to produce 1000.0 mol/h of methanol. The following reaction is taking place. (15,000 mol/h)

$$CH_4 + \frac{1}{2}O_2 \rightarrow CH_3OH$$

6.7 Propylene oxide (C_3H_6O) is produced by the partial oxidation of propane (C_3H_8). The fresh feed to the entire system is an equimolar stream of propane and oxygen. The reactor effluent stream is split into recycle stream and product stream. The ratio of the flow rate of recycle stream to product stream is 5.0. A total of 100 mol of product leaves the system. The mole fraction of propane in the product stream is 0.20. How much propane does enter the system in the fresh feed? What is the single pass of propane conversion? (100 mol/h, 0.2)

The following reaction is taking place:

$$C_3H_8 + O_2 \rightarrow C_3H_6O + H_2O$$

6.8 Raw natural gas flowing at 1000 mol/h contains 15.0 mol% CO_2, 5.00 mol% H_2S, and 1.41 mol% COS, and the balance is methane. The separations system shown in Problem Figure 6.8.1 is capable of handling a certain flow rate of raw natural gas. Any flow rate higher than this limit is sent into the bypass around the separation unit and then mixed with the treated gas to yield the final product. The final product is to contain no more than 1.00 mol% H_2S and no more than 0.300 mol% COS. Stream 6 leaving the separation unit contains H_2S and COS only. The separations unit maximum flow rate it can handle is 900 mol/h; any flow beyond this must go to the bypass system. Find out the flow rate of the treated natural gas stream and flow rate of the reject stream from the separator? (948.2, 51.8 mol/h)

6.9 Ethylene oxide (C_2H_4O) is formed by the partial oxidation of ethylene in a gas phase reaction. In this process, 1000 mol/h of pure ethylene is fed to the system. It is mixed with air. The feed rate of the air is such that the molar ratio of ethylene in the fresh feed to oxygen is 2:1. The ethylene single-pass conversion is 25%. The stream leaving the reactor contains ethylene, ethylene oxide, water, CO_2, and N_2 only; all of the oxygen is consumed. The reactor effluent goes to a separation unit in which the ethylene and nitrogen are separated from the other gases. A portion of the ethylene and nitrogen stream is purged, and the

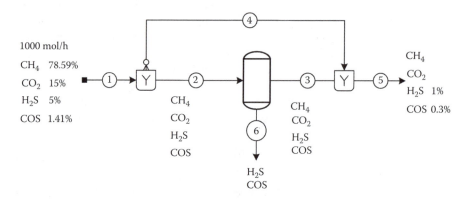

PROBLEM FIGURE 6.8.1
Separation of sulfur compounds from raw natural gas.

remainder is mixed with the fresh ethylene and air streams to form the reactor feed. The remaining gases (ethylene oxide, water, and CO_2) are sent to other units for further processing. A total of 50 mol/h of CO_2 appears in this stream. The gas phase reaction proceeds according to the following stoichiometry:

$$C_2H_4 + \frac{1}{2}O_2 \rightarrow C_2H_4O$$

$$C_2H_4 + 3O_2 \rightarrow 2H_2O + 2CO_2$$

Determine the amount of ethylene oxide is produced (Problem Figure 6.9.1). (850 mol/h)

6.10 Formaldehyde is produced via hot methanol vapor and air. Inside the reactor, hydrogen produced via the reaction reacts with oxygen in the air stream to produce water. The separator overhead steam was found to contain 667 mol/h of hydrogen, which leaves in the effluent gas stream. The complete conversion of methanol is achieved, and formaldehyde is separated from other gases. Formaldehyde is separated as a bottom product of separation column. Hydrogen, nitrogen, and water left the column in the overhead stream. Calculate the molar flow rate of the produced formaldehyde based on 1000 mol/h of methanol in fresh feed stream, and determine the composition of hydrogen in the overhead stream. (1000 mol/h, 41.0 mol% H_2)

Methanol reacts with oxygen to produce formaldehyde according to the following reactions:

$$CH_3OH \rightarrow HCHO + H_2$$

$$H_2 + \frac{1}{2}O_2 \rightarrow H_2O$$

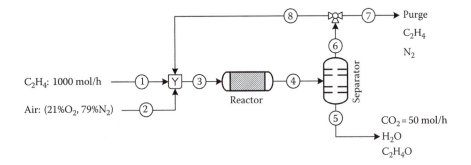

PROBLEM FIGURE 6.9.1
Production of formaldehyde.

References

1. Felder, R.M. and R.W. Rousseau (1999) *Elementary Principles of Chemical Processes*, 3rd edn., John Wiley & Sons, New York.
2. Reklaitis, G.V. (1983) *Introduction to Material and Energy Balances*, John Wiley & Sons, New York.
3. Himmelblau, D.M. and J.B. Riggs (2012) *Basic Principles and Calculations in Chemical Engineering*, 8th edn., Prentice-Hall, Upper Saddle River, NJ.

7

Single- and Multiphase Systems

The aim of this chapter is to introduce calculations of the physical properties, which are commonly encountered in chemical processes, of simple compounds and mixtures; it also introduces students to the fundamentals of operations in chemical processes, which involve changes in the phase of a material. These phase changes are used in separation and purification processes. The following learning objectives should be achieved by the end of this chapter.

Learning Objectives

1. Single-phase systems (Section 7.1).

2. Ideal gas equation of state (Section 7.2).

3. Real gas relationships (Section 7.3).

4. Multiphase systems (Section 7.4).

5. Vapor pressure estimation (Section 7.5).

6. Partial pressure (Section 7.6.)

7. Gibbs' phase rule (Section 7.7).

8. Boiling point, dew point, and critical point (Section 7.8).

7.1 Single-Phase Systems

Calculations in chemical process engineering depend on the availability of the physical properties of materials. There are three common ways to find these properties: obtained through experimental measurements, data available in the open literature, and estimated using physical models that generally come in the form of equations. Among these physical properties are the density, volume, pressure, enthalpy, etc. [1–3]

7.1.1 Liquid and Solid Densities

Density is defined as the mass per unit volume of a substance, and it is a physical property of matter. A physical property can be measured without changing the chemical identity of the substance. Since pure substances have unique density values, measuring the density of a substance can help identify that substance. Density is determined by dividing the mass of a substance by its volume:

$$\text{Density} = \frac{\text{Mass}}{\text{Volume}}$$

The effect of temperature on the density of solids and liquids are modest but sometimes important. In contrast, pressure dependence of the density is usually negligible for solid-state materials and incompressible fluids. Sometimes, additive volumes can be assumed while mixing liquids:

$$V = \sum V_i$$

where V_i is the volume of component i and V is the total volume of the mixture. For mixtures of similar liquid materials (e.g., ethanol and methanol), the density of a mixture may be approximated by averaging the specific volumes. This is termed volume additive rule:

$$\frac{1}{\rho} = \sum \frac{w_i}{\rho_i}$$

where
  w_i is the mass of component i
  ρ_i is the density of component
  ρ is the density of the mixture

Liquids may exhibit attraction or repulsion, so the volume occupied by the mixture can be either larger or smaller than the sum of the individual volumes of each liquid:

$$v = \sum x_i v_i + v^E$$

where v^E is the excess volume (per mole).
   This may occur near mixture critical conditions (high T and P) and for strongly interacting species (e.g., ethanol + water).

7.2 Ideal Gas Equation of State

The volume of pure component is the volume that the moles of pure gas component would occupy at the total pressure P and temperature T of the mixture. In an ideal gas mixture, each component satisfies the ideal gas law provided the partial pressure or pure component volumes are used! The ideal gas obeys the ideal gas law expressed as follows [4–6]:

$$PV = nRT$$

where
   P is the absolute pressure of the gas
   V is the total volume occupied by the gas
   n is the number of moles of the gas
   R is the ideal gas constant in appropriate units
   T is the absolute temperature of the gas

Sometimes, the ideal gas law is written as follows:

$$Pv = RT$$

where v is the specific volume (volume per mole or mass) of a gas (Table 7.1).

Example 7.1 Methane Volume at Standard Conditions

Problem
Calculate the volume, in cubic meters, occupied by 100 kg of methane (CH_4) at standard conditions. The standard temperature and pressure (STP) corresponds to a temperature of 273.15 K (0°C, 32°F) and an absolute pressure of 100 kPa (1 bar). At STP, 1 mol of ideal gas occupies 22.4 L.

TABLE 7.1

Common Standard Conditions for the Ideal Gas

System	T	P	v
SI	273.15 K	101.325 kPa	22.415 m³/kmol
Universal scientific	0.0°C	760 mmHg	22.415 L/mol
Natural gas	60.0°F	14.696 psia	379.4 ft³/lbmol
	(15.0°C)	101.325 kPa	
American engineering	32°F	1 atm	359.05 ft³/lbmol

Source: Himmelblau, D.M. and Riggs, J.B., *Basic Principles and Calculations in Chemical Engineering*, 8th edn., Prentice Hall, Upper Saddle River, NJ, 2012. With permission.

Solution

Known quantities: 100 kg of methane.

Find: Volume in cubic meters.

Analysis: Use standard molar volume.

Basis: 100 kg of carbon dioxide (CH_4):

$$\frac{100 \text{ kg } CH_4}{} \left| \frac{1 \text{ kmol } CH_4}{16 \text{ kg } CH_4} \right| \frac{22.42 \text{ m}^3 \text{ } CH_4}{\text{kmol } CH_4}$$

$$= 140.13 \text{ m}^3 \text{ } CH_4 \text{ at standard condition}$$

Example 7.2 Universal Gas Constant

Problem

For 1 mol of ideal gas when the pressure is in the units of atm, the volume is in cm^3, and the temperature in K, find the value for the universal gas constant R.

Solution

Known quantities: 1 mol of ideal gas.

Find: The value for the universal gas constant R.

Analysis: Use standard molar volume. At standard conditions, we will use the approximate molar volume value: $22{,}415$ cm^3/mol

$$R = \frac{Pv}{T} = \frac{1 \text{ atm}}{273.15 \text{ K}} \left| \frac{22{,}415 \text{ cm}^3}{1 \text{ mol}} \right. = 82.06 \frac{\text{cm}^3(\text{atm})}{(\text{K})(\text{mol})}$$

Example 7.3 Volume of Carbon Dioxide at Standard Conditions

Problem

Carbon dioxide is available under the following conditions: a pressure of 0.95 bar and a temperature of 288 K. Calculate the volume occupied by 100 kg of CO_2.

Solution

Known quantities: 100 kg of carbon dioxide.

Find: Volume in cubic meters.

Analysis: Use standard molar volume. Using the ideal gas law:

$$PV = nRT$$

To bypass the use of the ideal gas constant, R, divide by the standard case:

$$PV = nRT \quad \text{and} \quad P_sV_s = nRT_s$$

$$\frac{PV}{P_sV_s} = \frac{nRT}{nRT_s} = \frac{T}{T_s}$$

Rearranging, the volume occupied by CO_2 at the above conditions is

$$V = V_s \left(\frac{T}{T_s} \right) \left(\frac{P_s}{P} \right)$$

$$= \left(100 \text{ kg} \times \frac{1 \text{ kmol}}{44 \text{ kg}} \times \frac{22.415 \text{ m}^3}{1 \text{ kmol}} \right) \times \left(\frac{288 \text{ K}}{273.15 \text{ K}} \right) \left(\frac{1 \text{ bar}}{0.95 \text{ bar}} \right) = 56.57 \text{ m}^3$$

7.2.1 Gas Density

The density of a gas is defined as the mass per unit volume and can be expressed in kilograms per cubic meter, pounds per cubic foot, grams per liter, or other units.

Example 7.4 Nitrogen Density

Problem
What is the density of nitrogen gas in SI units? Nitrogen is available at a temperature of 300 K and a pressure of 1 bar.

Solution

Known quantities: Nitrogen at 300 K and 1 bar (100 kPa).

Find: Density of nitrogen.

Analysis: Use standard molar volume.

Basis: 1 m³ of N_2 at 300 K and 100 kPa.
 The specific molar volume is

$$v = v_s \left(\frac{T}{T_s} \right) \left(\frac{P_s}{P} \right) = \left(\frac{22.4 \text{ m}^3}{1 \text{ kmol}} \right) \left(\frac{300 \text{ K}}{273.15 \text{ K}} \right) \left(\frac{100 \text{ kPa}}{100 \text{ kPa}} \right) = 24.6 \text{ m}^3 / \text{kmol}$$

The density is calculated from the inverse of specific molar volume:

$$\text{Density} = \frac{1}{v} \times Mw = \frac{1}{24.6 \text{ m}^3 / \text{kmol}} \times 28 \frac{\text{kg}}{\text{kmol}} = 1.14 \text{ kg} / \text{m}^3$$

Example 7.5 Specific Gravity of Nitrogen

Problem
What is the specific gravity of 1 ft³ of N_2 at 540 R and 0.97 bar compared with 1 ft³ of air at 540 R and 0.97 bar?

Solution

Known quantities: Nitrogen is at 540 R and 0.97 bar.

Find: Specific gravity of N_2 and air.

Analysis: Use standard molar volume.

Basis: 1 ft³ of nitrogen at 540 R and 1 atm and the same conditions for air:

$$\frac{v}{v_s} = \frac{RT/P}{RT_s/P_s} = \left(\frac{T}{T_s}\right)\left(\frac{P_s}{P}\right)$$

The molar volume related to standard conditions is

$$v = v_s \frac{RT/P}{RT_s/P_s} = v_s \left(\frac{T}{T_s}\right)\left(\frac{P_s}{P}\right)$$

The density is

$$\rho_{N_2} = \frac{Mw}{v} = \frac{Mw}{v_s\left(\dfrac{T}{T_s}\right)\left(\dfrac{P_s}{P}\right)}$$

$$= \left(\frac{28\ \text{lb}}{\text{lbmol}}\right)\bigg/\left(\frac{359\ \text{ft}^3}{1\ \text{lbmol}}\right)\left(\frac{540\ \text{R}}{492\ \text{R}}\right)\left(\frac{1\ \text{bar}}{0.97\ \text{bar}}\right) = 0.069\ \text{lb/ft}^3$$

The density of air at 540 R and 0.97 bar is

$$\rho_{Air} = \frac{Mw}{v} = \frac{Mw}{v_s\left(\dfrac{T}{T_s}\right)\left(\dfrac{P_s}{P}\right)}$$

$$= \left(\frac{29\ \text{lb}}{\text{lbmol}}\right)\bigg/\left(\frac{359\ \text{ft}^3}{1\ \text{lbmol}}\right)\left(\frac{540\ \text{R}}{492\ \text{R}}\right)\left(\frac{1\ \text{bar}}{0.97\ \text{bar}}\right) = 0.071\ \text{lb/ft}^3$$

Specific gravity of nitrogen is equal to its density divided by the density of air:

$$\text{Specific gravity} = \frac{0.069}{0.071} = 0.97$$

7.3 Real Gas Relationships

At room temperature and pressure, many gases can be assumed to behave ideally. However, under conditions of high pressure, values of the gas properties that you might obtain using the ideal gas law would be at wide variance with the experimental data. In reality, the molecule of a gas does interact with each other, and all real gases are nonideal. Even though the ideal gas equation of state (EOS) may be a very good approximation at low pressure and high temperature, at higher pressures and/or lower temperatures, the impact of intermolecular interactions on gas behavior increases. In the next sections, the following five EOSs will be introduced:

1. Compressibility factor EOS
2. Virial EOS
3. van der Waals EOS
4. Soave–Redlich–Kwong (SRK) EOS
5. Kay's mixing rule for nonideal gas mixtures

7.3.1 Compressibility Factor (z)

One common way is to modify the ideal gas law by inserting an adjustable coefficient z, the compressibility factor, a factor that compensates for the nonideality of the gas. Thus, the ideal gas law becomes a real gas law or a generalized EOS:

$$PV = znRT$$

The use of the law of corresponding states for compressibility factors is very practical. According to simple (two-parameter) corresponding states approach, all fluids obey the same EOS when their properties are written in reduced form:

$$\frac{z}{z_c} = f\left(\frac{T}{T_c}, \frac{P}{P_c}\right)$$

$$z_r = f(T_r, P_r)$$

Since z_c is nearly a constant for many organic species (about 0.27), the most often used form is

$$z = f(T_r, P_r)$$

The following expression is useful when the molar volume is known:

$$V_r = V_r^{ideal} = \frac{v}{\left(\dfrac{RT_c}{P_c}\right)}$$

Applying the extended (three-parameter) corresponding states approach results in

$$z = f(T_r, P_r, \omega)$$

$$z = z^{(0)} + \omega z^{(1)}$$

7.3.2 Virial Equation of State

The virial EOS is an infinite power series in the inverse of the specific molar volume:

$$z = \frac{Pv}{RT} = 1 + \frac{B}{v} + \frac{C}{v} + \frac{D}{v} \cdots$$

$B, C, \ldots = $ 2nd, 3rd, $\ldots$ virial coefficients, which are functions of T.

For instance, B can be estimated from the following corresponding states correlations:

$$B_0 = 0.083 - \frac{0.422}{T_r^{1.6}}$$

$$B_1 = 0.139 - \frac{0.172}{T_r^{4.2}}$$

$$B = \frac{RT_c}{P_c}(B_0 + \omega B_1)$$

7.3.3 van der Waals Equation of State

Starting with the ideal EOS:

$$P = \frac{RT}{v}$$

The following modifications may account for a nonideal behavior:

$$P = \frac{RT}{v-b} - \frac{a}{v^2}$$

where v is in molar volume. The "$-b$" in the denominator accounts for the fact that real gas molecules do possess a volume, thus reducing the total volume available to the gas. The second term on the right, involves the parameter "a," which accounts for molecular interactions. The parameters a and b change from one gas to another but are independent of temperature. They can be expressed as functions of the critical temperature and pressure as

$$a = \frac{27R^2T_c^2}{64P_c}$$

$$b = \frac{RT_c}{8P_c}$$

Note: As the molar volume becomes large, $z \to 1$.

7.3.4 Soave–Redlich–Kwong Equation of State

A more accurate cubic EOS is the empirical SRK equation:

$$P = \frac{RT}{v-b} - \frac{\alpha a}{v(v+b)}$$

The parameters of the SRK EOS are calculated as follows:

$$R = 0.08205 \frac{L \cdot atm}{mol \cdot K}$$

$$\hat{V} = \frac{L}{mol}$$

$$a = 0.42747 \frac{R^2T_c^2}{P_c}$$

$$b = 0.08664 \frac{RT_c}{P_c}$$

$$\alpha = \left[1 + m\left(1 - \sqrt{T_r}\right)\right]^2$$

$$m = 0.48508 + 1.5517\omega - 0.1561\omega^2$$

Example 7.6 van der Waals Equation of State

Problem

A cylinder of 0.150 m³ in volume containing 22.75 kg of propane (C_3H_8) stands in the hot sun. A pressure gauge shows that the pressure is

4790 kPa gauge. What is the temperature of propane in the cylinder? Use the van der Waals equation.

Solution

Known quantities: Mass, volume, and pressure of propane in the cylinder is known.

Find: The temperature of propane in the cylinder.

Analysis: Use van der Waals EOS.

Basis: 22.75 kg of propane. The van der Waals constants for propane obtained from any suitable handbook:

$$a = 9.24 \times 10^6 \text{ atm} \left(\frac{\text{cm}^3}{\text{mol}} \right)^2$$

$$b = 90.7 \text{ cm}^3/\text{mol}$$

$$\left\{ P + \left(\frac{n^2 a}{V^2} \right) \right\} (V - nb) = nRT$$

$$P = (4790 + 101) \text{ kPa} \left| \frac{1 \text{ atm}}{101.3 \text{ kPa}} \right. = 48.3 \text{ atm}$$

$$R = 82.06 \frac{(\text{cm}^3)(\text{atm})}{\text{mol K}}$$

$$n = \frac{22.75 \text{ kg}}{44 \text{ kg/kmol}} = 0.517 \text{ kmol}$$

Substitute the values in this equation:

$$\left\{ 48.3 + \left(\frac{(517)^2 (9.24 \times 10^6)}{(0.15 \times 10^6)^2} \right) \right\} (0.15 \times 10^6 - (517)(90.7)) = (517)(82.06)T$$

$$T = 384 \text{ K}$$

Example 7.7 Soave–Redlich–Kwong Equation of State

Problem
Determine the hydrogen pressure in the tank using the SRK equation. The tank is at 40°C. The molar volume of hydrogen is 0.3 L/mol. Critical temperature and pressure of hydrogen are 33 K and 12.9 atm, respectively. Hydrogen acentric factor is −0.216.

Solution

Known quantities: Temperature and pressure of hydrogen gas in the cylinder and molar volume of hydrogen are known.

Find: The temperature of propane in the cylinder.

Analysis: Use SRK EOS. More accurate cubic is the empirical SRK equation:

$$P = \frac{RT}{v-b} - \frac{\alpha a}{v(v+b)}$$

The parameters of the SRK EOS are calculated from the following:

$$R = 0.08205 \frac{\text{L} \cdot \text{atm}}{\text{mol} \cdot \text{K}}$$

$$a = 0.42747 \frac{R^2 T_c^2}{P_c} = 0.42747 \frac{(0.082)^2 (33)^2}{(12.9)} = 0.243$$

$$b = 0.08664 \frac{RT_c}{P_c} = 0.08664 \frac{(0.082)(33)}{(12.9)} = 0.018$$

$$T_r = \frac{T}{T_c} = \frac{(40 + 273.15)\,\text{K}}{33\,\text{K}} = 9.49$$

$$\alpha = \left[1 + m\left(1 - \sqrt{T_r}\right)\right]^2 = \left[1 + 0.143\left(1 - \sqrt{9.45}\right)\right]^2 = 0.495$$

$$m = 0.48508 + 1.5517\omega - 0.1561\omega^2$$

$$m = 0.48508 + 1.5517(-0.216) - 0.1561(-0.216)^2 = 0.143$$

$$P = \frac{RT}{v-b} - \frac{\alpha a}{v(v+b)} = \frac{0.082(40 + 273.15)}{0.3 - 0.018}$$

$$-\frac{0.495 \times 0.243}{0.3(0.3 + 0.018)} = 91.06 - 1.26 = 89.8\ \text{atm}$$

7.3.5 Kay's Mixing Rules

Kay's rule is an empirical mixing rule that uses data from generalized compressibility charts. Kay's rule uses pseudocritical properties to calculate pseudoreduced quantities that are then used in generalized compressibility charts. The procedure for using Kay's rule is as follows:

$$T'_c = \sum y_i T_{c_i}$$

$$P'_c = \sum y_i P_{c_i}$$

Calculate the pseudoreduced pressure and the pseudoreduced temperature using

$$T'_r = \frac{T}{T'_c}$$

$$P'_r = \frac{T}{P'_c}$$

The compressibility factor for a nonideal mixture is defined as for a pure fluid:

$$Pv = Z_m RT$$

Example 7.8 Kay's Mixing Rule

Problem

A gas mixture contains 30% CO_2 and 70% CO at 202.73 K and 92.04 atm. The molecular weight, critical temperature, and pressure for CO_2 are 44 g/mol, 304 K, 72.9 atm, respectively. The molecular weight, critical temperature, and pressure for CO are 28 g/mol, 133 K, and 34.5 atm, respectively. Find the molar volume of the gas mixture using Kay's mixing rule.

Solution

Known quantities: The molecular weight, critical temperature, and pressure for CO_2 and CO are given. The gas composition is known.

Find: Find the molar volume of the gas mixture.

Analysis: Use Kay's mixing rule. The pseudocritical properties for the given mixture:

$$T_{c,m} = (0.3)(304) + (0.7)(133) = 184.3 \text{ K}$$

$$P_{c,m} = (0.3)(72.9) + (0.7)(34.5) = 46.02 \text{ atm}$$

$$\omega_m = (0.3)(0.239) + (0.7)(0.066) = 0.118$$

$$M_w = (0.3)(44) + (0.7)(28) = 32.8 \text{ g/mol}$$

The reduced temperature and pressure are

$$T_r = \frac{T}{T_{c,m}} = \frac{202.73}{184.3} = 1.1$$

$$P_r = \frac{P}{P_{c,m}} = \frac{92.04}{46.02} = 2.0$$

The compressibility factor of the mixture is determined by

$$z = z^0 + \omega z^1$$

where z^0 and z^1 depend on the reduced temperature and pressure:

$$z^0 = 1 + B^0 \frac{P_r}{T_r}$$

$$z^1 = B^1 \frac{P_r}{T_r}$$

$$B^0 = 0.083 - \frac{0.442}{T_r^{1.6}}$$

$$B^1 = 0.139 - \frac{0.172}{T_r^{4.2}}$$

Substituting values of reduced temperatures:

$$B^0 = 0.083 - \frac{0.442}{T_r^{1.6}} = 0.083 - \frac{0.442}{(1.1)^{1.6}} = -0.297$$

$$B^1 = 0.139 - \frac{0.172}{T_r^{4.2}} = 0.139 - \frac{0.172}{(1.1)^{4.2}} = 0.0237$$

Substitute calculated values of B^0 and B^1 in z^0 and z^1:

$$z^0 = 1 + B^0 \left(\frac{P_r}{T_r}\right) = 1 - 0.297 \left(\frac{2.0}{1.1}\right) = 0.46$$

$$z^1 = B^1 \frac{P_r}{T_r} = 0.0237 \left(\frac{2.0}{1.1}\right) = 0.04$$

The compressibility factor of the mixture is

$$Z_m = 0.46 + 0.118 \times 0.04 = 0.465$$

The compressibility factor was obtained from the table of compressibility factors in the tools section. The molar volume is then obtained from the definition of Z_m:

$$v = \frac{Z_m RT}{P} = \frac{(0.465)(0.08206 \text{ L} \cdot \text{atm/mol} \cdot \text{K})(202.73 \text{ K})}{(92.04 \text{ atm})} = 0.084 \text{ L/mol}$$

7.4 Multiphase Systems

Many flow systems in nature and industrial applications involve multiple phases. Multiphase systems are commonly encountered in chemical and biological systems. A multiphase system is one characterized by the simultaneous presence of several phases, the two-phase system being the simplest case. The individual phases in a multiphase system are commonly shown on a phase diagram. A phase diagram is a graphical representation of the physical states of a substance under different conditions of temperature and pressure. The term "phase" is sometimes used to refer to a set of equilibrium states distinguished in terms of state variables such as pressure and temperature by a phase boundary on a phase diagram.

7.4.1 Phase Diagram

One way to envision the behavior of a material is to use a phase diagram (Figure 7.1). The latter is the plot that shows the conditions under which a pure substance exists in a particular phase, that is, liquid, solid, or gas. It is often the case that we plot P vs. T in order to show the phase behavior of the material. Often, the y-axis indicates pressure and the x-axis the temperature. If vapor and liquid coexist, then setting either temperature or pressure determines the conditions of the state. On the vapor–liquid line, the temperature is called the boiling point, and the pressure is the vapor pressure. On the vapor–solid line, the temperature is called the sublimation point. On the liquid–solid line, the temperature is called the freezing/melting point.

7.4.2 Vapor–Liquid Equilibrium Curve

It is the locus of points for which liquid and vapor can coexist. For a point (T, P) on the vapor–liquid equilibrium (VLE) curve, T is the boiling point

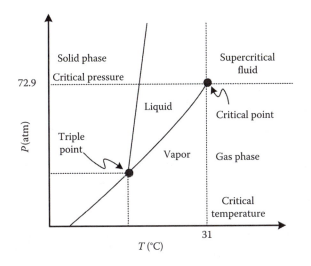

FIGURE 7.1
Phase diagram of CO_2.

of the substance at the pressure P. The normal boiling point is the boiling point temperature for $P=1$ atm. For a point (T, P) on the solid–liquid equilibrium curve, T is the freezing (equivalently, melting) temperature of the substance at the pressure P. Finally, for a point (T, P) on the solid–vapor equilibrium curve, T is the sublimation point of the substance at the pressure P.

7.5 Vapor Pressure Estimation

At a given temperature, the vapor pressure of a pure compound is the pressure at which vapor and liquid coexist at equilibrium. For a pure compound, there is only one vapor pressure at any temperature. In other words, vapor pressure is the pressure of the vapor when the vapor and liquid of a pure component are in equilibrium, also referred to as saturated pressure. Saturated temperature is the temperature at which the vapor and liquid coexist at a given pressure, commonly referred to as the boiling or condensation point. There are four ways to estimate the vapor pressure when data are not available at a temperature of interest: the Clapeyron equation, the Clausius–Clapeyron equation, Cox chart, and the Antoine equation.

7.5.1 Clapeyron Equation

The Clapeyron equation gives a rigorous quantitative relationship between vapor pressure and temperature:

$$\frac{dp^*}{dT} = \frac{\Delta \hat{H}_v}{T(v_g - v_l)}$$

where
$\Delta \hat{H}_v$ is the latent heat of vaporization
v_g, v_l are the specific molar volume of gas and liquid, respectively
T is the absolute temperature

Assuming ideal gas behavior of the vapor and neglecting the liquid volume, the Clapeyron equation can be simplified over a small temperature range to result in an expression known as the Clausius–Clapeyron equation.

7.5.2 Clausius–Clapeyron Equation

The Clausius–Clapeyron equation allows the calculation of the vapor pressure as a function of temperature. However, there are some additional parameters such as the heat of vaporization and the parameter B that must be calculated in order to use this equation. The Clausius–Clapeyron equation is expressed as follows:

$$\ln p^* = \frac{-\Delta H_v}{RT} + \beta$$

where
p^* is the vapor pressure
ΔH_v is the heat of vaporization
R is the ideal gas constant
T is the temperature in K
β is the constant

The Clausius–Clapeyron equation can be used to construct the entire vaporization curve. There is a deviation from experimental value; that is because the enthalpy of vaporization varies slightly with temperature. The Clausius–Clapeyron equation applies to any phase transition. The Clausius–Clapeyron equation can also be used to estimate the heat of phase transition from the vapor pressures measured at two temperatures. The use of the Clausius–Clapeyron equation is recommended over short temperature ranges in regions where the ideal gas law is valid.

Example 7.9 Clapeyron Equation

Problem
The normal boiling temperature of benzene is 353.25 K at a pressure of 101.325 kPa with a standard enthalpy of vaporization $\Delta H^\circ_{vap} = 30.8$ kJ/mol. Determine the vapor pressure of benzene at a temperature of 298.15 K.

Solution

Known quantities: Benzene boiling point.

Find: Benzene vapor pressure at 298.15 K.

Analysis: If we assume that the enthalpy of vaporization varies a little with temperature and pressure, we may use the integrated form of the Clapeyron equation. The normal boiling temperature is the tempera-ture at which the vapor pressure of benzene is 1 atm (101.325 kPa). We may therefore take $T_1 = 353.25$ K and $p_1 = 101{,}325$ Pa and substitute into the Clapeyron equation. This will allow us to determine a value for the vapor pressure p_2 at a temperature of $T_2 = 298.15$.

$$\ln p_2 - \ln p_1 = \frac{\Delta H_{vap}}{R}\left(\frac{1}{T_1} - \frac{1}{T_2}\right)$$

Substituting known values gives

$$\ln p_2 - \ln(101{,}325) = \frac{30.8 \times 10^3}{8.134}\left(\frac{1}{353.25} - \frac{1}{298.15}\right)$$

The vapor pressure p_2 at a temperature of $T_2 = 298.15$ is

$$p_2 = 14{,}600 \text{ Pa}$$

7.5.3 Cox Chart

Cox further improved the method of estimating the vapor pressure by plot-ting the logarithm of the vapor pressure of certain compounds against an arbitrary temperature scale. The vapor pressure/temperature plot forms a straight line mainly for petroleum hydrocarbons [7].

7.5.4 Antoine Equation

The Antoine equation is an empirical and accurate equation for correlating p^* versus T data. The Antoine equation is a simple three-parameter fit to experimental vapor pressures measured over a restricted temperature range:

$$\ln(p^*) = A - \frac{B}{T + C} \quad \text{or} \quad p^* = \exp\left(A - \frac{B}{T + C}\right)$$

where A, B, and C are "Antoine coefficients" that vary from substance to substance. The units of p^* and T are mmHg and K, respectively

TABLE 7.2

Constants for the Antoine Equation for
Vapor Pressure of Pure Components

Component	A	B	C
Acetic acid	16.8080	3405.57	−56.34
Acetone	16.6513	2940.46	−35.93
Ammonia	16.9481	2132.50	−32.98
Benzene	15.9008	2788.51	−52.36
Ethyl acetate	16.1516	2790.50	−57.15
Ethanol	18.5242	3578.91	−50.50
Methanol	18.5875	3626.55	−34.29
n-heptane	15.8737	2911.32	−56.51
n-hexane	15.8366	2697.55	−48.78
n-pentane	15.8333	2477.07	−39.94
Toluene	16.0137	3096.52	−53.67
Water	18.3036	3816.44	−46.13

Source: Himmelblau, D.M. and Riggs, J.B., *Basic Principles and Calculations in Chemical Engineering*, 8th edn., Prentice Hall, Upper Saddle River, NJ, 2012. With permission.

(Table 7.2). Antoine parameters can also be found in the literature in the following form where T is in degree Celsius, and values of A, B, and C are different than those in Table 7.2:

$$\text{Log}_{10}p^* (\text{mmHg}) = A - \frac{B}{T(°C)+C} \quad \text{or} \quad p^* (\text{mmHg}) = 10^{\left(A - \frac{B}{T(°C)+C}\right)}$$

Accordingly, use the values of the parameters belonging to each correlation.

Example 7.10 Vapor Pressure of Ethanol

Problem
Use the Antoine equation to calculate the vapor pressure of ethanol at 90°C.

Solution

Known quantities: Ethanol at 90°C.

Find: Vapor pressure.

Analysis: Use the Antoine equation.

From Table 7.2, the Antoine coefficients for ethanol are
$A = 18.5242$
$B = 3578.91$
$C = -50.50$

$$\ln\left(p^*\right) = A - \frac{B}{T+C}$$

$$\ln\left(p^*\right) = 18.5242 - \frac{3578.91}{(90+273) - 50.50}$$

$$p^* = 1178.14 \text{ mmHg}$$

Example 7.11 Vapor Pressure of *n*-Butane Using Cox Chart and the Second Form of Antoine Equation

Problem
Use the Antoine equation and Cox vapor pressure chart to calculate the vapor pressure of *n*-butane at 80°C.

Solution

Known quantities: Butane at 80°C.

Find: Vapor pressure.

Analysis: Use Cox chart and the Antoine equation. Antoine coefficients for *n*-butane are as follows [2]:
$A = 6.808$
$B = 935.77$
$C = 238.789$

$$\text{Log}_{10}p^* = 6.808 - \frac{935.77}{80 + 238.789}$$

$$p^* = 7458 \text{ mmHg } (745.8 \text{ cmHg})$$

Using Cox chart, the approximate value of vapor pressure is estimated to be around 715 cmHg as shown in Example Figure 7.11.1. The figure is extracted from Cox vapor pressure chart for paraffin [6,7].

Example 7.12 Molar Composition of Gas Phase

Problem
Air and liquid water are at equilibrium in a closed chamber at 75°C and 760 mmHg. Calculate the molar composition of the gas phase. The vapor pressure of water at 75°C is $p^*_{H_2O} = 289$ mmHg.

Solution

Known quantities: Air and liquid water are contained at equilibrium in a closed chamber at 75°C and 760 mmHg.

Find: The molar composition of the gas phase.

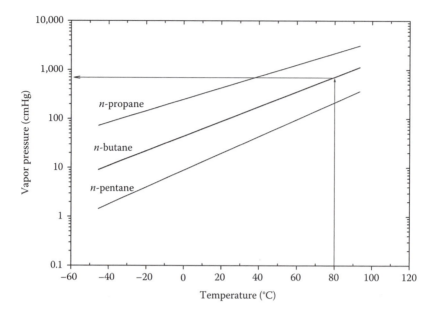

EXAMPLE FIGURE 7.11.1
Cox vapor pressure chart for *n*-butane. (From Cox, E.R., *Ind. Eng. Chem.*, 15(6), 592, 1923. With permission.)

Analysis: Raoult's law may be applied since the gas and liquid are in equilibrium:

$$y_{H_2O} = \frac{p_i^*}{P} = \frac{289 \text{ mmHg}}{760 \text{ mmHg}} = 0.38$$

$$y_{dry\ air} = 1 - y_{H_2O} = 1 - 0.38 = 0.62$$

7.6 Partial Pressure

Partial pressure is a way of describing how much of gas is present in a gas mixture. It is defined as the pressure that any one gas would exert on the walls of a container, if it were the only gas present. This relationship is known as Dalton's law of partial pressures.

7.6.1 Dalton's Law of Partial Pressures

Dalton's law of partial pressures states that the total pressure of the mixture is the sum of the partial pressures of all the gases present. The law of partial pressures is expressed as follows:

$$P_{\text{Total}} = P_1 + P_2 + P_3 + \cdots$$

where

P_{Total} is the total pressure of the mixture
$P_1, P_2, P_3, \ldots$ are the partial pressures of the gases present in the mixture

The partial pressure of component A in an ideal gas mixture is given by

$$p_A = Py_A$$

where p_A is the partial pressure of component A. It is equal to the pressure that would be exerted if A alone occupied the container only for ideal gases.

7.6.2 Raoult's Law for a Single Condensable Species

Raoult's law states that if i is the condensable component, then at equilibrium the partial pressure of i in the gas mixture must equal the vapor pressure of the pure i liquid. It is applicable for processes where the gas and the liquid are in equilibrium, and it is given by the following equation [6]:

$$y_i = \frac{p_i^*}{P}$$

where

y_i is the mole fraction of component i in the vapor phase
p_i^* is the vapor pressure of component i
P is the total pressure

With one or more volatile components i, the partial pressure P_i in a gas that is in equilibrium with the liquid is given by

$$p_i = y_i P = x_i p_i^*$$

where

p_i is the partial pressure of component i
y_i is the mole fraction of component i in the gas phase
P is the total pressure
x_i is the mole fraction of i in the liquid phase
p_i^* is the vapor pressure of pure liquid i

Example 7.13 Partial Pressure Estimation

Problem
A flue gas effluent from a combustion chamber was analyzed and found to contain the following gases: 14.0% CO_2, 6.0% O_2, and 80.0% N_2.

The gas is at 400°F and 765.0 mmHg. Calculate the partial pressure of each component.

Solution

Known quantities: Flue gas composition.

Find: Partial pressure of each component.

Analysis: Use Dalton's laws.

Basis: 1.00 kmol flue gas.

The partial pressure of each component is

$$P_{CO_2} = P_{tot} y_{CO_2} = 765 \text{ mmHg} \times 0.14 = 107.1 \text{ mmHg}$$

$$P_{O_2} = P_{tot} y_{O_2} = 765 \text{ mmHg} \times 0.06 = 45.9 \text{ mmHg}$$

$$P_{N_2} = P_{tot} y_{N_2} = 765 \text{ mmHg} \times 0.80 = 612 \text{ mmHg}$$

Example 7.14 Dalton's Law

Problem
Dry air at 1 atm contains 78% nitrogen, 21% oxygen, and 1.0% argon. Calculate the partial pressure of each component.

Solution

Known quantities: Dry air contains 78% nitrogen, 21% oxygen, and 1.0% argon.

Find: The partial pressure of each component.

Analysis: Calculate the partial pressure of each gas by using Dalton's law, which states that each partial pressure is the same percent of the total pressure as the percent each gas is of the total volume.

$$p_{N_2} = y_{N_2} \times P_{Tot} = 0.78 \times 1 \text{ atm} = 0.78 \text{ atm}$$

$$p_{O_2} = y_{O_2} \times P_{Tot} = 0.21 \times 1 \text{ atm} = 0.21 \text{ atm}$$

$$p_{Ar} = y_{Ar} \times P_{Tot} = 0.01 \times 1 \text{ atm} = 0.01 \text{ atm}$$

Example 7.15 Ideal Gas Law

Problem
Nitrogen gas is in contact with water, the gas had a volume of 6.523 L at 299 K and a total pressure of 0.98 atm. In any gas sample saturated with water vapor at 299 K, the partial pressure of the water is 0.03 atm. How many moles of nitrogen did the nitrogen contain?

Solution

Known quantities: Gas volume and temperature is known.

Find: Moles of nitrogen did the sample contain?

Analysis: The solution to this problem requires the use of the ideal gas equation in the form:

$$n = \frac{PV}{RT}$$

The total pressure of the gas sample is the sum of the partial pressure of the nitrogen and the partial pressure of the water vapor:

$$P_{Total} = P_{N_2} + P_{H_2O}$$

Rearranging this equation gives

$$P_{N_2} = P_{Total} - P_{H_2O} = 0.98 - 0.03 = 0.95 \text{ atm}$$

Substituting these values into the ideal gas equation gives

$$n_{N_2} = \frac{P_{N_2}V}{RT} = \frac{(0.95 \text{ atm})(6.523 \text{ L})}{(0.082 \text{ L} \cdot \text{atm}/\text{mol} \cdot \text{K})(299 \text{ K})} = 0.252 \text{ mol N}_2$$

Example 7.16 Partial Pressure of a Gas Mixture

Problem
A gas mixture which is 5 mol% H_2 and 95 mol% N_2 is present in a container at 200 atm. What is the partial pressure of each gas?

Known quantities: A gas mixture which is 5 mol% H_2 and 95 mol% N_2 is present in a container at 200 atm.

Find: The partial pressure of each gas.

Analysis: Use Dalton's law to calculate the partial pressure of any gas in a mixture if we know the total pressure and the number of moles of each gas.

$$p_{H_2} = y_{H_2} P_{tot} = 0.05 \times 200 \text{ atm} = 10 \text{ atm}$$

$$p_{N_2} = y_{N_2} P_{tot} = 0.95 \times 200 \text{ atm} = 190 \text{ atm}$$

Henry's law: Henry's law is similar in form to Raoult's law, except that it applies to VLE when the volatile species i is dilute in the liquid solution. Henry's law states that for a liquid solution with a volatile component i, the partial pressure of i in a gas mixture in equilibrium with the liquid is given by

$$P_i = y_i P = x_i H_i$$

where
H_i is Henry's law constant of species i. Like p_i^*, the vapor pressure of i, Henry's law constant is also a function of temperature. In contrast to

Raoult's law, Henry's law applies when x_i is close to 0, that is, when the volatile i component is dilute in the liquid phase.

Example 7.17 Estimation of the Fraction of Dissolved Ethane in Water

Problem
A gas containing 1.00 mol% ethane is in direct contact with water at 20.0°C and 20.0 atm. Estimate the mole fraction of dissolved ethane. Henry's law constant for ethane in water at 20°C is 2.63×10^4 atm/mol fraction.

Solution

Known quantities: The gas contains 1.00 mol% ethane; the gas is in contact with water at 20.0°C and 20.0 atm.

Find: The mole fraction of dissolved ethane.

Analysis: Use Henry's law.
  Henry's law is applied in this case since hydrocarbons are relatively insoluble in water.

$$x_{C_2H_6} H_{C_2H_6} = y_{C_2H_6} P$$

Mole fraction of dissolved ethane in water is

$$x_{C_2H_6} = \frac{y_{C_2H_6} P}{H_{C_2H_6}} = \frac{0.01 \times 20\ \text{atm}}{2.63 \times 10^4\ \text{atm/mol fraction}} = 7.6 \times 10^{-6}$$

Example 7.18 Partial Pressure of Component in a Mixture

Problem
An equimolar liquid mixture of benzene (B) and toluene (T) is in equilibrium with its vapor at 30.0°C. Using the Antoine equation it was estimated that at 30.0°C, the vapor pressures of benzene and toluene are 119 and 36.7 mmHg, respectively. What are the system pressure and composition of the vapor phase?

Solution

Known quantities: 50 mol% benzene and 50% toluene is in equilibrium at 30.0°C.

Find: The system pressure and composition mole fraction of vapor.

Analysis: Apply Raoult's law. The partial pressures of benzene and toluene are

$$P_B = x_B P_B^* = 0.5\,(119\ \text{mmHg}) = 59.5\ \text{mmHg}$$

$$P_T = x_T P_T^* = 0.5\,(36.7\ \text{mmHg}) = 18.35\ \text{mmHg}$$

The total pressure is the summation of benzene and toluene partial pressures:

$$P = P_B + P_T = 59.5 + 18.35 = 77.9 \text{ mmHg}$$

Mole fraction of benzene and toluene in the vapor phase:

$$y_B = \frac{P_B}{P} = \frac{59.5 \text{ mmHg}}{77.9 \text{ mmHg}} = 0.764$$

$$y_T = \frac{P_T}{P} = \frac{18.35 \text{ mmHg}}{77.9 \text{ mmHg}} = 0.236$$

Example 7.19 Clausius–Clapeyron Equation

Problem
Methanol is stored at the temperature of 60°C. The vapor pressure of methanol is 13.2 kPa at 20°C and 347 kPa at 100°C. Determine the vapor pressure of methanol at 60°C.

Solution

Known quantities: Methanol vapor pressure is 13.2 kPa at 20°C and 347 kPa at 100°C.

Find: Vapor pressure of methanol.

Analysis: Use Clausius–Clapeyron equation. The Clapeyron equation is

$$\ln p_1^* = -\frac{\Delta H_v}{RT_1} + \beta$$

$$\ln p_2^* = -\frac{\Delta H_v}{RT_2} + \beta$$

Subtract 1 from 2

$$\ln p_2^* - \ln p_1^* = -\frac{\Delta H_v}{R}\left[\frac{1}{T_2} - \frac{1}{T_1}\right]$$

Rearranging:

$$\ln p_2^*/p_1^* = -\frac{\Delta H_v}{R}\left[\frac{1}{T_2} - \frac{1}{T_1}\right]$$

Substitute values in the previous equation:

$$\ln 13.2 \text{ kPa}/347 \text{ kPa} = -\frac{\Delta H_v}{R}\left[\frac{1}{293.15} - \frac{1}{373.15}\right]$$

$$\frac{\Delta H_v}{R} = 4470 \text{ K}$$

$$\Delta H_v = R \times 4{,}470 \text{ K} = \left(8.314 \frac{\text{J}}{\text{mol} \cdot \text{K}} \right) \times 4{,}470 = 37{,}164 \frac{\text{J}}{\text{mol}}$$

To calculate β

$$\ln p_1^* = -\frac{\Delta H_v}{R} \frac{1}{T_1} + \beta$$

$$\ln 13.2 = -4470 \text{ K} \frac{1}{293.15} + \beta$$

$$\beta = 17.83$$

The Clausius–Clapeyron equation for methanol at 60°C is

$$\ln p^* = -\frac{4470}{60 + 273.15} + 17.83$$

$$p^* = 42.49 \text{ kPa}$$

Example 7.20 Composition of a Saturated Gas–Vapor System

Problem

Air, at a temperature of 80°C and a total pressure of 2.5 bar, enters a humidifier where it is brought in direct contact with water. The air composition is 15 mol% oxygen and 85 mol% nitrogen on a dry basis. The value of the vapor pressure of water at the temperature of 80°C is 0.47 bar. Calculate the composition of the exhaust gases.

Solution

Known quantities: Air inlet temperature and pressure.

Find: The composition of exit gases.

Analysis: Subtitle the vapor pressure and the absolute pressure values into Raoult's law gives

$$y_{H_2O} = \frac{p_{H_2O}^*}{P} = \frac{0.47 \text{ bar}}{2.5 \text{ bar}} = 0.19 \frac{\text{mol } H_2O}{\text{mole humid air}}$$

The sum of the molar fractions of water and dry air is 1, since only air and water are present in the exhaust gases:

$$1 = y_{H_2O} + y_{\text{dry-air}}$$

Substitute values of water mole fraction:

$$1 = 0.19 + y_{\text{dry-air}}$$

The mole fraction of dry air is

$$y_{\text{dry-air}} = 1 - 0.19 = 0.81 \frac{\text{mole dry air}}{\text{mole humid air}}$$

Example 7.21 Dew Point Calculation

Problem

A stream of air exiting a condenser at a temperature of 87°C and a pressure of 1.5 bar contains 25 mol% of water. Determine the dew point and the degrees of superheat of the air.

Solution

Known quantities: Air inlet temperature and pressure.

Find: Dew point and the degree of superheat of the air.

Analysis: For this problem, we will assume that the air and the water are in equilibrium. Thus, we can use Raoult's law to compare the partial pressure of water to the vapor pressure at the dew point temperature. First, the partial pressure of water in the air entering the condenser must be determined. This can be done using the definition of partial pressure:

$$p_{H_2O} = y_{H_2O}\, P = 0.25(1.5 \text{ bar}) = 0.375 \text{ bar}$$

If we are assuming equilibrium between the two phases, the partial pressure of water must be equal to the vapor pressure at the dew point. This can be written as follows:

$$p_{H_2O} = p^*_{H_2O}(T_{\text{dp}})$$

Using the steam table, the dew point temperature at 0.375 bar is 74.66°C.
The degrees of superheat of the air entering the condenser can be obtained by subtracting the dew point from the temperature of the air:

$$\text{Degree superheat} = 87°C - 74.66°C = 12.3°C$$

7.7 Gibbs' Phase Rule

The Gibbs' phase rule defines the number of independent intensive properties that must be specified in order to define the state of a system, and which determine all remaining intensive properties. It is given by the following equation:

$$F = m + 2 - \pi - r$$

where

F is the degree of freedom = the number of phase equilibrium variables (T, P, mole fraction compositions in each phase) that can be independently specified

m is the number of components

π is the number of phases

r is the number of reactions

Example 7.22 Phase Rule

Problem

A system contains water and carbon dioxide and exists in two phases: vapor and liquid. Calculate the degrees of freedom.

Solution

Known quantities: The number of components and two phases no reaction.

Find: Degrees of freedom.

Analysis: Use Gibbs' phase rule.

The phase rule must be modified to become

$$F = m + 2 - \pi - r$$

$m = 2$: Two components: water and carbon dioxide

$\pi = 2$: Two phases: liquid and vapor

If r is the number of independent chemical reactions at equilibrium in the system, then $r = 0$, since no reaction is taking place.

$$\text{Degrees of freedom: } F = 2 + 2 - 2 - 0 = 2$$

Two of the phase equilibrium variables must be given ($F = 2$) to fix all the others; for example, P and T can be fixed, or P and x_A, or T and y_B.

Variables: P, T, y_A (*Note:* $y_B = 1 - y_A$; and $x_A = 1$, pure water assuming carbon dioxide is insoluble in water). Thus, fix any two variables and find the third from the equilibrium relationship.

7.8 Bubble Point, Dew Point, and Critical Point

The bubble point is the temperature at which saturation occurs in the liquid phase. In the case of a liquid solution, the composition of the vapor in the bubbles that form in the liquid is not the same as that of the liquid. The dew point is the temperature at which saturation occurs in the gas phase for a given pressure. The degree of superheat refers to the difference in temperature between the actual temperature and the dew point. At higher temperatures and/or higher pressures, the difference between a gas and a liquid eventually disappears, and a supercritical fluid is formed. The point at which this happens is called the critical point.

Example 7.23 Ammonia Gas Absorber

Problem

Ammonia gas is absorbed from an air stream by liquid water as a solvent. Conditions in the absorber are 100°F and 50 psia. The liquid solution entering the top of the tower is 10.0 mol% ammonia. What is the composition of the gas stream leaving the top of the absorber if it can be assumed to be in equilibrium with the liquid phase? If the inlet gas contained 30.0 mol% ammonia, the equilibrium partial pressures of the two liquid components over a 10.0 mol% solution would be 1.98 psia for ammonia and 0.45 psia for water. Determine the fraction of ammonia that was removed in the absorber.

Solution

Basis: 1 mol of inlet gas.

The mole fractions of ammonia in exit gas are calculated from partial pressure to total pressure as follows:

$$y_{NH_3} = \frac{P_{NH_3}}{P_{tot}} = \frac{1.98}{50} = 0.040$$

$$y_{H_2O} = \frac{P_{H_2O}}{P_{tot}} = \frac{0.45}{50} = 0.01$$

$$y_{air} = 1 - y_{NH_3} - y_{H_2O} = 1 - 0.04 - 0.01 = 0.95$$

Moles of exit air

$$0.7 \times 1 = 0.95 \times n_{Air}$$

$$n_{Air} = \frac{0.7}{0.95}$$

Moles of ammonia in exit air = $n_{NH_3} = 0.04 \times 0.7/0.95 = 0.0295$

The fraction of ammonia removed can be found by (moles of ammonia in–moles of ammonia out)/moles of ammonia in:

The fraction of ammonia removed is

$$\frac{0.30 - 0.0295}{0.30} = 0.902$$

Homework Problems

7.1 Calculate the volume, in liters, occupied by 100 g of ethane (C_2H_6) at standard conditions. (74 L)

7.2 Find the value for the universal gas constant R for the following combinations of units: For 1 mol of ideal gas when the pressure is in atm, the volume is in m^3, and the temperature is in K. (0.082 L·atm/mol·K)

7.3 A tank contains 100 g of CO_2 under a pressure of 0.90 atm and a temperature of 300 K. Calculate the volume occupied by this CO_2 at these conditions. (62.18 L)

7.4 Helium gas (He) is stored in a 1 m^3 tank at 300 K and 100 kPa. What is the density of helium under these conditions? (0.16 g/L)

7.5 Determine the specific gravity of argon at 550 R and 0.9 atm compared with air under same temperature and pressure assuming a volume of 1 ft^3 gas. (1.38)

7.6 A gas cylinder 10 L in volume and containing 40 g of hydrogen H_2(20 mol) is kept in a store. The store is not air-conditioned. A pressure gauge shows that the pressure is 5 atm. What is the temperature of the methane in the cylinder? Use van der Waals EOS. The van der Waals constants for hydrogen are $a = 0.2444$ $L^2 \cdot atm/mol^2$, $b = 0.02661$ L/mol. (352 K)

7.7 Determine the nitrogen pressure in a tank using the SRK equation. The tank is at 40°C, and the molar volume of nitrogen is 0.9 L/mol. The critical temperature and pressure of nitrogen are 126.2 K and 34 atm, respectively. Nitrogen acentric factor is 0.039. (41.43 atm)

7.8 A gas mixture contains 30% CO_2 and 70% N_2 at 202.73 K and 92.04 atm. The molecular weight, critical temperature, critical pressure, and acentric factor for CO_2 are 44 g/mol, 304 K, 72.9 atm, and 0.239, respectively. The molecular weight, critical temperature, critical pressure, and acentric factor for N_2 are 28 g/mol, 126 K and 34 atm, and 0.039, respectively. Find the molar volume of the gas mixture using Kay's mixing rule. (0.086 L/mol)

7.9 Typically, a firefighter sucks in air 15 times per minute, and each mouthful of air consists of a volume of 450.0 cm^3. The inhalation device will continue to deliver air until the pressure in the tank reaches 1.10 atm (abs). Assuming the tank volume is 15.0 L, what is the pressure inside the tank at 300.0 K, if it is desired to provide 30.0 min of breathing, determine the pressure in the tank to hold this amount of air. (15.9 atm)

7.10 It is desired to eliminate acetone from a stream containing air and acetone by cooling the stream and condensing some of the acetone. The original stream is at 104°F and 1.00 atm (absolute) and contains a relative saturation of acetone of 0.900. The vapor pressure of acetone at 104°F is 424.5 mmHg. Based on 1 mol of air, to what temperature must this stream be cooled so that 95% of the acetone in the feed stream is condensed? (−11.7°C)

7.11 Butanol is being promoted as a potential biofuel. Suppose you have a gaseous mixture that contains 10 mol% 1-butanol and 90 mol% air at 760 mmHg (absolute) and 75°C. If the pressure remains constant at 760 mmHg (absolute), what is the dew point temperature? If the temperature remains constant at 75°C, at what pressure does the first drop of 1-butanol form? If the pressure remains constant at 760 mmHg (absolute), to what temperature must the gas be cooled in order to condense 90% of the 1-butanol? (29.2°C)

7.12 A tank contains vapor that is 65 mol% benzene, and the balance toluene is in equilibrium with a liquid mixture, which contains benzene and toluene. The absolute pressure in the container is 150 mmHg. Estimate the composition of liquid and system temperature. (48.2°C) The Antoine coefficients are as follows:

Substance	A	B	C
Benzene	6.893	1203.531	219.888
Toluene	6.958	1346.773	219.693

7.13 Benzene and toluene enter a reboiler at a molar flow rate of 100 mol/h. The reboiler is operating at 500 mmHg and 90°C. The vapor and liquid are split equally (i.e., the vapor is 50% of the feed stream). Estimate the liquid and vapor mole fraction of benzene and toluene. ($x_B = 0.15$, $y_B = 0.31$)

7.14 It is desired to prepare a mixture of *n*-hexane ($T_{bp} = 68.74°C$) and *n*-octane ($T_{bp} = 125.5°C$) at a pressure of 760 mmHg and a temperature of 90°C in a sealed container. Under these conditions, both vapor and liquid are present. At 90°C, the vapor pressure of *n*-hexane is 1417 mmHg, and the vapor pressure for *n*-octane is 250.8 mmHg. What is the composition of the liquid? What is the composition of the vapor? If there is 0.5 mol of *n*-hexane and 0.5 mol of *n*-octane total in the container, how many moles exist in the liquid phase? (0.832 mol)

7.15 A mixture consists of two components, A and B, which are both in their liquid state. When there are as many moles of A as there are of B in the mixture, the first bubbles of vapor form when the pressure is 900 mmHg at 30°C. When there are three times as many moles of B as there are of A in the mixture, the first bubbles of vapor form when the pressure is 1050 mmHg at 30°C. Which of the two components is the most volatile? How do you know which component is the most volatile? If the first bubbles of vapor are observed at 760 mmHg and 30°C, what is the composition of the liquid phase? (0.733)

7.16 Determine the percent error if the Antoine equation is used to estimate the normal boiling point of benzene. From the literature, the normal boiling point of benzene is 353.26 K. For benzene, values of the constants for the Antoine equation are as follows: $A = 15.90$, $B = 2788.51$, and $C = -52.36$. (0.01%)

7.17 The vapor pressure of benzene is measured at two temperatures, with the following results: $T_1 = 280.8$ K, $p_1^* = 40$ mmHg, $T_2 = 288.6$ K, $p_2^* = 60$ mmHg. Calculate the latent heat of vaporization and the parameter β in the Clausius–Clapeyron equation and then estimate p^* at 40°C using this equation. (188.22 mmHg)

7.18 The vapor pressure of water is 1.0 atm at 373 K, and the enthalpy of vaporization is 40.7 kJ/mol. Estimate the vapor pressure at a temperature of 383 K. (1.409 atm)

7.19 The vapor pressure of pure hexane as a function of temperature is 54.04 kPa at 50°C and 188.76 kPa at 90°C. Estimate the vapor pressure of hexane at 100°C using the Clausius–Clapeyron equation. (248.73 kPa)

7.20 Humid air flowing at a molar flow rate of 100 mol/h contains 25% water, and the balance is dry air. The air at 87°C and 1.5 bar enters a partial condenser and exits it at 60°C and 1.5 bar. Determine the composition of the gas exiting the condenser and mole of water condensed. The vapor pressure of water at 60°C is 0.2 bar. (14 mol/h)

References

1. Perry, R.H. and D.W. Green (1997) *Perry's Chemical Engineering Handbook*, 7th edn., McGraw-Hill, New York.
2. Haynes, W.M., Ed. (2013–2014) *Handbook of Chemistry and Physics*, 94th edn., CRC Press, New York.
3. Reid, R.C., J.M. Praustniz, and T.K. Sherwood (1977) *The Properties of Gases and Liquids*, 3rd edn., McGraw Hill, New York.
4. Himmelblau, D.M. and J.B. Riggs (2012) *Basic Principles and Calculations in Chemical Engineering*, 8th edn., Prentice Hall, Upper Saddle River, NJ.
5. Richard Elliott, J. and C.T. Lira (2012) *Introductory Chemical Engineering Thermodynamics*, 2nd edn., Prentice Hall, Upper Saddle River, NJ.
6. Felder, R.M. and W.R. Rousseau (1999) *Elementary Principles of Chemical Processes*, 3rd edn., John Wiley & Sons, New York.
7. Cox, E.R. (1923) Pressure-temperature chart for hydrocarbon vapors. *Ind. Eng. Chem.* 15 (6), 592–593.

8

Energy and Energy Balances

This chapter illustrates the fundamental theory of energy balance without reactions. The concept of energy conservation as expressed by an energy balance equation is essential to chemical engineering calculations. Similar to material balances studied in previous chapters, a balance of energy is important to solving many problems. The chapter begins with definitions of the first law of thermodynamics, each term in the first law, and application for closed and open systems. Next, the three forms of energy, that is, kinetic, potential, and internal, are explained. Mechanical energy balance and Bernoulli's equations are also covered in this chapter. The following items outline the principal learning objectives of this chapter.

Learning Objectives

1. Calculate energy balance for closed and open systems (Section 8.1).
2. Write mechanical energy balance for a nonreacting system (Section 8.2).
3. Use Bernoulli's equation to solve mechanical energy problems involving flowing fluids with no work input/output (Section 8.3).
4. Use heat capacities to calculate enthalpy changes (Section 8.4).
5. Use latent heats within energy balances for systems involving phase changes (Section 8.5).
6. Use psychrometric charts (Section 8.6).

8.1 Energy Balance for Closed and Open Systems

A *system* is an object or a collection of objects that an analysis is carried out on. The system has a definite boundary, called the system boundary, which is chosen and specified at the beginning of the analysis. Once a system is defined, through the choice of a system boundary, everything external to it is called the surroundings. All energy and material that are transferred out of the system enter the surroundings, and vice versa.

An *isolated system* is a system that does not exchange heat, work, or material with the surroundings.

A *closed system* is a system in which heat and work are exchanged across its boundary, but material is not.

An *open system* can exchange heat, work, and material with the surroundings.

8.1.1 Forms of Energy: The First Law of Thermodynamics

Energy is often categorized as kinetic energy, potential energy, and internal energy. The first law of thermodynamics is a statement of energy conservation. Although energy cannot be created or destroyed, it can be converted from one form to another. Energy can also be transferred from one point to another or from one body to another one. Energy transfer can occur by flow of heat, by transport of mass, or by performance of work [1]. The general energy balance for a thermodynamic process can be expressed in words as the accumulation of energy in a system equals the input of energy into the system minus the output of energy from the system.

8.1.2 Energy Balance for a Closed System

Energy can cross the boundaries of a closed system in the form of heat and work (Figure 8.1).

The energy balance of a *system* is used to determine the amount of energy that flows into or out of each process unit, calculate the net energy requirement for the process, and assess ways of reducing energy requirements in

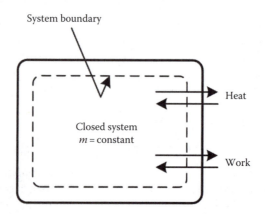

FIGURE 8.1
Energy balance for a closed system.

order to improve process profitability and efficiency [2]. The energy balance for a closed system takes the form

$$Q - W = \Delta U + \Delta KE + \Delta PE \qquad (8.1)$$

where heat (Q), work (W), internal energy (U), kinetic energy (KE), and potential energy (PE) are defined as follows.

Heat is the energy that flows due to a temperature difference between the system and its surroundings and always flows from regions at high temperatures to regions at low temperatures. By convention, heat is defined to be positive if it flows to a system (i.e., gained). For systems with no significant heat exchange with the surroundings, $Q = 0$. Such a system is said to be *adiabatic*. The absence of any heat transfer can be due to perfect thermal insulation or the fact that the system and surroundings are at the same temperature.

Work is the energy that flows in response to any driving force (e.g., applied force, torque) other than temperature, and is defined as positive if it flows *from* the system (i.e., work done by the system). In chemical processes, work may, for instance, come from pumps, compressors, moving pistons, and moving turbines. Heat or work only refers to energy that is being transferred to or from the system. If there is no motion along the system boundary, then $W = 0$.

Internal Energy is all the energy associated with a system that does not fall under the earlier definitions of kinetic or potential energy. More specifically, internal energy is the energy due to all molecular, atomic, and subatomic motions, and interactions. Usually, the complexity of these various contributions means that no simple analytical expression is available from which internal energy can be readily calculated. An *isothermal* system is one where the temperature does not change with time and in space. This does not mean that no heat crosses the boundaries.

Kinetic Energy is associated with directed motion of the system. Translation refers to straight line motion. If the system is not accelerating, then $\Delta KE = 0$.

Potential Energy of a system is due to the position of the system in a potential field. There are various forms of potential energy, but only gravitational potential energy will be considered in this course. If the system is not experiencing a displacement in the direction of the gravitational field, then $\Delta PE = 0$.

8.1.2.1 Kinetic Energy

Kinetic energy is the energy carried by a moving system because of its velocity. The kinetic energy KE of a moving object of mass m, traveling with speed v, is given by

$$KE = \frac{1}{2}\dot{m}v^2 \Rightarrow \left(\frac{kg}{s}\right)\left(\frac{m}{s}\right)^2 \left| \frac{N}{kg\, m/s^2} \right| \frac{J}{N \cdot m} \left| \frac{W}{J/s} \right| = W \qquad (8.2)$$

KE has units of energy, $\dot{m}$ has units of mass flow rate (mass/time), and v has units of velocity (length/time).

Example 8.1 Kinetic Energy Calculations

Problem

Water flows from a large lake into a process unit through a 0.02 m inside diameter pipe at a rate of 2.0 m³/h. Calculate the change in kinetic energy for this stream in joules per second.

Solution

Known quantities: Pipe diameter (0.02 m), water volumetric flow rate (2.0 m³/h), density of water (1000 kg/m³).

Find: Change in kinetic energy.

Analysis: First, calculate the mass flow rate from the density and volumetric flow rate, and, next, determine the velocity as the volumetric flow rate divided by the pipe inner cross-sectional area. The rate of change in kinetic energy is calculated by

$$\Delta KE = \frac{1}{2}\dot{m}\Delta v^2 = \frac{1}{2}\dot{m}(v_2^2 - v_1^2) \tag{8.3}$$

The mass flow rate, $\dot{m}$, is the density (ρ) multiplied by volumetric flow rate ($\dot{V}$):

$$\dot{m} = \rho\dot{V} = \frac{1000\ \text{kg}}{\cancel{\text{m}^3}}\left|\frac{2\ \cancel{\text{m}^3}}{\cancel{\text{h}}}\right|\frac{\cancel{\text{h}}}{3600\ \text{s}} = 0.56\ \text{kg/s}$$

The water exit velocity (v_2) is calculated from the volumetric flow rate ($\dot{V}$) divided by pipe inner cross-sectional area of the exit of the pipe (A). The surface of the lake being large, the water surface can be assumed to be almost stagnant. Accordingly, the initial velocity is negligible ($v_1 = 0$):

$$v_2 = \frac{\dot{V}}{A = \dfrac{\pi D^2}{4}} = \left(\frac{2.00\ \dfrac{\text{m}^3}{\text{h}}\left|\dfrac{\text{h}}{3600\ \text{s}}\right.}{\dfrac{3.14 \times (0.02\ \text{m})^2}{4}}\right) = 1.77\ \text{m/s}$$

Substituting the values of mass flow rate and velocities in the kinetic energy equation,

$$\Delta KE = \frac{1}{2}\dot{m}(v_2^2 - v_1^2) = \frac{1}{2}\left(0.56\ \frac{\text{kg}}{\text{s}}\right)\left(\left(1.77\ \frac{\text{m}}{\text{s}}\right)^2 - 0\right)\left|\frac{1\ \text{N}}{\dfrac{\text{kg m}}{\text{s}^2}}\right.$$

$$\times\left(\frac{1\ \text{J}}{1\ \text{N m}}\right) = 0.88\ \text{J/s}$$

8.1.2.2 Potential Energy

Potential energy is the energy due to the position of the system in a potential field (e.g., earth's gravitational field, $g = 9.81$ m/s^2). The gravitational potential energy (ΔPE) of an object of mass m at an elevation z in a gravitational field, relative to its gravitational potential energy at a reference elevation z_0, is given by

$$\Delta PE = mg(z - z_0) \Rightarrow m(\text{kg}) g(\text{m/s}^2) \Delta z(\text{m}) = \text{N} \cdot \text{m} = \text{J} \tag{8.4}$$

To calculate the change in the rate of potential energy (ΔPE), often, the earth's surface is used as the reference, assigning $z_0 = 0$:

$$\Delta PE = \dot{m}g(z - z_0) \tag{8.5}$$

The unit of the change in transport rate of potential energy is obtained as follows:

$$\Delta PE = \dot{m}(\text{kg/s}) g(\text{m/s}^2) \Delta z(\text{m}) = \text{N} \cdot \text{m/s} = \text{J/s} = \text{W} \tag{8.6}$$

Example 8.2 Potential Energy Calculation

Problem
Water is pumped at a rate of 10.0 kg/s from a point 200.0 m below the earth's surface to a point 100.0 m above the ground level. Calculate the rate of change in potential energy.

Solution

Known quantities: Water mass flow rate (10.0 kg/s), initial location of water below the earth's surface (−200.0 m), and final location of water above the earth's surface (100 m).

Find: The rate of change in potential energy.

Analysis: Use the definition of potential energy.
   Taking the surface of the earth as a reference, the distance below the earth's surface is negative ($z_1 = -200.0$) and above the surface is positive ($z_2 = 100$):

$$\Delta PE = \dot{m}g(z_2 - z_1)$$

Substituting the values of the mass flow rate, gravitational acceleration, and change in inlet and exit pipe elevation from the surface of the earth,

$$\Delta PE = \left(10.0 \, \frac{\text{kg}}{\text{s}}\right) \times \left(9.81 \, \frac{\text{m}}{\text{s}^2}\right) \times \left(100.0 - (-200.0)\right) \text{m} \left|\frac{\text{J}}{\text{kg} \cdot \text{m}^2/\text{s}^2}\right. = 29,430 \, \text{J/s}$$

The rate of change in potential energy $\Delta PE = 29.43 \, \dfrac{\text{kJ}}{\text{s}} = 29.43$ kW.

Example 8.3 Internal Energy Calculation

Problem

A cylinder fitted with a movable piston is filled with gas. An amount of 2.00 kcal of heat is transferred to the gas to raise the gas temperature 100°C higher. The gas does 68 J of work in moving the piston to its new equilibrium position. Calculate the change in internal energy of the system (Example Figure 8.3.1).

Solution

Known quantities: The difference in gas temperature (100°C), work done by the system (+68 J), and heat added to the system (+2.00 kcal).

Find: Change in internal energy.

Analysis: Use the energy balance equation for a closed system.

System: Gas in the system, closed system

$$\Delta U + \Delta KE + \Delta PE = Q - W$$

Assumption: No change in kinetic and potential energy; accordingly, both are set to zero.

   The equation is reduced to

$$\Delta U = Q - W$$

Substitute the values of Q and W to calculate the change in internal energy (make sure units are consistent). The heat is added to the system (positive value) and the work is done by the system (positive value as well):

$$\Delta U = (2.0 \text{ kcal}) \left[\frac{1000 \text{ cal}}{\text{kcal}} \frac{1 \text{ J}}{0.239 \text{ cal}} \right] - 68 \text{ J} = 8300 \text{ J}$$

The change in internal energy $\Delta U = 8.30$ kJ.

   The specific enthalpy ($h = H/m$) can be calculated using the following equation:

$$h = u + Pv \qquad\qquad (8.7)$$

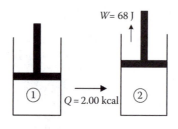

EXAMPLE FIGURE 8.3.1
Heat added to a cylinder fitted with a piston.

Substituting the values of specific internal energy ($u = U/m$), pressure (P), and specific volume (v) in the earlier equations gives the specific enthalpy h.

Example 8.4 Enthalpy from Internal Energy

Problem

The specific internal energy of helium at 25°C and 1 atm is 3.80 kJ/mol, and the specific molar volume under the same conditions is 25 L/mol. Calculate the specific enthalpy of helium at this temperature and pressure, and the rate at which enthalpy is transported by a stream with a molar flow rate of 250 kmol/h.

Solution

Known quantities: Internal energy, pressure, temperature, molar volume, molar flow.

Find: Specific molar enthalpy (h), rate of enthalpy transport ($\dot{H}$).

Analysis: Follow the specific enthalpy definition. The specific enthalpy of helium is given by

$$h = u + Pv$$

Substituting the values of specific internal energy, pressure (P), and specific volume (v) in the earlier equations,

$$h = \left(3800 \, \frac{J}{mol}\right) + (1 \, atm)\left(25 \, \frac{L}{mol}\right)\left[\frac{1 \, m^3}{1000 \, L} \, \frac{1.01325 \times 10^5 \, \frac{N}{m^2}}{1 \, atm} \, \frac{J}{N \cdot m}\right]$$

$$= 6333 \, J/mol$$

The enthalpy transport rate ($\dot{H}$) is calculated by multiplying the molar flow rate ($\dot{n}$) with the specific molar enthalpy (h):

$$\dot{H} = \dot{n} \times h$$

Substitute the values of molar flow rate ($\dot{n}$) and specific enthalpy (h) to find the enthalpy transport rate ($\dot{H}$):

$$\dot{H} = \left(250 \, \frac{kmol}{h}\right) \times \left(6333 \, \frac{J}{mol}\right)\left[\frac{1000 \, mol}{kmol} \, \frac{kJ}{1000 \, J}\right] = 1.58 \times 10^6 \, kJ/h$$

8.1.3 Energy Balance for an Open System

In open systems, material crosses the system boundary as the process occurs (e.g., continuous process at steady state). In an open system, work must be

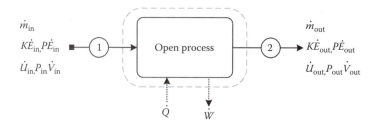

FIGURE 8.2
Energy balance for an open system.

done on the system to push input fluid streams at a pressure P_{in} into the system, and work is done on the surroundings to push output fluid streams at a pressure P_{out} out of the system, as shown in the schematic diagram in Figure 8.2 [3].

Net rate of work done by the system is

$$\dot{W}_f = \dot{W}_{out} - \dot{W}_{in} = P_{out}\dot{V}_{out} - P_{in}\dot{V}_{in} \tag{8.8}$$

For several input and output streams,

$$\dot{W}_f = \sum_{output} P_j \dot{V}_j - \sum_{input} P_j \dot{V}_j \tag{8.9}$$

The total rate of work $\left(\dot{W}\right)$ done by a system on its surroundings is divided into two parts:

$$\dot{W} = \dot{W}_s + \dot{W}_f \tag{8.10}$$

where shaft work ($\dot{W}_s$) is the rate of work done by the fluid on a moving part within the system (e.g., piston, turbine, and rotor), and flow work ($\dot{W}_f$) is the rate of work done by the fluid at the system outlet minus the rate of work done on the fluid at the system inlet. The general balance equation for an open continuous system (Figure 8.2) under steady state in the absence of generation/consumption term is

$$\text{Energy input} = \dot{U}_{in} + \dot{KE}_{,in} + \dot{PE}_{,in} + P_{in}\dot{V}_{in} \tag{8.11}$$

$$\text{Energy output} = \dot{U}_{out} + \dot{KE}_{,out} + \dot{PE}_{,out} + P_{out}\dot{V}_{out} \tag{8.12}$$

$$\text{Energy transferred} = \dot{Q} - \dot{W}_s \tag{8.13}$$

Under steady state, the accumulation term is set to zero and the following equation is valid:

$$\text{Energy input} = \text{Energy output}$$

$$\dot{Q} - W_s = \Delta \dot{U} + \Delta \dot{KE} + \Delta \dot{PE} + \Delta \left(P\dot{V} \right) \tag{8.14}$$

Enthalpy (H) is the sum of the *internal energy* (U) of fluid volume added to the system plus the *flow work* (PV) performed on the system in order to push the fluid in/out of the system:

$$H = U + PV \tag{8.15}$$

The change in enthalpy transport rate is given by

$$\Delta \dot{H} = \Delta \dot{U} + \Delta \left(P\dot{V} \right) \tag{8.16}$$

Rearranging the earlier equations leads to the first law of thermodynamics for an open system under steady state:

$$\dot{Q} - \dot{W}_s = \Delta \dot{H} + \Delta \dot{KE} + \Delta \dot{PE} \tag{8.17}$$

Steam turbine is an example of an open system. Electrical generating plants operate by generating steam at elevated temperatures and pressures, then reducing the pressure in a turbine. As the pressure is reduced, the high pressure, high temperature steam expands (and cools down), driving the turbine. The shaft work produced by the turbine is transferred to a generator to produce electricity. One limitation on steam turbines is that they cannot tolerate small amounts of water in its liquid state in the gases passing through the turbine. If the liquid content of the steam is above the threshold limit (a few percentage points), the liquid droplets damage the turbine blades and lead to failure of the turbine. The steam tables in Appendix A.3 are used to determine the temperature, specific internal energy, and specific enthalpy of saturated steam and superheated steam.

Example 8.5 Energy Balance for an Open System: The Steam Turbine

Problem

Steam flowing at a rate of 10 kg/h enters a steam turbine at a velocity of 50 m/s and leaves at a point 5 m below the inlet at a velocity of 300 m/s. The heat loss from the turbine is estimated to be 10 kW, and the turbine delivers shaft work at a rate of 70 kW. Calculate the change in enthalpy transport rate of the process.

Solution

Known quantities: Steam flow rate, inlet and exit velocity, heat loss, and work delivered.

Find: Change in enthalpy transport rate.

Analysis: Use the general energy balance equation for an open system.

System: Steam turbine as open system
The energy balance for an open system has been derived as

$$\dot{Q} - \dot{W}_s = \Delta \dot{H} + \Delta KE + \Delta P\dot{E}$$

In this example, heat is lost (negative value) from the system:

$$Q = -10 \text{ kW} = -10 \text{ kJ/s}$$

The shaft work is delivered (positive value) by the system:

$$W_s = 70 \text{ kW} = 70 \text{ kJ/s}$$

The change in kinetic energy

$$\Delta KE = \frac{1}{2}\dot{m}\left(v_2^2 - v_1^2\right)$$

Substitute the values of mass flow rate ($\dot{m}$), inlet (v_1), and exit (v_2) velocities, and use conversion factors (make sure units are consistent):

$$\Delta KE = \frac{1}{2}\left(10 \ \frac{\text{kg}}{\text{h}} \ \frac{\text{h}}{3600 \text{ s}}\right)\left(300^2 - 50^2\right)$$

$$\times \left(\frac{\text{m}}{\text{s}}\right)^2 \left| \frac{\text{N}}{\text{kg m/s}^2} \right| \frac{\text{J}}{\text{N}\cdot\text{m}} \left| \frac{\text{kJ}}{1000 \text{ J}} \right. = 0.12 \text{ kJ}$$

Change in potential energy

$$\Delta PE = \dot{m}g\left(z_2 - z_1\right)$$

Substitute the values of mass flow rate ($\dot{m}$), and inlet and exit heights from the surface of the earth (z_1, z_2):

$$\Delta PE = 10 \ \frac{\text{kg}}{\text{h}} \ \frac{\text{h}}{3600 \text{ s}} \times 9.81 \ \frac{\text{m}}{\text{s}^2}$$

$$\times (-5 - 0) \text{m} \left| \frac{\text{N}}{\text{kg m/s}^2} \right| \frac{\text{J}}{\text{N}\cdot\text{m}} \left| \frac{\text{kJ}}{1000 \text{ J}} \right. = -0.00014 \text{ kJ}$$

The change in potential energy is almost negligible compared to the magnitudes of heat and work.

Substitute the values of Q, W_s, and changes in kinetic and potential energies in the energy balance equation for an open system:

$$\Delta \dot{H} + \Delta KE + \Delta PE = Q - W_s$$

$$\Delta \dot{H} + 0.12 \, \frac{kJ}{s} - 0.00014 \, \frac{kJ}{s} = -10 \, \frac{kJ}{s} - \left(70 \, \frac{kJ}{s} \right)$$

The change in enthalpy transport rate is

$$\Delta \dot{H} = -80.12 \, kJ/s$$

Example 8.6 Use of a Steam Table

Problem

Use steam tables in the appendix to determine the temperature, specific internal energy, and specific enthalpy of saturated steam at 3.0 bar. What is the state of the steam at 10 bar and 400°C? (i.e., is it saturated or superheated steam?)

Solution

Known quantities: *Case 1*: 3 bar, saturated steam, *Case 2*: 10 bar, 400°C.

Find: Specific enthalpy (h) and specific internal energy (u), specific volume (v). The state of steam at 10 bar and 400°C.

Analysis: Two properties are needed to be able to use saturated steam table and superheated steam table in the appendix.

Case 1: At 3 bar, steam is saturated: use saturated steam table (Appendix A.3). The temperature is 133.5°C, specific enthalpy is 2724.7 kJ/kg, and specific internal energy is 2543 kJ/kg.

Case 2: At 10 bar and 400°C: At 10 bar the saturated temperature is 179.9°C, and since the steam is at 400°C, this temperature is higher than the saturated temperature at 10 bar. Therefore, the state of water is superheated steam, and hence, the superheated steam table (Table A.5) is used.

Specific enthalpy is 3264 kJ/kg, specific internal energy is 2958 kJ/kg, and specific volume is 0.307 m³/kg.

8.1.4 Steam Turbine

Steam turbines are open systems used to generate electricity; in most cases, the turbine operates adiabatically. The exit pressure of turbine is lower than the inlet pressure. Turbines produce work; by contrast, work should be provided to a compressor or a pump. The following examples explain the possible operations for a steam turbine.

Example 8.7 Steam Table and Turbine Work

Problem
Steam at a rate of 1500 kg/s enters a turbine at 40 bar and 400°C. It comes out of the turbine as wet steam at 4 bar. The turbine operates adiabatically and produces 1000 MW of work. What is the temperature of the steam leaving the turbine? What is the mass fraction of vapor in the stream leaving the turbine?

Solution

Known quantities: Steam mass flow rate (1500 kg/s), inlet conditions (40 bar and 400°C), exit steam conditions (4 bar, wet steam).

Find: Mass fraction of vapor in the stream leaving the turbine.

Assumptions: No change in kinetic and potential energy.

Analysis: Use steam tables to find inlet and exit enthalpy and the first law for an open system. Inlet and exit steam enthalpies: Saturated steam (Table A.4), superheated steam (Table A.5).

Inlet steam conditions: at 40 bar and 400°C: the enthalpy of the incoming steam is 3216 kJ/kg (Table A.5).

Exit steam conditions: at 4 bar: steam is either wet or saturated (Table A.4). Since the steam leaving the turbine is a vapor–liquid mixture, it must be saturated. From Table A.4, for saturated steam at 4 bar the enthalpies of the liquid and vapor are 604.7 and 2737.6 kJ/kg, respectively, and the temperature is 143.6°C.

The general energy balance applied to this process, after neglecting the potential and kinetic energy terms and bearing in mind that the turbine is adiabatic, can be expressed as

$$\Delta \dot{H} = -\dot{W}_\text{s}$$

Rearranging the earlier equation,

$$-\dot{W}_\text{s} = \Delta \dot{H} = \dot{H}_\text{out} - \dot{H}_\text{in} = \dot{m}(h_\text{out} - h_\text{in})$$

Substituting known values of shaft work, mass flow rate, and inlet specific enthalpy, since the turbine is producing work, the sign of W_s is +:

$$-\dot{W}_\text{s} = -1000 \text{ MW} = -1 \times 10^6 \, \frac{\text{kJ}}{\text{s}} = 1500 \, \frac{\text{kg}}{\text{s}} (h_\text{out} - 3216) \, \frac{\text{kJ}}{\text{kg}}$$

The specific enthalpy of the exit steam is

$$h_\text{out} = 2549.3 \text{ kJ/kg}$$

Let x be the mass fraction of the steam that is in the vapor phase, then

$$h_\text{out} = 2549.3 \, \frac{\text{kJ}}{\text{kg}} = h_\text{f} + x h_\text{fg} = 604.7 \, \frac{\text{kJ}}{\text{kg}} + x \left(2133.0 \text{ kJ/kg} \right)$$

The steam quality is $x = 0.912 \rightarrow$ The wet steam is 91.2 wt% vapor. The wet contains 91.2% water vapor and 8.80 wt% liquid water.

Example 8.8 Steam Turbine

Problem

Steam enters a turbine at a pressure of 10.0 bar (absolute) and a temperature of 600°C. The steam leaving the turbine is at 1 atm (absolute) pressure and is of 90% quality (90 wt% steam, 10 wt% liquid). How much steam has to go into the turbine to yield 1.5 × 10⁶ kW of shaft work?

Solution

Known quantities: Steam inlet conditions (10 bar, 600°C), exit steam conditions (1 atm, 90% quality), shaft work is 1.5 × 10⁶ kW.

Find: Amount of steam that has to go into the turbine.

Assumptions: No change in kinetic and potential energy, turbine is adiabatic.

Analysis: Use steam tables to find inlet and exit enthalpy and the first law for an open system. From the first law,

$$\Delta H + \Delta KE + \Delta PE = Q - W_s$$

After applying the earlier assumptions, the equation is reduced to

$$\Delta \dot{H} = \dot{m}\left(h_{out} - h_{in}\right) = -W_s$$

To find the enthalpy of the steam leaving the turbine, use Table A.4. At 1 atm the enthalpies of saturated water and steam are 419.1 and 2676.0 kJ/kg, respectively.

Thus, the enthalpy of the steam leaving the turbine is

$$h_{out} = h_f + x h_{fg} = 419.1 + 0.9\left(2676.0 - 419.1\right) = 2450.3 \text{ kJ/kg}$$

The enthalpy of the input steam can be found from Table A.5 to be 3697 kJ/kg.

Substitute the values of inlet and outlet specific enthalpy and shaft work in the first law:

$$\dot{m}\left(2450.3 - 3697\right) = -\left(1.5 \times 10^6 \text{ kJ/s}\right)$$

The required steam mass flow rate is

$$\dot{m} = 1.20 \times 10^3 \text{ kg/s}$$

8.1.5 Heaters and Coolers

Heaters and coolers such as shell and tube heat exchangers are open systems employed to cool down or heat up certain fluid streams. In most cases, the external surface of heaters and coolers is insulated and heat is just transferred between the cold and hot streams across the walls of the exchanger tubes. The following example illustrates the use of heat exchangers for cooling and heating purposes.

Example 8.9 Heat Exchanger

Problem
Steam at a rate of 60 kg/h, at 200°C, and 1 bar enters the tube side of a shell and tube heat exchanger. The steam is used to heat cold water flowing on the shell side; the steam leaves as saturated liquid. Neglect pressure drop of the steam on the tube side and the water on the shell side of the heat exchanger. How much heat must be transferred from the steam to the water side?

Solution

Known quantities: Mass flow rate (60 kg/h), inlet temperature and pressure (200°C, 1 bar), exit conditions (saturated water, 1 bar).

Find: Heat transfer rate from steam to water.

Assumptions: Pressure drop across the boiler is neglected, so exit pressure is at 1 bar.

Analysis: Use steam tables to find inlet and exit enthalpy.

Basis: 60 kg/h of feed steam. The schematic diagram of the problem is shown in Example Figure 8.9.1.

From the superheated steam table (Table A.5),

Inlet: (1 bar, 200°C): $h_1 = 2875$ kJ/kg
Using saturated steam table (Table A.4),
Outlet: (1 bar, saturated water): $h_2 = h_f|_{at\,1\,bar} = 417.5$ kJ/kg
No change in steam mass flow rate: $\dot{m} = \dot{m}_{in} = \dot{m}_{out} = 60$ kg/h

The general energy balance equation for an open system is

$$\Delta \dot{H} + \Delta KE + \Delta PE = \dot{Q} - \dot{W}_s$$

The following simplifying assumptions for the condenser are used:
No shaft work: $W_s = 0$.
No change in elevation. The inlet and outlet lines are at the same level: $\Delta PE = 0$.

Since we do not know anything about the diameters of the inlet and exit pipes, same pipe diameters are used for inlet and exit streams; accordingly, there is no change in velocity, and change in kinetic energy is negligible: $\Delta KE = 0$.

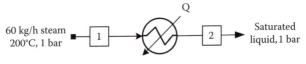

EXAMPLE FIGURE 8.9.1
Schematic of a heat exchanger system.

The simplified form of the energy balance is therefore

$$\dot{Q} = \Delta\dot{H} = \dot{H}_{out} - \dot{H}_{in}$$

The rate of enthalpy transport ($\dot{H}$) as a function of specific enthalpy (h),

$$\Delta\dot{H} = \dot{m}\Delta h$$

Replacing enthalpy change rate ($\Delta\dot{H}$) with specific enthalpy (Δh) at constant mass flow rate,

$$\dot{Q} = \dot{m}\Delta h = \dot{m}(h_{out} - h_{in})$$

Substituting the values of mass flow rate and exit and inlet specific enthalpy in the earlier equation,

$$Q = \left(60\ \frac{kg}{h} \right)(417.5 - 2875)\frac{kJ}{kg} = -147,450\ kJ/h$$

The value of heat transfer is negative; that is, heat is transferred from the system (steam) to the surrounding (cold water).

8.1.6 Compressors

Compressors are open systems utilized to raise the pressure of gas steams. The exit pressure is higher than the inlet pressure. Work is required for the compressor to operate. The following example illustrates the use of a compressor to pressurize a vapor steam.

Example 8.10 Compressor

Problem
The feed to a compressor is superheated steam at 300°C and 20 bar absolute pressure. It enters the compressor at a velocity of 20 m/s. The pipe inlet inside diameter is 0.10 m. The discharging pipe, after the compressor, has a smaller inside diameter and the discharge velocity is 170 m/s. The exit of the compressor is superheated steam at 350°C and 60 bar absolute. Heat loss from the compressor to the surroundings is 5 kW. Determine the compressor horsepower.

Solution

Known quantities: Inlet and exit steam temperature and pressure, inlet and exit velocities, heat loss from the compressor, inlet pipe diameter.

Find: Compressor horsepower.

Assumption: The system is located on a horizontal plane and no change in elevation between inlet and exit of the compressor; hence change in potential energy is negligible.

Analysis: The compressor is an open system. The steady-state energy balance can be used to describe the compressor system:

$$\Delta \dot{H} + \Delta KE + \Delta PE = \dot{Q} - W_s$$

The general energy balance reduces to

$$\Delta H + \Delta KE = \dot{Q} - W_s$$

Determination of the specific enthalpy and specific volume from the superheated steam table (Appendix A.3):

Inlet stream at $P_1 = 20$ bar, $T_1 = 300°C$, $h_1 = 3025$ kJ/kg, $v_1 = 0.125\,m^3/kg$
Exit stream at $P_2 = 60$ bar, $T_2 = 350°C$, $h_2 = 3046$ kJ/kg, $v_2 = 0.0422\,m^3/kg$

Mass flow rate of the inlet steam is equal to density multiplied by volumetric flow rate; the steam density is the inverse of steam specific volume:

$$\dot{m} = \rho \times \dot{V} = \rho \times (v \times A) = \rho \times \left(v \times \frac{\pi D^2}{4} \right) = \frac{1}{v} \times \left(v \times \frac{\pi D^2}{4} \right)$$

Substitute the values of density, velocity, and diameter:

$$\dot{m} = \frac{1}{0.125\,m^3/kg} \times \left(\frac{20\,m}{s} \times \frac{\pi (0.1m)^2}{4} \right) = 1.25\,kg/s$$

The change in enthalpy transport rate $\Delta \dot{H}$ is given by

$$\Delta \dot{H} = \dot{m}(h_2 - h_1) = 1.25\,\frac{kg}{s}(3046 - 3025)\,\frac{kJ}{kg} = 26.5\,\frac{kJ}{s} = 26.25\,kW$$

The change in kinetic energy is

$$\Delta KE = \frac{1}{2} \dot{m} \left(v_2^2 - v_1^2 \right)$$

Substitute the values of mass and inlet and exit velocity:

$$\Delta KE = \frac{1}{2} \times 1.25\,\frac{kg}{s} \left\{ \left(\frac{170\,m}{s} \right)^2 - \left(\frac{20\,m}{s} \right)^2 \right\}$$

$$\times \frac{N}{kg\,m/s^2} \left| \frac{J}{N \cdot m} \right| \frac{kJ}{1000\,J} = 18\,kJ/s$$

The change in the kinetic energy is

$$\Delta KE = 18\,kW$$

The heat loss from the system to the surroundings is 5 kW. Since heat is transferred from the system to the surroundings, $Q = -5$ kW.

The general energy balance equation reduces to

$$\Delta H + \Delta E_k = Q - W_s$$

Substituting the values of change in enthalpy, kinetic energy, and heat loss,

$$26.25 \text{ kW} + 18 \text{ kW} = -5 \text{ kW} - W_s$$

Rearranging and solving for the shaft work,

$$W_s = -49.25 \text{ kW}$$

$$\text{Power} = 49.25 \text{ kW} \left(\frac{1.341 \text{ hp}}{1 \text{ kW}} \right) = 66.04 \text{ hp}$$

The sign of the shaft work is negative since work is done on the system by compressor blades. To convert the shaft work to horsepower, use the proper conversion factor.

8.2 Mechanical Energy Balance

The mechanical energy balance is most useful for processes in which changes in the potential and kinetic energies are of primary interest, rather than changes in internal energy or heat associated with the process. Thus, the mechanical energy balance is mainly used for purely mechanical flow problems—that is, problems in which heat transfer, chemical reactions, or phase changes are not present. First, we assume the steady-state condition so that all terms on the left-hand side become zero. Second, we assume that the system has only a single inlet and a single outlet. Moreover, steady state implies that the inlet mass flow rate must equal the outlet mass flow rate, in order to avoid accumulation of material in the system. Let us start with the general energy balance equation:

$$\text{Energy transferred} = \text{Energy out} - \text{Energy in}$$

$$\dot{Q} - \dot{W}_s = \left(\dot{U}_{out} + \dot{KE}_{,out} + \dot{PE}_{,out} + P_{out}\dot{V}_{out} \right) - \left(\dot{U}_{in} + \dot{KE}_{,in} + \dot{PE}_{,in} + P_{in}\dot{V}_{in} \right) \quad (8.17)$$

Rearrange the earlier equation by taking the mass flow rate ($\dot{m}$) as a common factor. In this case, the internal energy and volumetric flow rate will become specific internal energy and specific volumetric flow rate, respectively:

$$\dot{Q} - \dot{W}_s = \dot{m} \left(u_{out} + \frac{v_{out}^2}{2} + gz_{out} + P_{out}v_{out} - u_{in} - \frac{v_{in}^2}{2} - gz_{in} - P_{in}v_{in} \right) \quad (8.18)$$

In this equation, subscript "in" refers to the inlet section, and subscript "out" to the outlet port. Now, we divide the entire equation by $\dot{m}$, and express the specific volume (volume/mass) as $v = 1/\rho$, where ρ is the density (mass/volume) of the flowing material. Assuming incompressible flow rate, so that the density is constant, $v_{in} = v_{out} = 1/\rho$. Also, we define $\Delta u = u_{out} - u_{in}$ and $\Delta P = P_{out} - P_{in}$. With these changes, the general energy balance equation becomes

$$\frac{-W_s}{\dot{m}} = \frac{\Delta P}{\rho} + \frac{\Delta v^2}{2} + g\Delta z + \Delta u - \frac{\dot{Q}}{\dot{m}} \qquad (8.19)$$

The term $(\Delta u - \dot{Q}/\dot{m})$ in the absence of chemical reactions, phase changes, or other sources of large amounts of heat transfer will generally represent heat generated due to the viscous friction in the fluid. In such situations, this term is called the friction loss and we will write it as F. With this last change, the general energy balance represents the usual form of the mechanical energy balance:

$$\frac{-\dot{W}_s}{\dot{m}} = \frac{\Delta P}{\rho} + \frac{\Delta v^2}{2} + g\Delta z + F \qquad (8.20)$$

where $\dot{W}/\dot{m}$ is the shaft work performed by the system on the surroundings, per unit mass of material passing through the system. The following example illustrates the use of the mechanical energy balance equation.

Example 8.11 Mechanical Energy Balance Equation

Problem
A water supply tank is capable of delivering 0.3 m³/s of water for fire-fighting purposes in a chemical plant. The water supply is to come from a lake, the elevation of the surface of the lake is 800 m and the elevation of the factory is 852 m from sea level. The water discharge pipe is located at a depth of 100 m from the surface of the lake. The frictional losses in the water line to the plant are given by the relation (0.01 m/s^2) L, where L is the length of the pipe line. The water line to the supply tank has an inner diameter of 0.15 m and a length of 8000 m. How much energy must a pump deliver to the water?

Solution

Known quantities: Discharge line volumetric flow rate, initial and final elevation, friction losses, length and diameter of the pipe.

Find: Pump horsepower.

Assumption: Pressure drop is neglected because the pressure at both ends of line is atmospheric.

Analysis: Use the mechanical energy balance equation:

$$\frac{\Delta P}{\rho} + \frac{\Delta v^2}{2} + g\Delta z + F = \frac{-\dot{W}_s}{\dot{m}}$$

The pressure at both ends of the line is atmospheric, so $\Delta P = 0$. The velocity at the inlet of the lake is zero but the velocity out of the discharge end of the pipe is

$$v_2 = \dot{V} \times \frac{1}{\frac{\pi D^2}{4}} = \left(0.3 \frac{m^3}{s}\right) \times \frac{1}{\frac{\pi (0.15 m)^2}{4}} = 17 \text{ m/s}$$

The mass flow rate

$$\dot{m} = \dot{V} \times \rho = \frac{0.3 \text{ m}^3}{s} \times \frac{1000 \text{ kg}}{m^3} = 300 \text{ kg/s}$$

The elevation change is from 800 m (800–100 to the lake) to 852 to the factory, or the difference is equivalent to 152 m. So the mechanical energy balance becomes

$$\frac{0}{\rho} + \frac{\left(17 \frac{m}{s}\right)^2 - 0}{2} + 9.81 \frac{m}{s^2}(152 \text{ m}) + 0.01 \frac{m}{s^2}(8000 \text{ m}) = \frac{-\dot{W}_s}{300 \text{ kg/s}}$$

Solving for shaft work,

$$-\dot{W}_s = 514{,}686 \text{ W} \times \frac{hp}{746} = 690 \text{ hp}$$

The minus sign indicates that the energy is going into the system.

Example 8.12 Fire Extinguishment Process

Problem
A large tank filled with water and open to atmosphere is used for fire extinguishment in an ethylene production plant. The water is taken from the tank, passed through a pump, and then delivered to hoses. It is desired to deliver 1890 L of water per minute at a pressure of 15 bar (gauge). If there is a negligible elevation change between the water level in the tank and the discharge of the pump, no changes in the diameter of the pipes and hoses, and if the pump has an efficiency of 70.0%, how much work must be supplied to the pump in order to meet the pressure and discharge rate specifications?

Solution

Known quantities: Discharge line volumetric flow rate, initial and final elevation, friction losses, length and diameter of the pipe.

Find: Pump horsepower.

Assumption: Pressure drop is neglected because the pressure at both ends of the line is atmospheric.

Analysis: Use the mechanical energy balance to solve this problem:

$$\frac{\Delta P}{\rho} + \frac{\Delta v^2}{2} + g\Delta z + F = \frac{-\dot{W}_s}{\dot{m}}$$

Because there is no change in elevation or velocity (no change in pipe/hose diameter) and no frictional losses are given, the earlier equation reduces to

$$\frac{\Delta P}{\rho} = \frac{-\dot{W}_s}{\dot{m}}$$

The water mass flow rate is

$$\dot{m} = \dot{V} \times \rho = 1890\frac{L}{min} \times \frac{1\,min}{60\,s} \times \frac{1\,kg}{L} = 31.5\,kg/s$$

The discharge pressure is given as 15 bar (gauge). This means that the absolute pressure is this pressure plus the ambient pressure. Substitute the known values to get

$$\frac{\left[(15\,bar + P_{ambient}) - P_{ambient}\right]\frac{10^5\,Pa}{bar}\left|\frac{1\,N/m^2}{Pa}\right.}{1000\,\frac{kg}{m^3}} = \frac{-\dot{W}_s}{31.5\,\frac{kg}{s}}$$

Simplifying,

$$-\dot{W}_s = 47{,}250\frac{N\cdot m}{s}\left|\frac{J}{N\cdot m}\right|\frac{kJ}{1000\,J} = 47.25\frac{kJ}{s} = 47.25\,kW$$

The theoretical shaft work in horsepower is

$$-\dot{W}_s = 47.25\,kW\left|\frac{1\,hp}{0.754\,kW}\right. = 62.7\,hp$$

The pump has an efficiency of 70.0%; accordingly, the actual work ($\dot{W}_{s.a}$) that must be supplied to the pump in order to meet the pressure and discharge rate specifications is

$$-\dot{W}_{s.a} = \frac{62.7\,hp}{0.7} = 89.5\,hp$$

The sign of the work is negative, which means that the work is done on the system. The actual work that must be supplied to the pump in order to meet the pressure and discharge rate specifications is higher than the theoretical work.

8.3 Bernoulli's Equation

In many instances, the amount of energy lost to viscous dissipation in the fluid is small compared to magnitudes of the other terms in the general energy balance equation. In such a case, $F=0$. Moreover, many common flows such as fluid flow through a pipe do not have any appreciable shaft work associated with them; accordingly, $\dot{W} = 0$. For such frictionless flows with no shaft work, the mechanical energy balance simplifies to *Bernoulli's equation*:

$$\frac{\Delta P}{\rho} + \frac{\Delta v^2}{2} + g\Delta z = 0 \tag{8.21}$$

Bernoulli's equation has a wide range of applications, despite its simplified assumptions. The following example illustrates the use of Bernoulli's equation.

Example 8.13 Bernoulli's Equation

Problem
The pressure difference between the underside of the wing and the top of the wing that is necessary to lift the weight of an aircraft is 0.08 atm. At an elevation of approximately 10,000 m, the aircraft velocity is 275 m/s and the density of air is 0.45 kg/m³. Assume that the velocity of the air on the underside of the wing is the plane velocity of 275 m/s. What is the velocity of the air on the topside of the wing, which is necessary to generate the pressure difference needed to lift the plane?

Solution

Known values: Pressure drop around the wing, velocity of air on the underside of the wing

Find: Velocity of air on the topside of the wing

Analysis: Use Bernoulli's equation around the wing (1: topside of the wing, 2: underside of the wing). Use Bernoulli's Equation to relate the pressure difference to a velocity difference so that

$$\frac{P_2 - P_1}{\rho} + \frac{v_2^2 - v_1^2}{2} + g(z_2 - z_1) = 0$$

Neglect the effect of wing thickness on change in potential energy. The equation is reduced to

$$\frac{P_2 - P_1}{\rho} + \frac{v_2^2 - v_1^2}{2} + 0 = 0$$

Substituting the values of pressure drop, air density, and velocity under the wing,

$$\frac{0.08\ \text{atm} \left| \dfrac{101{,}325\ \text{Pa}}{1\ \text{atm}} \right.}{0.45\ \text{kg/m}^3} + \frac{275^2 - v_1^2}{2} = 0$$

Solving for velocity on the topside of the wing,

$$v_1 = 334\ \frac{\text{m}}{\text{s}}$$

The velocity on the topside of the wing is higher than that on the underside of the wing.

8.4 Enthalpy Calculations

Change in enthalpy can occur because of change in temperature, change in phase, or mixing of solutions and reactions.

8.4.1 Enthalpy Change as a Result of Temperature

Sensible heat is the heat transferred to raise or lower the temperature of a material in the absence of phase change. In the energy balance calculations, sensible heat change is determined by using a property of matter called the heat capacity at constant pressure, or just heat capacity (C_P). Units for C_P are (J/mol/K) or (cal/g/°C). Appendix A.2 lists C_P values for several organic and inorganic compounds. There are several methods for calculating enthalpy change using C_P values. When C_P is constant, the change in the enthalpy of a substance due to change in temperature at constant pressure is given by

$$\Delta H = mC_P(T - T_{ref}) \tag{8.22}$$

Heat capacities for most substances vary with temperature where the values of C_P vary for the range of the change in temperature. Heat capacities are tabulated as polynomial functions of temperature such as

$$C_P = a + bT + cT^2 + dT^3 \tag{8.23}$$

Coefficients a, b, c, and d for a number of substances are given in Appendix A.2. In this case, the enthalpy change is

$$\Delta \dot{H} = \dot{m} \int_{T_{ref}}^{T} C_P \, dT = \dot{m} \int_{T_{ref}}^{T} \left(a + bT + cT^2 + dT^3 \right) dT \qquad (8.24)$$

Sometimes, you need an estimate of specific enthalpy, specific internal energy, or specific volume at a temperature and a pressure that is between tabulated values. In this case, one can use a linear interpolation.

The following example demonstrates the determination of internal energy from heat capacity.

Example 8.14 Internal Energy and Heat Capacity

Problem

A closed rigid vessel that contains 200 kg of a fluid is heated from 20°C to 150°C. Calculate the heat required for this purpose. The constant volume heat capacity of the fluid is given by the following relation:

$$C_v \left(\frac{kJ}{kg \cdot {}^{\circ}C} \right) = a + bT = 0.855 + 9.42 \times 10^{-4} T$$

Solution

Known quantities: Mass of fluid, initial (20°C) and final temperature (150°C), heat capacity at constant volume as a function of temperature.

Find: Heat required to heat the content of the closed vessel.

Analysis: Use the general energy balance for a closed system, no change in kinetic and potential energies as the system is a rigid vessel:

$$Q - W = \Delta U$$

$W = 0.0$ (rigid vessel; no moving part), the change in internal energy is

$$Q = \Delta U$$

The change in internal energy is a function of heat capacity at constant volume; since the heat capacity is a function of temperature and mass, we multiply mass by heat capacity as follows:

$$\Delta U = m \int_{T_1}^{T_2} C_V \, dT$$

Substitute the heat capacity at constant volume:

$$\Delta U = m \int_{T_1}^{T_2} (0.855 + 9.42 \times 10^{-4} T) dT$$

Integrating the earlier equation as a function of initial and final temperature, we obtain

$$\Delta U = m \left[0.855(T_2 - T_1) + 9.42 \times 10^{-4} \left(\frac{T_2^2 - T_1^2}{2} \right) \right]$$

Substituting the values of initial (20°C) and final temperature (150°C),

$$Q = \Delta U = 200 \text{ kg} \left[0.855(150 - 20) + 9.42 \times 10^{-4} \frac{(150^2 - 20^2)}{2} \right]$$

$$\times \left[\frac{kJ}{kg \cdot {}^\circ C} \right] = 24,312 \text{ kJ}$$

Example 8.15 Use of Data from Tables and Reference State

Problem
The data shown in Table E8.15 are for a saturated fluid. Calculate Δh and Δu for the transition of saturated vapor from 10°C to –20°C.

Solution

Known quantities: Enthalpy, pressure, and temperature.

Find: Change of specific enthalpy and specific internal energy.

Analysis: The reference is liquid at –40°C, because the enthalpy at this temperature is zero. Change in specific enthalpy (Δh) and change in specific internal energy (Δu) for the transition of saturated CH_3Cl vapor from 10°C to –20°C can be calculated as

$$\Delta h = h_{-20^\circ C} - h_{10^\circ C} = 456 - 470 = -14 \text{ kJ/kg}$$

The change in specific internal energy starts using $h = u + Pv$ and $\Delta h = \Delta u + \Delta(Pv)$.
  Rearranging for Δu,

$$\Delta u = \Delta h - \Delta(Pv) = \Delta h - \left\{ (Pv)_{-20} - (Pv)_{10} \right\}$$

TABLE E8.15

Properties of Saturated Methyl Chloride

State	T (°C)	P (atm)	v (m³/kg)	h (kJ/kg)
Liquid	–40	0.47	0.001	0.00
Vapor	–20	1.30	0.310	456
Vapor	10	3.54	0.120	470

To calculate the change in internal energy,

$$\Delta u = -14\,\frac{\text{kJ}}{\text{kg}} - \{1.30 \times 0.312 - 3.54 \times 0.12\}\left(\text{atm} \times \frac{\text{m}^3}{\text{kg}}\right)$$

$$\times \left(\frac{101.325\ \text{kN/m}^2}{1\ \text{atm}}\right)\left(\frac{\text{kJ}}{\text{kN} \cdot \text{m}}\right)$$

The rounded result of change in internal energy is

$$\Delta u = -12\ \text{kJ/kg}$$

8.4.2 Constant Heat Capacity

Keeping P constant and letting T change, we can get the expression for the constant P part as $\Delta h = \int C_P dT \approx C_P \Delta T$ (at constant P). It is not necessary to know the reference state to calculate ΔH for the transition from one state to another. Δh from state 1 to state 2 equals $h_2 - h_1$ regardless of the reference state upon which h_1 and h_2 were based. If different tables are used, one must make sure they have the same reference state. h and u are state properties; their values depend only on the state of the species, temperature, and pressure and not on how the species reached its state. When a species passes from one state to another, both Δu and Δh for the process are independent of the path taken from the first state to the second one.

Example 8.16 Constant Heat Capacity

Problem
What is the change in the enthalpy of 100 g/s acid heated in a double pipe heat exchanger from 20°C to 80°C, if the average heat capacity at constant pressure is 0.50 cal/g°C?

Solution

Known quantities: Mass of acid, constant heat capacity, initial and final temperatures.

Find: Change in enthalpy.

Analysis: Use change in enthalpy with const heat capacity.
The change in enthalpy as a function of specific heat is given by

$$\Delta \dot{H} = \int_{T_1}^{T_2} \dot{m} C_P\, dT$$

Since the heat capacity (C_P) is constant, the earlier equation is simplified to

$$\Delta \dot{H} = \dot{m} C_P (T_2 - T_1)$$

Substitute the values of mass flow rate, heat capacity at constant pressure, and difference in temperature (the reference temperature is 20°C):

$$\Delta \dot{H} = \left(100 \, \frac{g}{s} \right) \left(0.5 \frac{cal}{g°C} \right) (80 - 20)°C = 3000 \, cal/s$$

The change in enthalpy transport rate is

$$\Delta \dot{H} = 3.0 \, kcal/s$$

Example 8.17 Heat Added to a Boiler

Problem

Liquid water is fed to a boiler at 23°C under a pressure of 10 bar, and is converted at constant pressure to saturated steam. Calculate Δh for this process and the heat input required for producing 15,000 m³/h of steam at the exit conditions. Assume that the inlet velocity of liquid entering the boiler is negligible and that the steam is discharged through a 0.15 m ID (inner diameter) pipe (Example Figure 8.17.1). Inlet and exit pipes are at the same level.

Solution

Known quantities: Water inlet conditions (23°C under a pressure of 10 bar), exit steam conditions (10 bar, saturated steam), exit steam volumetric flow rate (15,000 m³/h), exit pipe diameter (0.15 m).

Find: Change in specific enthalpy (Δh).

Analysis: The reboiler is an open system, and the general energy balance equation is

$$Q - W_s = \dot{m}\Delta h + \Delta KE + \Delta PE$$

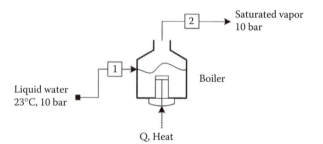

EXAMPLE FIGURE 8.17.1
Production of saturated steam.

Since the reboiler does not deliver shaft work, no change is seen in elevation between inlet and exit steams (change in potential energy is zero);
the energy balance equation reduces to

$$Q_s = \dot{m}\Delta h + \Delta KE$$

The change in specific enthalpy:
   Since no value of specific enthalpy is available at 23°C and 10 bar, the
value is taken at 23°C (saturated water):

$$h_1\big|_{\text{at } 23°C,\, 10 \text{ bar}} = 96.2 \text{ kJ/kg}$$

The specific enthalpy value for the exit conditions at 10 bar, saturated
steam is

$$h_2\big|_{\text{at } 10 \text{ bar, sat'd steam}} = 2776.2 \text{ kJ/kg}$$

The change in specific enthalpy is

$$\Delta h = h_2 - h_1$$

Substitute the values of inlet and exit specific enthalpy:

$$\Delta h = 2776.2\,\frac{\text{kJ}}{\text{kg}} - 96.2\,\frac{\text{kJ}}{\text{kg}} = 2680 \text{ kJ/kg}$$

The discharge mass flow rate ($\dot{m}_2$) is calculated at the exit steam because
exit steam volumetric flow rate and diameter of discharge pipe are given.
The density is calculated from the inverse of specific volume ($\rho = 1/v$).
The specific volume (v) at 10 bar, saturated steam is 0.1943 m³/kg (used
saturated steam table, Appendix A.3):

$$\dot{m}_2 = \rho \times \dot{V} = \frac{1}{0.1943 \text{ m}^3/\text{kg}} \times 15,000\,\frac{\text{m}^3}{\text{h}} \times \frac{\text{h}}{3600 \text{ s}} = 21.45 \text{ kg/s}$$

The inlet velocity is negligible as given in the problem statement. The exit
velocity is calculated from the discharge volumetric flow rate divided by
pipe cross sectional area:

$$v_2 = \frac{\dot{V}_2}{\dfrac{\pi D^2}{4}} = \frac{15,000 \text{ m}^3/\text{h}}{\dfrac{\pi (0.15)^2}{4} \text{ m}^2} \times \frac{\text{h}}{3600 \text{ s}} = 235.79 \text{ m/s}$$

The simplified general energy balance equation becomes

$$Q = \dot{m}\Delta h + \Delta KE = \dot{m}\Delta h + \frac{1}{2}\dot{m}(v_2^2 - v_1^2)$$

Substitute the values of mass flow rate, specific enthalpy, and velocity:

$$Q = 21.45 \frac{kg}{s} \times \left(2680 \frac{kJ}{kg} \right) + \frac{1}{2} \times 21.45 \frac{kg}{s} \left\{ \left(235.79 \frac{m}{s} \right)^2 - 0 \right\}$$

$$\times \frac{kJ}{1000 \text{ J}} = 58{,}082 \text{ kJ/s}$$

The sign of the heat transfer across system boundaries is positive; that is, heat is transferred from the surroundings to the system.

8.5 Enthalpy Calculations with Phase Changes

The state of a system can be changed, for example, by increasing its temperature or changing its composition. Properties of the system whose change depends only on the initial and final states of the system but not on the manner used to realize the change from the initial to the final state are referred to as state properties [4].

Phase changes, such as evaporation and melting, are accompanied by relatively large changes in internal energy and enthalpy, as bonds between molecules are broken and reformed. Heat transferred to or from a system, causing change of phase at constant temperature and pressure, is known as latent heat. The types of latent heats are latent heat of vaporization, which is the heat required to vaporize a liquid; latent heat of fusion, which is the heat required to melt a solid; and latent heat of sublimation, which is the heat required to directly vaporize a solid. Heat is released during condensation, and heat is required to vaporize a liquid or melt a solid. Table A.1 reports these two latent heats for substances at their normal melting and boiling points (i.e., at a pressure of 1 atm). Sensible heat refers to heat that must be transferred to raise or lower the temperature of a substance without change in phase as defined earlier. The quantity of sensible heat required to produce a temperature change in a system can be determined from the appropriate form of the first law of thermodynamics. The heat capacity at constant pressure, C_P, for most incompressible liquids and solids is equal; $C_P \approx C_V$ and for ideal gases, $C_P = C_V + R$.

Example 8.18 Enthalpy of Phase Change

Problem
Steam at a rate of 100 kg/h is used to heat a stream of gas flowing on the tube side of a heat exchanger. The steam enters the shell side of the heat exchanger as saturated vapor at 10 bar of 90% quality, and exits as saturated liquid water at 10 bar. Calculate the rate of heat transfer to the gas side.

Solution

Known quantities: Inlet (10 bar, 90% quality) and exit (10 bar, saturated water) steam conditions.

Find: The change in enthalpy transport rate.

Assumption: No change in potential and kinetic energy, no shaft work.

Analysis: Use the general energy balance equation for an open system around the heat exchanger. The simplified energy balance is obtained as follows.

Energy balance for an open system is

$$\Delta \dot{H} + \Delta KE + \Delta PE = \dot{Q} - \dot{W}_s$$

After including the assumptions, the equation is reduced to

$$\Delta \dot{H} = Q$$

Setting enthalpy transport rate ($\dot{H}$) in terms of specific enthalpy h,

$$\Delta \dot{H} = \dot{m}_s (h_2 - h_1) = Q$$

The change in specific enthalpy of steam is

$$\Delta h_s = h_{s,2} - h_{s,1}$$

The inlet steam specific enthalpy ($h_{s,1}$) of saturated vapor at 10 bar and 90% quality is

$$h_{s,1}\big|_{16\ bar,\ x=0.9} = h_f + x h_{fg} = 762.6 + 0.9 \times 213.6 = 2574.84\ \text{kJ/kg}$$

The exit steam specific enthalpy at 10 bar, saturated water is

$$h_{s,2}\big|_{10\ bar,\ sat'd\ water} = 762.6\ \text{kJ/kg}$$

Substituting the values of specific enthalpies of steam,

$$\Delta h_s = h_{s,2} - h_{s,1} = 762.6 - 2574.84 = -1812.24\ \text{kJ/kg}$$

The rate of heat transfer from condensed steam to gas stream is

$$Q = m_s \Delta h_s = 100\ \frac{\text{kg}}{\text{h}} \left(-1812.24\ \frac{\text{kJ}}{\text{kg}} \right) \frac{\text{h}}{3600\ \text{s}} = -50.34\ \text{kJ/s}$$

The sign of Q value is negative; that is, heat is transferred from the condensed steam to gas stream.

8.5.1 Energy Balance for Open Systems with Multiple Inputs and Multiple Outputs

The general energy balance for an open system is

$$Q - \dot{W}_s = \Delta \dot{H} + \Delta KE + \Delta PE \qquad (8.25)$$

The change in the rate of enthalpy for multiple streams is

$$\Delta \dot{H} = \sum \dot{H}_{out} - \sum \dot{H}_{in} \qquad (8.26)$$

Setting enthalpy transport rate ($\dot{H}$) in terms of specific enthalpy h,

$$\Delta \dot{H} = \sum \dot{m}_{out} h_{out} - \sum \dot{m}_{in} h_{in} \qquad (8.27)$$

Example 8.19 Enthalpy Change of Mixtures and Phase Change

Problem
Thousand kilomoles per hour of a liquid mixture of 70 mol% acetone and 30 mol% benzene is heated from 10°C to 50°C in a shell-and-tube heat exchanger using steam as the heating medium. The steam enters the heat exchanger in the shell as a saturated vapor at 16 bar of 90% quality, and exits as saturated liquid water at 16 bar. Calculate the mass flow rate of the inlet steam required for this purpose.

Solution

Known quantities: Inlet mixture flow rate and composition, inlet and exit temperature, steam inlet and outlet conditions.

Find: The mass flow rate of inlet steam.

Assumptions: The boiler is adiabatic, no shaft work, no change in kinetic and potential energy, inlet and exit pipe is at the same diameter and level.

Analysis: Use energy balance for an open system around the heat exchanger. Energy balance for an open system is given by

$$\Delta \dot{H} + \Delta KE + \Delta PE = \dot{Q} - \dot{W}_s$$

After including the assumptions, the equation is reduced to

$$\Delta \dot{H} = 0$$

Since the system is of multiple inputs and multiple outputs, the change in enthalpy around the heat exchanger is

$$\Delta \dot{H} = 0 = \sum \dot{H}_{out} - \sum \dot{H}_{in}$$

Setting the enthalpy transport rate ($\dot{H}$) in terms of specific enthalpy h,

$$\Delta\dot{H} = 0 = \sum \dot{m}_{out} h_{out} - \sum \dot{m}_{in} h_{in}$$

In more detail,

$$\Delta\dot{H} = 0 = \left\{ \dot{m}_{s,out} h_{s,out} + \dot{m}_{a,out} h_{a,out} + \dot{m}_{b,out} h_{b,out} \right\}$$
$$- \left\{ \dot{m}_{s,in} h_{s,in} + \dot{m}_{a,in} h_{a,in} + \dot{m}_{b,in} h_{b,in} \right\}$$

where

$m_{s,in}$, $m_{s,out}$ are the inlet and exit mass flow rates of steam which are equal

$\dot{m}_{a,in}$, $\dot{m}_{a,out}$ are the inlet and exit mass flow rates of acetone

$\dot{m}_{b,in}$, $\dot{m}_{b,out}$ are the inlet and exit mass flow rates of benzene

Rearranging the earlier equation,

$$\Delta\dot{H} = 0 = \dot{m}_s \left(h_{s,out} - h_{s,in} \right) + \dot{m}_a \left(h_{a,out} - h_{a,in} \right) + \dot{m}_b \left(h_{b,out} - h_{b,in} \right)$$

where

$\dot{m}_a = \dot{m}_{a,in} = \dot{m}_{a,out}$ is the mass flow rate of acetone

$\dot{m}_b = \dot{m}_{b,in} = \dot{m}_{b,out}$ is the mass flow rate of benzene

Rearranging,

$$\Delta\dot{H} = 0 = \dot{m}_s \Delta h_s + \dot{m}_a \Delta h_a + \dot{m}_b \Delta h_b$$

where

Δh_s is the change in the specific enthalpy of steam

Δh_a is the change in the specific enthalpy of acetone

Δh_b is the change in the specific enthalpy of benzene

Since the mixture contains 70% acetone and 30% benzene, the mixture mass flow rate and change of mixture enthalpy can be written as

$$m_{mix} = 0.7\dot{m}_a + 0.3\dot{m}_b$$

The change in mixture specific enthalpy is given by

$$\Delta h_{mix} = 0.7\Delta h_a + 0.3\Delta h_b$$

The change in the specific enthalpy of steam, Δh_s, is given by

$$\Delta h_s = h_{s,2} - h_{s,1}$$

The inlet steam specific enthalpy ($h_{s,1}$) of saturated vapor at 16 bar and 90% quality is

$$h_{s,1}\big|_{16\,bar,\,x=0.9} = h_f + xh_{fg} = 858.6 + 0.9 \times 1933.2 = 2598.5 \text{ kJ/kg}$$

The exit steam specific enthalpy at 16 bar, saturated water is

$$h_{s,2}\big|_{16\,bar,\,sat'd\,water} = 858.6 \text{ kJ/kg}$$

Substituting the values of the specific enthalpies of steam,

$$\Delta h_s = h_{s,2} - h_{s,1} = 858.6 - 2598.5 = -1740 \text{ kJ/kg}$$

The change in specific enthalpy of acetone and benzene mixture, Δh_{mix}, is given by

$$\Delta h_{mix} = 0.7\Delta h_a + 0.3\Delta h_b = \int_{10°C}^{50°C} C_{P,mix}\, dT$$

The specific heat capacity of the mixture is given by

$$C_{P,mix} = \sum y_i C_{Pi} = 0.7 C_{P,acetone} + 0.3 C_{P,benzene}$$

The heat capacity at constant pressure as a function of temperature:

$$\text{Acetone (liquid): } C_{Pa}\left(\frac{J}{mol\,°C}\right) = 123 + 0.186\,T$$

$$\text{Benzene (liquid): } C_{Pb}\left(\frac{J}{mol\,°C}\right) = 126.5 + 0.234\,T$$

Substitute the heat capacities of acetone and benzene:

$$C_{P,mix} = \{0.7(123) + 0.3(126.5)\} + \{0.7(0.186) + 0.3(0.234)\}T$$

Rearranging,

$$C_{P,mix} = 124 + 0.20\,T$$

Substituting the mixture heat capacity,

$$\Delta h_{mix} = \int_{10°C}^{50°C} C_{P,mix}\, dT = \int_{10°C}^{50°C} (124 + 0.20\,T)\, dT$$

Integrating,

$$\Delta h_{mix} = \int_{10°C}^{50°C} (124 + 0.20T)dT = \left(124T + 0.20T^2\right)\Big|_{10}^{50}$$

The change in enthalpy of the acetone–benzene mixture, Δh_{mix}, is given by

$$\Delta h_{mix} = 124(50 - 10) + \frac{0.20}{2}\left(50^2 - 10^2\right) = 5200 \text{ J/mol}$$

Substituting the values of change in steam enthalpy and mixture enthalpy,

$$0 = \dot{m}_s \Delta h_s + \dot{m}_{mix} \Delta h_{mix}$$

$$= \dot{m}_s\left(-1740 \frac{\text{kJ}}{\text{kg}}\right) + 1000 \frac{\text{kmol}}{\text{h}}\left(\frac{1000 \text{ mol}}{\text{kmol}}\right)\left(5200 \frac{\text{J}}{\text{mol}}\Big|\frac{\text{kJ}}{1000 \text{ J}}\right)$$

Solving for $\dot{m}_s$, $\dot{m}_s\left(1740 \frac{\text{kJ}}{\text{kg}}\right) = \left(5.20 \times 10^6 \text{ kJ/h}\right)$.

The rounded value of the steam mass flow rate is $\dot{m}_s = 2990 \text{ kg/h}$.
The amount of steam required for heating the acetone–benzene mixture is 2990 kg/h.

8.5.2 Enthalpy Change because of Mixing

The thermodynamic property of an ideal mixture is the sum of the contributions from the individual compounds. The following example illustrates the thermodynamic property of an ideal mixing.

Example 8.20 Mixing

Problem
Hundred kilograms per hour of a saturated steam at 1 bar is mixed with superheated steam available at 400°C and 1 bar to produce superheated steam at 300°C and 1 bar. Calculate the amount of superheated steam produced at 300°C, and the required mass flow rate of the 400°C steam.

Solution

Known quantities: Stream 1: mass flow rate, saturated steam, 1 bar. Stream 2: 400°C and 1 atm. Stream 3: superheated steam produced at 300°C, 1 bar.

Find: Volumetric flow rate of stream 2.

Assumptions: No change in kinetic and potential energy, no shaft work.

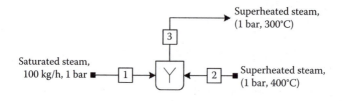

EXAMPLE FIGURE 8.20.1
Mixing of saturated and superheated steam.

Analysis: Use open system energy balance with multiple inputs, single output. The process flow sheet is shown in Example Figure 8.20.1.

The general energy balance for an open system after applying the assumptions is reduced to

$$\Delta H = 0$$

For two inputs, single output,

$$\Delta \dot{H} = \dot{H}_3 - \dot{H}_1 - \dot{H}_2 = 0$$

Putting the equation in terms of mass flow rate and specific enthalpy,

$$\Delta \dot{H} = \dot{m}_3 h_3 - \dot{m}_1 h_1 - \dot{m}_2 h_2 = 0$$

Overall mass balance for the mixing system is

$$\dot{m}_1 + \dot{m}_2 = \dot{m}_3 \Rightarrow 100\,\frac{\text{kg}}{\text{h}} + \dot{m}_2 = \dot{m}_3$$

The specific enthalpy of stream 1 is

$$h_1\big|_{1\,\text{bar, sat'd steam}} = 2675.4 \text{ kJ/kg}$$

The specific enthalpy of stream 2 is

$$h_2\big|_{1\,\text{bar, }400°C} = 3278 \text{ kJ/kg}$$

The specific enthalpy of stream 3 is

$$h_3\big|_{1\,\text{bar, }300°C} = 3074 \text{ kJ/kg}$$

The general energy balance for the mixing process is

$$\dot{m}_1 \hat{H}_1 + \dot{m}_2 \hat{H}_2 = \dot{m}_3 \hat{H}_3$$

Substituting the values,

$$100 \frac{kg}{h}\left(2675.4 \frac{kJ}{kg}\right) + \dot{m}_2\left(3278 \frac{kJ}{kg}\right) = \dot{m}_3\left(3074 \text{ kJ/kg}\right)$$

From the material balance equation,

$$\dot{m}_2 = \dot{m}_3 - 100$$

Substitute the value of $\dot{m}_2$ in the earlier equation:

$$100 \text{ kg/h}\left(2675.4 \text{ kJ/kg}\right) + \left(\dot{m}_3 - 100\right)\left(3278 \text{ kJ/kg}\right) = \dot{m}_3\left(3074 \text{ kJ/kg}\right)$$

Solving for $\dot{m}_3$,

$$100 \text{ kg/h}\left(2675.4 - 3278\right) \text{ kJ/kg} = \dot{m}_3\left(3074 \text{ kJ/kg}\right) - m_3\left(3278 \text{ kJ/kg}\right)$$

Rearranging,

$$\frac{100 \times \left(2675.4 - 3278\right) \dfrac{kJ}{h}}{\left(3074 - 3278\right) \dfrac{kJ}{kg}} = \dot{m}_3$$

The rounded values of the mass flow rates of streams 3 and 2 are

$$\dot{m}_3 = 295 \text{ kg/h and } \dot{m}_2 = 195 \text{ kg/h}$$

8.5.3 Energy Balance for Bioprocesses

Bioprocesses are unlike many chemical processes. Bioprocesses are not particularly energy intensive. Fermentation and enzyme reactors are operated at temperatures and pressures close to ambient conditions, and energy input for downstream processing is minimized to avoid damaging heat-labile products. Nevertheless, energy effects are important because biological catalysts are very sensitive to heat and changes in temperature. In large-scale processes, heat released during biochemical reactions can cause cell death of enzymes, if heat is not properly removed. The law of conservation of energy means that an energy accounting system can be set up to determine the amount of steam or cooling water required to maintain optimum process temperature [4].

Examples 8.21 Cooling of Fatty Acids

Problem
A liquid at the rate of 2000 kg/h fat (20 wt% acid, 80 wt% water) at 90°C is to be cooled to 7°C. Cooling is achieved by heat exchange with

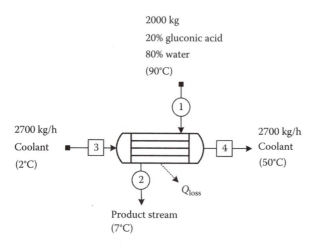

EXAMPLE FIGURE 8.21.1
Shell and tube heat exchanger.

2700 kg/h coolant fluid initially at 2°C. The final temperature of the coolant liquid is 50°C. The fat is flowing on the shell side and coolant liquid is on the tube side. The heat exchanger is not adiabatic, so part of the heat is lost through the exchanger walls and the rest to coolant fluid. What is the rate of heat loss from the acid solution to the surroundings? Assume the heat capacity of acid is 1.463 (kJ/kg°C). The process flow sheet is shown in Example Figure 8.21.1.

Solution

Known quantities: Inlet and exit conditions of cooling water, inlet and exit temperature liquid stream.

Assumptions: No change in kinetic and potential energy, no shaft work.

Find: Heat loss through surroundings.

Analysis: Use the first law for an open system:

$$Q - W_s = \Delta \dot{H} + \Delta KE + \Delta PE$$

The simplified equation is

$$Q - 0 = \Delta \dot{H} + 0 + 0$$

For multiple input and multiple outputs, this equation is written as follows:

$$Q = \Delta \dot{H} = \sum_{\text{out}} \dot{m}_i h_i - \sum_{\text{in}} \dot{m}_i h_i$$

Let the subscript "c" be for coolant, "a" be for acid, and "w" be for water associated with the fat:

$$Q_{loss} = \left\{ m_c h_{c,4} + m_a h_{a,2} + m_w h_{w,2} \right\}_{out} - \left\{ m_c h_{c,3} + m_a h_{a,1} + m_w h_{w,1} \right\}_{in}$$

The mass flow rates of coolant, acid, and water are constant. Rearranging,

$$Q_{loss} = m_c \left(h_{c,4} - h_{c,3} \right) + m_a \left(h_{a,2} - h_{a,1} \right) + m_w \left(h_{w,2} - h_{w,1} \right)$$

The enthalpy of water is found from the steam table as saturated liquid water. The specific enthalpy of coolant medium at 2°C is 8.124 kJ/kg and at 50°C is 209.5 kJ/kg. Use the saturated steam table (Table A.4) to find the specific enthalpy of the water associated with the fat:

$$h_{w,1}\big|_{@\,90°C} = 376.8 \ \frac{kJ}{kg}, \quad h_{w,1}\big|_{@\,7°C} = 29.3 \ \frac{kJ}{kg}$$

Substituting the values of mass flow rates and specific enthalpies of coolant, and water associated with steam and acid,

$$Q = 2700 \ \frac{kg}{h} \times \frac{h}{3600 \ s} \left(209.5 \ \frac{kJ}{kg} - 8.124 \ \frac{kJ}{kg} \right) + \left(2000 \ \frac{kg}{h} \ \frac{h}{3600 \ s} \right)$$

$$\times \left(29.3 - 376.8 \ \frac{kJ}{kg} \right) + \left(400 \ \frac{kg}{h} \ \frac{h}{3600 \ s} \right) \left(1.463 \ \frac{kJ}{kg°C} \right) (7°C - 90°C)$$

Solving for the heat loss Q,

$$Q = 151 \ kJ/s - 193 \ kJ/s - 13.49 \ kJ/s = -55.49 \ kJ/s$$

The sign for the heat transfer through the exchanger's wall is negative; that is, heat is lost from the fatty acid to the surroundings.

8.6 Psychrometric Chart

The psychrometric chart (Figure 8.3) displays the relationship between dry-bulb, wet-bulb, and dew point temperatures and specific and relative humidity. Given any two properties, the others can be calculated. To use the chart, take the point of intersection of the lines of any two known factors (interpolate if necessary), and, from that intersection point, follow the lines

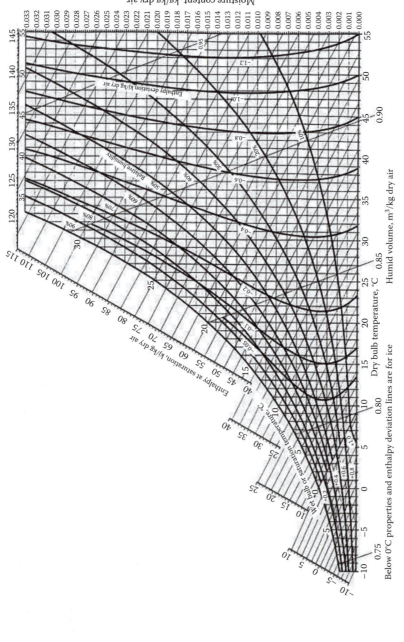

FIGURE 8.3
Psychrometric chart.

of the unknown factors to their numbered scales to obtain the corresponding values. The thermo-physical properties found on most psychrometric charts are as follows.

Dry-Bulb Temperature
Dry-bulb temperature is the temperature of an air sample, as determined by an ordinary thermometer, the thermometer bulb being dry. It is typically the abscissa or horizontal axis of the graph. The SI unit for temperature is Celsius, the other unit is Fahrenheit.

Wet-Bulb Temperature
Wet-bulb temperature is the temperature of an air sample after it has passed through a constant-pressure, ideal, adiabatic saturation process, that is, after the air has passed over a large surface of liquid water in an insulated channel. In practice, this is the reading of a thermometer whose sensing bulb is covered with a wet sock evaporating into a rapid stream of the air sample. The wet-bulb temperature is the same as the dry-bulb temperature when the air sample is saturated with water.

Dew Point Temperature
Dew point temperature is that temperature at which a moist air sample at the same pressure would reach water vapor saturation. At this saturation point, water vapor would begin to condense into liquid water fog.

Relative Humidity
Relative humidity is the ratio of the mole fraction of water vapor to the mole fraction of saturated moist air at the same temperature and pressure. Relative humidity is dimensionless, and is usually expressed as a percentage.

Humidity Ratio
Humidity ratio, also known as moisture content, mixing ratio, or specific humidity, is the proportion of mass of water vapor per unit mass of dry air at the given conditions. For a given dry-bulb temperature, there will be a particular humidity ratio for which the air sample is at 100% relative humidity. Humidity ratio is dimensionless, but is sometimes expressed as grams of water per kilogram of dry air.

Specific Enthalpy
Specific enthalpy, also called heat content per unit mass, is the sum of the internal (heat) energy of the moist air in question, including the heat of the air and water vapor within. In the approximation of ideal gases, lines of constant enthalpy are parallel to lines of constant wet-bulb temperature.

Specific Volume
Specific volume, also called inverse density, is the volume per unit mass of the air sample. The SI unit is cubic meters per kilogram of air; the other unit is cubic feet per pound of dry air.

Example 8.22 Psychrometric Chart

Problem
Humid air at 28°C has a dew point of 8°C. Using the psychrometric chart provided, determine the following: relative humidity, absolute humidity, wet-bulb temperature, dry-bulb temperature, humid volume, specific enthalpy, and mass of air that contains 2 kg of water, and volume occupied by air that contains 2 kg of water.

Solution

Known quantities: Humid air at 28°C has a dew point of 8°C. Air contains 2 kg of water, and volume occupied by air that contains 2 kg of water.

Find: Relative humidity, absolute humidity, wet-bulb temperature, dry-bulb temperature, humid volume, specific enthalpy.

Analysis: Use psychrometric chart. Humid air at 28°C has a dew point of 8°C. Using the psychrometric chart provided, determine the following:

1. Relative humidity = 30% (Example Figure 8.22a.1).
2. Absolute humidity = 0.007 kg water/kg dry air (Example Figure 8.22b.1).
3. Wet-bulb temperature = 16.5°C (Example Figure 8.22c.1). Follow the constant enthalpy line from the intersection of the dry-bulb and dew point temperatures.
4. Dry-bulb temperature = 28°C.
5. Humid volume = 0.86 m³/kg (Example Figure 8.22d.1).
6. Specific enthalpy = 46 kJ/kg − 0.3 kJ/kg = 45.7 kJ/kg (Example Figure 8.22e.1).
7. Mass of air that contains 2 kg of water:

$$2\text{ kg H}_2\text{O}\frac{\text{kg dry air}}{0.007\text{ kg H}_2\text{O}} = 285.7\text{ kg dry air}$$

8. Volume occupied by air that contains 2 kg of water:

$$\frac{0.86\text{ m}^3}{\text{kg dry air}}\frac{\text{kg dry air}}{0.007\text{ kg H}_2\text{O}}2\text{ kg H}_2\text{O} = 245.7\text{ m}^3$$

Example 8.23 Relative Humidity

Problem
Air flowing at a volumetric flow rate of 0.80 m³/h under a temperature of 25°C and a pressure of 3 bar is fed to a process. The air has a relative humidity of 80%. The molar composition of dry air is 21% oxygen and 79% nitrogen. Determine the molar flow rates of water, dry air, and oxygen entering the process.

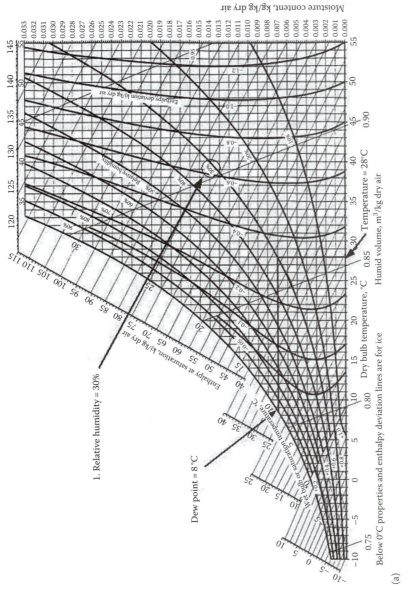

EXAMPLE FIGURE 8.22.1

(a) Estimation of relative humidity.

(Continued)

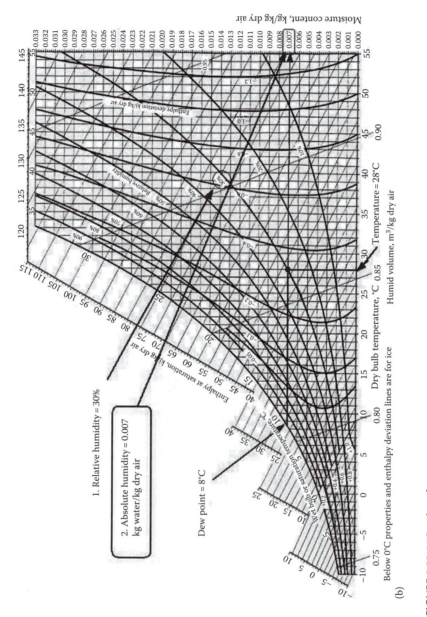

EXAMPLE FIGURE 8.22.1 (Continued)
(b) Absolute humidity of air.

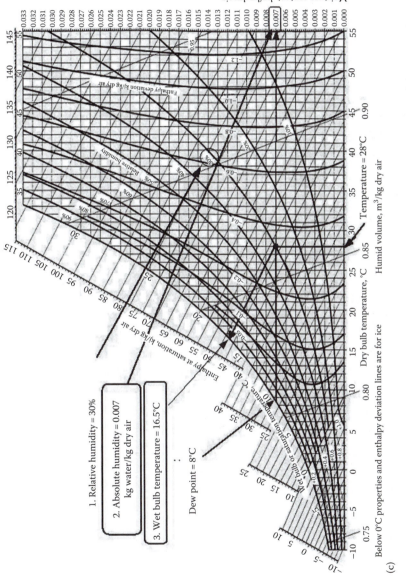

EXAMPLE FIGURE 8.22.1 (*Continued*)
(c) Wet-bulb temperature estimation

(*Continued*)

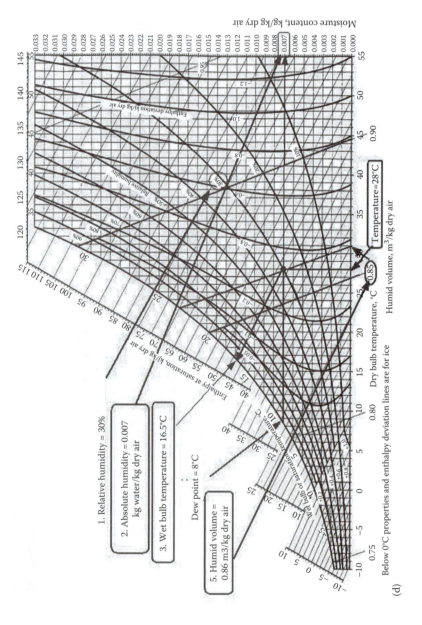

EXAMPLE FIGURE 8.22.1 (*Continued*)
(d) Determination of humid volume.

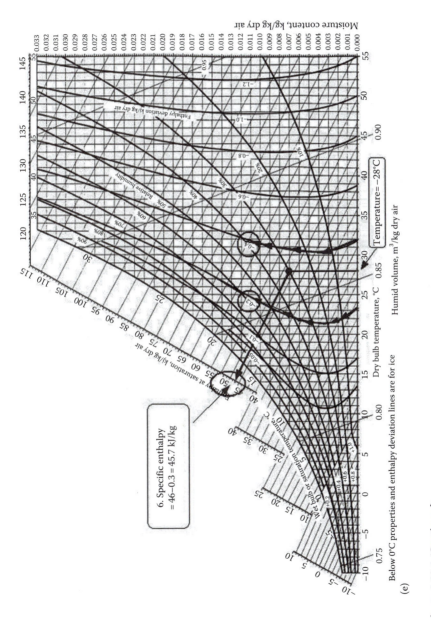

Example Figure 8.22.1 (*Continued*)
(e) Specific enthalpy calculation.

Solution

Known quantities: Inlet air flow rate temperature and pressure.

Find: The molar flow rates of water, dry air, and oxygen entering the process.

Analysis: First, we calculate the partial pressure of water by using the definition of relative humidity:

$$0.8 = \frac{p_{H_2O}}{p_{H_2O}^*}$$

The vapor pressure at 25°C is 0.03 bar:

$$0.8 = \frac{p_{H_2O}}{0.03}$$

The partial pressure of water (p_{H_2O}) is 0.025 bar:

$$p_{H_2O} = y_{H_2O}P$$

Substitute values of partial pressure and total pressure:

$$0.025 = y_{H_2O}(3.0)$$

The mole fraction of water is 0.008 mol water/mole humid air, and the mole fraction of dry air is 0.992. The molar flow rate of air can be obtained using ideal gas equation of state using the volumetric flow rate given in the problem statement. After substituting the known information into this equation, we have

$$n = \frac{PV}{RT} = \frac{(3\text{ bar})(0.8\text{ m}^3/\text{h})}{\left(8.314 \times 10^{-5}\dfrac{\text{m}^3 \cdot \text{bar}}{\text{mol K}}\right)(298\text{ K})} = 96.87\text{ mol/h}$$

The number of moles of dry air is

$$n_{\text{dry-air}} = y_{\text{dry-air}} \times \dot{n} = 0.992 \times 96.87\ \frac{\text{mol}}{\text{h}} = 96.1\text{ mol/h}$$

Example 8.25 Relative and Absolute Humidity

Problem

The dry-bulb temperature is measured as 20°C and the wet-bulb temperature as 15°C. If the total pressure is 1 atm, what are the relative humidity and the absolute humidity? Use the humidity chart.

Solution

Known quantities: Dry-bulb and wet-bulb temperatures of air.

Find: Relative and absolute humidity.

Analysis: Use the psychrometric chart.
Relative humidity = 59%
Absolute humidity = 0.0087 kg water/kg dry air

8.7 Summary

The first law of thermodynamics for a closed system is

$$\Delta U + \Delta E_k + \Delta E_p = Q - W$$

The first law of thermodynamics for an open system at steady state (i.e., continuous) is

$$\Delta \dot{H} + \Delta \dot{E}_k + \Delta \dot{E}_p = \dot{Q} - \dot{W}_s$$

Procedure for energy balance calculations:

1. Draw and completely label a process flowchart.
2. Perform all material balance calculations.
3. Write the appropriate form of the energy balance equation and remove any negligible terms.
4. Choose a reference state (phase-gas/liquid, T, P) for each species involved. If using enthalpy tables, use reference state to generate table. If no tables are available, choose one inlet or outlet condition as the reference state for the species.
5. Construct an inlet–outlet enthalpy table.
6. Calculate all required values of u_i or h_i and insert the values into the table.
7. Calculate ΔU or ΔH (e.g., $\Delta H = \Sigma m_i h_i - \Sigma m_i h_i$).
8. Calculate any other terms in the energy balance equation (i.e., W, ΔE_k, ΔE_p).
9. Solve for the unknown quantity in the energy balance equation.

Homework Problems

8.1 Liquid methanol at 25°C is heated and vaporized for use in a chemical reaction. How much heat is required to heat and vaporize 10 mol/s of methanol to 600°C. (744 kJ/s)

8.2   Propane gas at 40°C and 250 kPa enters an adiabatic heat exchanger and exits the heat exchanger at 240°C. The flow rate of propane is 100 mol/min. The saturated steam with a flow rate of 6 kg/min at 5 bar (absolute) enters the heat exchanger. The discharge wet steam (contains vapor and liquid) is at 3 bar. Calculate the exit steam quality and temperature. (0.86, 133.5°C)

8.3   Wet steam at 20 bar with 97% quality is passed through a throttling valve and expands to 1 bar pressure. Calculate the temperature of the steam that departs from the adiabatic expansion valve. (132.25°C)

8.4   Hundred kilograms per hour of wet steam at 20 bar with 97% quality is passed through a throttling valve and expands to 10 bar pressure. Calculate the temperature of the steam that departs from the adiabatic expansion valve and the quality of steam. (179.9°C, 0.982)

8.5   Methanol is heated by condensing steam in a concentric double pipe heat exchanger as depicted in Problem Figure 8.5.1. Methanol flowing through the inner pipe at 64.0 kg/s enters at 20°C and exits at 60°C. Steam enters the outer pipe at 350°C and 80 bar (absolute) and leaves the heat exchanger as saturated water at 1.0 bar. Assume that the outer pipe is well insulated and no heat is lost to the surroundings. Determine the mass flow rate of the steam. (2.57 kg/s)

8.6   Calculate the heat rate required to heat 32.04 kg/s of liquid methyl alcohol (CH_3OH) at 5°C and 1 atm to vapor at 500°C and 1 atm. The heat of vaporization of methanol at 64.7°C and 1 atm is 35.27 kJ/mol. (69.0 MW)

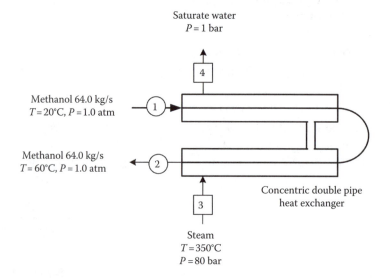

PROBLEM FIGURE 8.5.1
Schematic of double pipe heat exchanger.

8.7 Determine the total amount of heat required to convert 2.00 mol of liquid *n*-hexane (C_6H_{14}) at 10°C to vapor at 55°C in a closed container. Assume that hexane vapor behaves as an ideal gas at the system pressure. Neglect any effect of a change in pressure on the liquid enthalpy. The heat of vaporization of hexane at 68.74°C and 1 atm is 28.85 kJ/mol. (63.26 kJ)

8.8 A volume of 734 cm³ of liquid acetone is contained in a closed cylinder fitted with a movable frictionless piston at 10°C. The acetone is heated via heating coil inserted inside the cylinder to vapor at 500°C. The piston area is 50.0 cm², and the piston weighs 200 kg. The heat of vaporization of acetone at its normal boiling point (56.0°C) is 30.2 kJ/mol. Assume the cylinder is perfectly insulated and no heat is lost to the surroundings. Calculate the heat transferred from the heating process. (929 kJ)

8.9 A volume of 734 cm³ of liquid acetone is contained in a closed (vertical) cylinder fitted with a movable frictionless piston at 10°C. The acetone is heated via heating coil inserted inside the cylinder to vapor at 500°C. The piston area is 50.0 cm², and the piston weighs 200 kg. The heat of vaporization of acetone at its normal boiling point (56.0°C) is 30.2 kJ/mol. Assume the cylinder is perfectly insulated and no heat is lost to the surroundings. If the heat is provided by superheated steam at 550°C and 1.0 bar, the final condition of the steam is saturated at 100°C. How much steam is needed? (1.0 kg)

8.10 A hydroelectric project has a volumetric flow rate of 1.2 m³/s. The water flowing in the river at atmospheric pressure and 20.4°C falls vertically for 300 m and then passes through a turbine. The water exits the turbine at atmospheric pressure and 19°C. What is the power output of the turbine? (10.1 kW)

8.11 Air at 100 kPa and 10°C enters a compressor and is brought to 1000 kPa and 50°C. The constant pressure heat capacity of air is 1.01 kJ/kg K. If 15 kg/min of air are to be compressed, determine the power requirement of the compressor. State your assumptions. (12.625 kW)

8.12 A gasoline engine has an efficiency of 25%. If the engine consumes 0.75 L/h of gasoline with a heating value of 3.0×10^4 kJ/L, how much power does it provide? Express the answer in kilowatts. (1.56 kW)

8.13 A liquid stream (10 kg/min) flows through a heat exchanger in which it is heated from 25°C to 80°C. The liquid specific heat is 4.18 kJ/kg K. The inlet and outlet pipes have the same diameter, and there is no change in elevation between these points. Calculate the heat required. (38.3 kW)

8.14 Water (100 kg/s) passes through the gate of a dam and falls on a turbine 10 m below, which turns a shaft connected to a generator. The fluid velocity on both sides of the dam is negligible, and the water undergoes insignificant pressure and temperature changes between the inlet and outlet. Calculate the work generated by the turbine. (9.81 kW)

8.15 Crude oil is to be pumped at 1000 kg/min through a pipeline 2 km in length. The pipe inlet is 200 m below the outlet, the pipe diameter is constant. Neglect the pipe frictional losses. Calculate the work required by the pump. (32.7 kW)

8.16 A cylinder is fitted with a frictionless floating piston and contains 24.8 L of air at 25°C and 1 bar. The system is then heated to 250°C. ($C_v = 5R/2$). How much work was done by the system on the surroundings? (1.871 kJ)

8.17 A power plant (800 MW) burns natural gas to boil water producing saturated steam (100% quality) at 70 bar. This steam is expanded in a turbine to steam at 100°C and 1 bar. The steam enters the turbine at 10 m/s and exits 5 m below the entrance point level at 100 m/s. The turbine is connected to an electrical generator by a shaft. The efficiency of the turbine is 60%. What is the mass flow rate of steam to the turbine (kg/h)? (5.05×10^7 kg/h)

8.18 A large tank that is filled with water is open to the atmosphere. Water is taken from the tank, passed through a pump, and then delivered to the hoses of a firefighting extinguisher in a chemical factory. It is desired to deliver 69.5 lbm of water per second at a pressure of 200 psi (gauge). If there is a negligible elevation change between the water level in the tank and the discharge of the pump, no changes in the diameter of the pipes and hoses, and if the pump has an efficiency of 65.0%, how much work must be supplied to the pump in order to meet the pressure and discharge rate specifications? (89.7 hp)

8.19 Suppose you are operating a steam turbine where the steam leaving the turbine is at 5 bar (absolute). This steam contains 95 wt% vapors. If the shaft work produced by the turbine is 1100 kJ/kg and the high pressure, high temperature steam enters the turbine at 100 bar (abs), what is the temperature of the steam entering the turbine? If water at 10°C is supplied to the steam boiler to generate steam, how much heat is required per kilogram of steam produced? You may assume that heat losses from the turbine are negligible. (3700 kJ/kg)

8.20 Steam at 60 bar and 500°C enters an adiabatic turbine at a steady flow rate of 1 kg/s; the turbine outlet stream is at 1 bar and 400°C. The inlet and exit streams of the turbine are at the same height and the pipes have the same diameter of 0.15 m. How much work can be obtained from an adiabatic, continuous-flow turbine? (−144 kJ/s)

8.21 Consider taking 1000.0 kg of outside air, which is then heated to make your apartment comfortable in the winter. The outside air has a dry-bulb temperature of 10.0°C and a wet-bulb temperature of 5°C. You want the air in your apartment to be at a dry-bulb temperature of 25°C and a relative humidity of 60.0%.

   (a) How much water must be added to the 1000.0 kg of outside air to reach the desired humidity level? (8.57 kg)

(b) If the water you are using to alter the humidity of the air is coming from a tap (assume its temperature is 10°C), how much heat must be added to just the water to attain the desired temperature and humidity level in your apartment? (2.15×10^4kJ)

8.22 For healthy air quality, it is recommended that a 200 m^2 house have an air exchange rate of 60.0 ft^3 bone-dry air/min with the outside. It is also suggested that the relative humidity in the house be 65%. Suppose the outside air is at 30°F and a relative humidity of 40%. If this air is brought into the house and heated to 75°F without addition of water, what is the relative humidity? If you desire to maintain a relative humidity of 65%, how much water must be added to the air inside the house (g/min)? (24.3 g/min)

References

1. Reklaitis, G.V. (1983) *Introduction to Material and Energy Balances*, John Wiley & Sons, New York.
2. Felder, R.M. and R.W. Rousseau (1999) *Elementary Principles of Chemical Processes*, 3rd edn., John Wiley, New York.
3. Himmelblau, D.M. (1974) *Basic Principles and Calculations in Chemical Engineering*, 3rd edn., Prentice-Hall, Englewood Cliffs, NJ.
4. Whirwell, J.C. and R.K. Toner (1969) *Conservation of Mass and Energy*, Blaisdell, Waltham, MA.

9

Energy Balance with Reaction

The large changes in enthalpy and internal energy throughout a chemical reaction require significant heat transfer (heating or cooling) from the reactor in order to eventually maintain the reactor under optimum operating conditions. This chapter demonstrates how to calculate the heat of reaction at a specific temperature and illustrates how to estimate the heat of reaction from the heat of formation. The energy balances for a reacting system using two methods are defined. Problems that involve the application of combined material and energy balances are addressed. Finally, a few applications, specifically, combustion, bioprocesses, and membrane reactors, are discussed at length to practice the concepts further. The following items outline the principal learning objectives of this chapter.

Learning Objectives

1. Calculate the heat of reaction (Section 9.1).
2. Estimate the heat of reaction from heats of formation (Section 9.2).
3. Establish the energy balance for a reacting system (Section 9.3).
4. Write simultaneous material and energy balances for a reacting system (Section 9.4).
5. Write the appropriate balances for combustion processes (Section 9.5).
6. Apply the energy balance to bioprocesses (Section 9.6).
7. Perform material and energy balances for a membrane reactor system (Section 9.7).

9.1 Heat of Reaction

The heat of reaction, $\Delta H_{Rx}(T,P)$, is the enthalpy change for a process in which stoichiometric quantities of reactants at temperature T and pressure P react completely to form products at the same temperature and pressure. Consider the following reaction:

$$aA + bB \rightarrow cC + dD$$

The standard heat of reaction (ΔH_{Rx}^o) is calculated as the difference between the product and reactant enthalpies when both reactants and products are at standard conditions, that is, at 25°C and 1 atm [1]. The symbol "o" denotes standard conditions. Therefore,

$$\Delta H_{Rx}^o [kJ/mol] = H_{products} - H_{reactants}$$

$$= c\Delta H_{f,C}^o + d\Delta H_{f,D}^o - a\Delta H_{f,A}^o - b\Delta H_{f,B}^o$$

$$= \sum v_i \Delta H_i^o$$

where ΔH_f^o is the standard heat of formation. The reported ΔH_{Rx} applies to stoichiometric quantities of each species. Consider the following example:

$$A + 2B \rightarrow 3C, \quad \Delta H_{Rx}(100°C, 1\ atm) = -150\ kJ/mol$$

The enthalpy change for the given reaction is

$$\frac{-150\ kJ}{1\ mol\ A\ consumed} = \frac{-150\ kJ}{2\ mol\ B\ consumed} = \frac{-150\ kJ}{3\ mol\ C\ generated}$$

If 150 mol/s of C was generated at 100°C and 1 atm, then

$$\Delta \dot{H} = \left(\frac{-150\ kJ}{3\ mol\ C\ generated} \right) \left(\frac{150\ mol\ C\ generated}{s} \right) = -7500\ kJ/s$$

If $\Delta H_{Rx}(T)$ is negative, the reaction is exothermic; that is, energy must be removed from the reactor to prevent the temperature from increasing. If $\Delta H_{Rx}(T)$ is positive, the reaction is endothermic; that is, energy must be added to the reactor to prevent the temperature from decreasing. The heat of reaction ($\Delta H_{Rx}(T,P)$) is nearly independent of pressure. The value of the heat of reaction depends on how the stoichiometric equation is written and on the phase of the reactants and products.

9.2 Heats of Formation and Heat of Combustion

The standard heat of reaction (ΔH_{Rx}^o) can be calculated from the standard heat of formation (ΔH_f^o). The standard heat of formation is the enthalpy change associated with the formation of 1 mol of the compound at 25°C and 1 atm. The values of ΔH_f^o for many compounds can be obtained from tabulated data (Table A.2). The standard heat of reaction (ΔH_{Rx}^o) from the heat of formation ($\Delta H_{f,i}^o$) of any reaction can be calculated as

$$\Delta H_{Rx}^o = \sum_i v_i \Delta H_{f,i}^o$$

where
v_i is the stoichiometric coefficient of reactant or product species i
$\Delta H_{f,i}^o$ is the standard heat of formation of species i

The standard heats of formation of all elemental species are zero (H_2, O_2, N_2). The standard heat of reaction (ΔH_{Rx}^o) of any reaction involving only oxygen and a combustible species can be calculated as

$$\Delta H_{Rx}^o = -\sum_i v_i (\Delta H_c^o)_i$$

This is the reverse of determining the heat of reaction from heats of formation, where v_i is the stoichiometric coefficient of reactant or product species i. $(\Delta H_c^o)_i$ is the standard heat of combustion of species i. If any reactants or products are combustion products (i.e., CO_2, H_2O, SO_2), their heats of combustion are equal to zero. For many substances, it is much easier to measure the standard heat of combustion (ΔH_c^o) than measuring the standard heat of formation (ΔH_f^o). Consider the formation of pentane:

$$5C\,(s) + 6H_2\,(g) \rightarrow C_5H_{12}\,(l) \quad \Delta H_{Rx}^o = ?$$

Carbon, hydrogen, and pentane can all be burned, and their standard heats of combustion can be determined experimentally. Therefore,

$$\Delta H_{Rx}^o = 5\Delta H_{c,C(s)}^o + 6\Delta H_{c,H2\,(g)}^o - \Delta H_{c,C_5H_{12}\,(l)}^o$$

The standard enthalpy of combustion is the enthalpy change when 1 mol of a reactant completely burns in excess oxygen under standard thermodynamic conditions. The standard heat of combustion of a species i, $\Delta H_{c,i}^o$, is the enthalpy change associated with the complete combustion of 1 mol of species i with oxygen at 25°C and 1 atm such that all the carbon forms CO_2 (g), all the hydrogen forms H_2O (l), all the sulfur forms SO_2 (g), and all the nitrogen forms NO_2 (g). The same value of standard heat of reaction can be used to measure the standard heat of formation of pentane:

$$\Delta H_{Rx}^o = \Delta H_{f,C_5H_{12}\,(l)}^o - 5\Delta H_{f,C(s)}^o - 6\Delta H_{f,H2\,(g)}^o$$

Since carbon and hydrogen are atoms, the magnitude of their standard heats of formation is zero. Accordingly,

$$\Delta H^\circ_{Rx} = \Delta H^\circ_{f, C_5H_{12} (l)} - 0 - 0$$

Example 9.1 Heat of Reaction from Heats of Formation

Problem

Consider the combustion of liquid ethanol as shown in the following reaction scheme:

$$C_2H_5OH \; (l) + 3O_2 \; (g) \rightarrow 2CO_2 \; (g) + 3H_2O \; (l)$$

Use heat of formation and heat of combustion to determine the standard heat of reaction.

Solution

Known quantities: Reaction stoichiometry.

Find: Standard heat of reaction.

Analysis: Values for standard heat of combustion and standard heat of formation are available in the appendix.

$$\Delta H^\circ_{Rx} = 3\Delta H^\circ_{f, H_2O (l)} + 2\Delta H^\circ_{f, CO_2} - 0 - \Delta H^\circ_{f, C_2H_5OH (l)}$$

Substitute the values of the standard heat of formation:

$$\Delta H^\circ_{Rx} (kJ/mol) = 3(-285.84) + 2(-393.51) - 0 - (-277.63) = -1366.9 \; kJ/mol$$

The standard heat of reaction is calculated from the standard heat of combustion as

$$\Delta H^\circ_{Rx} = \Delta H^\circ_{c, C_2H_5OH (l)} + 3\Delta H^\circ_{c, O_2} - 3\Delta H^\circ_{c, H_2O (l)} - 2\Delta H^\circ_{c, CO_2 (g)}$$

Substitute the values of the standard heat of combustion, knowing that the magnitudes of the standard heat of combustion of oxygen, water, and carbon dioxide are zero:

$$\Delta H^\circ_{Rx} (kJ/mol) = -1366.91 + 0 - 0 - 0 = -1366.9 \; kJ/mol$$

Results reveal that both values of standard heat of reactions are identical.

Example 9.2 Standard Rate of Change in Enthalpy

Problem

If 240 mol/s of CO_2 is produced in the following reaction where reactants and products are all at 25°C, and no CO_2 is present in the feed stream,

$$2C_4H_{10(g)} + 13O_{2(g)} \rightarrow 8CO_{2(g)} + 10H_2O_{(l)}$$

what is the standard rate of change in enthalpy?

Solution

Known quantities: Molar flow rate of effluent CO_2 and reaction temperature.

Find: Standard rate of change in enthalpy.

Analysis: The standard heat of reaction from the heat of formation is as follows:

$$\Delta H_{Rx}^{\circ}(25°C, 1\,atm) = \sum v_i \Delta H_{f,i}^{\circ}$$

The heat of reaction is given by

$$\Delta H_{Rx}^{\circ}(25°C, 1\,atm) = 10\Delta H_{f,H_2O\,(l)}^{\circ} + 8\Delta H_{f,CO_2}^{\circ} - 2\Delta H_{f,C_4H_{10}}^{\circ} - 13\Delta H_{f,O_2}^{\circ}$$

$$\Delta H_{Rx}^{\circ}(25°C, 1\,atm) = 10(-285.84) + 8(-393.5) - 2(-124.7) - 13(0)$$

$$\Delta H_{Rx}^{\circ}(25°C, 1\,atm) = -5757\ kJ/mol$$

The extent of reaction is calculated using the mole balance of CO_2; note that no carbon dioxide is present in the feed stream before the reaction takes place:

$$n_{CO_2} = 0 + 8\xi = 240 = 0 + 8\xi$$

$$\xi = \frac{240}{8} = 30\ mol/s$$

The change in enthalpy transfer rate is

$$\Delta \dot{H} = \xi \times \Delta H_{Rx}^{\circ} = 30 \times -5757\ kJ/mol = -1.727 \times 10^5\ kJ/s$$

Example 9.3 Butane Combustion

Problem
The reaction stoichiometry and standard heat of the reaction on n-butane vapor is shown here:

$$C_4H_{10}(g) + \frac{13}{2}O_2\,(g) \rightarrow 4CO_2\,(g) + 5H_2O\,(l),\ \Delta H_{Rx}^{\circ} = -2900\ kJ/mol$$

Assume that 40 mol/s of CO_2 is produced in this reaction and the reactants and products are all at 25°C. The fresh feed to the reactor contains 20 mol/s of CO_2. Calculate the rate of change in enthalpy $\Delta \dot{H}$ (kJ/s).

Solution

Known quantities: Inlet and exit CO_2 molar flow rate and reaction temperature.

Find: The rate of change in enthalpy.

Analysis: Since inlet and exit streams' temperature is equal, the change in the sensible heat is irrelevant. Accordingly, the rate of change in enthalpy is only due to heat of reaction and is calculated as $\Delta \dot{H} = \dot{\xi}\Delta H^{\circ}_{Rx}(T, P)$ at 25°C, $\Delta \dot{H} = \dot{\xi}\Delta H^{\circ}_{Rx}$, where ΔH°_{Rx} is the standard heat of reaction.

The extent of reaction is calculated using the mole balance of CO_2:

$$\dot{n}_{CO_2} = \dot{n}_{CO_2,\,feed} + 4\dot{\xi}$$

Substitute the values of inlet and exit molar flow rates of carbon dioxide:

$$40 = 20 + 4\dot{\xi}$$

The extent of reaction is $\dot{\xi} = \dfrac{40 - 20}{4} = \dfrac{20}{4} = 5\ mol/s$

The standard rate of change in enthalpy, $\Delta \dot{H}$, is

$$\Delta \dot{H} = \dot{\xi}\Delta H^{\circ}_{Rx} = \left(5\ \frac{mol}{s}\right)\left(\frac{-2900\ kJ}{mol}\right) = -1.45 \times 10^4\ kJ/s$$

9.2.1 Extent of Reaction

If n_{Ar} is the moles of A generated or consumed by a reaction at a temperature T and pressure P, and v_A is the stoichiometric coefficient of the reactant or product, the associated enthalpy change is

$$\Delta \dot{H} = \dot{\xi}\Delta H_{Rx}(T, P)$$

The extent of reaction, ξ, is a measure of how far a reaction has proceeded:

$$(\dot{n}_i)_{out} = (\dot{n}_i)_{in} + v_i\dot{\xi}$$

Rearranging, the extent of reaction is expressed as

$$\dot{\xi} = \frac{(\dot{n}_i)_{out} - (\dot{n}_i)_{in}}{v_i}$$

9.2.2 Reactions in Closed Processes

If the reaction is taking place in a closed system of constant volume, the change in the internal energy of reaction, ΔU_{Rx}, is given as

$$\Delta U_{Rx} = \Delta H_{Rx} - (\Delta n)RT$$

The change in the number of moles is equivalent to the change in the number of stoichiometric coefficients between the product and the reactant. Substituting $\Delta n = \sum v_i$,

$$\Delta U_{Rx} = \Delta H_{Rx} - RT \sum v_i$$

where v_i is the stoichiometric coefficient of the gaseous reactant or product component (+for product, −for reactant).

Example 9.4 Heat of Reaction from Internal Energy

Problem
Calculate the standard heat of the reaction of the following reaction:

$$C_2H_4 \, (g) + 2Cl_2 \, (g) \rightarrow C_2HCl_3 \, (l) + H_2 \, (g) + HCl \, (g)$$

The internal energy of reaction at standard conditions (25°C, 1 atm) is $\Delta U_{Rx} = -418$ kJ/mol.

Solution

Known quantities: Internal energy of reaction.

Find: The standard heat of reaction.

Analysis: Use the following equation for closed system:

$$\Delta U_{Rx}(T) = \Delta H_{Rx}(T) - RT \sum v_i$$

Substitute known quantities:

$$-418 \text{ kJ/mol} = \Delta H_{Rx}^o - \frac{8.314 \text{ J}}{\text{mol K}} \frac{1 \text{ kJ}}{1000 \text{ J}} \times 298.15 \text{ K} \times (1+1+0-2-1)$$

$$\Delta H_{Rx}^o = -420.5 \text{ kJ/mol}$$

Note that v_i is the stoichiometric coefficient (+for product, −for reactant) of the gaseous reactant or product component only (not liquids or solids components). If a set of reactions can be manipulated through a series of algebraic operations to yield the desired reaction, then the desired heat of reaction can be obtained by performing the same algebraic operations on the heats of reaction of the manipulated set of reactions (Hess' law).

Example 9.5 Standard Heat of Reaction

Problem

Calculate the heat of combustion for C_2H_6 from the following reactions:

$$C_2H_4 + 3O_2 \rightarrow 2CO_2 + 2H_2O \quad \Delta H^\circ_{Rx1} = -1409.5 \text{ kJ/mol}$$

$$C_2H_4 + H_2 \rightarrow C_2H_6, \quad \Delta H^\circ_{Rx2} = -136.7 \text{ kJ/mol}$$

$$H_2 + \frac{1}{2}O_2 \rightarrow H_2O \quad \Delta H^\circ_{Rx3} = -285.5 \text{ kJ/mol}$$

Solution

Known quantities: Standard heat of reaction.

Find: Standard heat of reaction for the combustion of ethane.

Analysis: Use Hess' law.

The reaction for combustion of ethane is as follows:

$$C_2H_6 + 3.5O_2 \rightarrow 2CO_2 + 3H_2O \quad \Delta H^\circ_{Rx1} = ?$$

The first reaction minus the second reaction based on one mole reacted

$$C_2H_4 + 3O_2 \rightarrow 2CO_2 + 2H_2O, \qquad -1409.5 \text{ kJ}$$
$$\underline{C_2H_6 \rightarrow C_2H_4 + H_2, \qquad\qquad +136.7 \text{ kJ}}$$

$$C_2H_6 + 3O_2 \rightarrow 2CO_2 + 2H_2O + H_2 \quad -1272.8 \text{ kJ}$$

The resulting reaction is added with the third one:

$$C_2H_6 + 3O_2 \rightarrow 2CO_2 + 2H_2O + H_2 \quad -1272.8 \text{ kJ}$$
$$\underline{H_2 + \frac{1}{2}O_2 \rightarrow H_2O \qquad\qquad\qquad -285.5 \text{ kJ}}$$

$$C_2H_6 + 3.5O_2 \rightarrow 2CO_2 + 3H_2O \qquad -1558.3 \text{ kJ}$$

Accordingly, the heat of combustion of C_2H_6 is −1558.3 kJ

Example 9.6 Dehydrogenation of Ethane

Problem

Calculate the standard heat of reaction from the dehydrogenation of ethane using the standard heats of combustion:

$$C_2H_6 \rightarrow C_2H_4 + H_2$$

Solution

$$\Delta H^\circ_{Rx} = \Delta H^\circ_{c,\,C_2H_6} - \Delta H^\circ_{c,\,C_2H_4} - \Delta H^\circ_{c,\,H_2}$$

Substituting the values of standard heat of combustion (from the appendix) yields

$$\Delta H^\circ_{Rx}(kJ/mol) = -1559.9 - (-1410.99) - (-285.84) = 136.93\ kJ/mol$$

9.3 Energy Balance for Reactive Processes

For energy balances with reaction, we have two methods for solving these types of problems: the heat of reaction method (extent of reaction) and the heat of formation method (element balance). These two methods differ in the choice of the reference state [2,3].

9.3.1 Heat of Reaction Method

The heat of reaction method is ideal when there is a single reaction for which ΔH°_{Rx} is known. This method requires calculation of the extent of reaction, ξ. The extent of reaction can be obtained by performing material balance for any reactant or product for which the feed and product flow rates are known. The reference state is such that all reactant and product species are at 25°C and 1 atm in the states for which the heat of reaction is known (Figure 9.1).

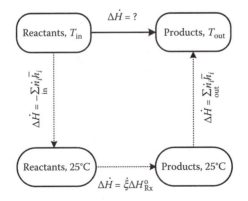

FIGURE 9.1
Rate of change in enthalpy for a reactive process.

For a single reaction at a reference state of 25°C and 1 atm while reactant and product are at different inlet and exit temperatures [4],

$$\Delta\dot{H} = \dot{\xi}\Delta H_{Rx}^{o} + \sum_{out}\dot{n}_i\bar{h}_i - \sum_{in}\dot{n}_i\bar{h}_i$$

where $\bar{h}_i$(J/mol) is the specific molar enthalpy of a definite component. For multiple reactions, where the reference state is 25°C and 1 atm and the inlet and exit streams are at temperatures other than the reference states,

$$\Delta\dot{H} = \sum_{reactions}\dot{\xi}_j\Delta H_{Rx,j}^{o} + \sum_{out}\dot{n}_i\bar{h}_i - \sum_{in}\dot{n}_i\bar{h}_i$$

A reference temperature other than 25°C can be considered in this case, and the heat of reaction should be calculated at the new reference state:

$$\Delta\dot{H} = \sum_{reactions}\dot{\xi}_j\Delta H_{Rx,j} + \sum_{out}\dot{n}_i\bar{h}_i - \sum_{in}\dot{n}_i\bar{h}_i$$

The heat of reaction at any temperature

$$\Delta H_{Rx} = \Delta H_{Rx}^{o}\,(25°C) + \int_{25°C}^{T}\Delta C_p dT$$

where $\Delta C_p = \sum v_i C_{p,i}$

9.3.2 Heat of Formation or Element Balance Method

In the heat of formation method, the heats of reaction terms (ΔH_{rxn}^{o}) are not required as they are implicitly included when heats of formation of the reactants are subtracted from the products [5]. For single and multiple reactions,

$$\Delta\dot{H} = \sum_{out}\dot{n}_i\bar{h}_i - \sum_{in}\dot{n}_i\bar{h}_i$$

where $\bar{h}_i$ accounts for the change in molar enthalpy with T and phase $+\Delta H_f^{o}$ (Figure 9.2).

In this case we find the enthalpy of all of the compounds relative to the elements at 25°C. No heat of reaction needs to be calculated at all. We then plug these enthalpies directly into the energy balance expression. In the absence of kinetic and potential energy, the energy balance equation is

$$Q - W = \Delta\dot{H} = \sum_{out}\dot{n}_i\bar{h}_i - \sum_{in}\dot{n}_i\bar{h}_i$$

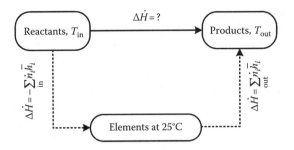

FIGURE 9.2
Heat of formation or element balance method.

where the specific molar enthalpy of component i in the inlet streams is

$$\bar{h}_{i,\text{in}} = \int_{25}^{T_{\text{in}}} C_{p,i} dT + \Delta H^o_{f,i}$$

Specific molar enthalpy of component i in the exit streams is

$$\bar{h}_{i,\text{out}} = \int_{25}^{T_{\text{out}}} C_{p,i} dT + \Delta H^o_{f,i}$$

9.4 Simultaneous Material and Energy Balances

Material balances could be written on either compound that requires the extent of reaction or their elements, which requires only balances without generation terms for each element. Similarly, we can also write down energy balances using either compounds or elements. From material balances with reaction that we had discussed, there are three methods of analyzing these types of reactive processes: atomic species balances, extents of reaction, and component balances. For energy balances with reaction, we have two methods for solving these types of problems: the heat of reaction method and the heat of formation method or element balance method.

Example 9.7 Gas Phase Reaction

Problem
Gas phase reaction is taking place in a continuous reactor. Stoichiometric proportions of CO_2 and H_2 are fed to the reactor at 400°C. The reaction proceeds to 80% completion. Given 1 mol of carbon dioxide, estimate the heat that must be provided or removed, if the gas exit steams are to

be kept at 500°C. Perform the energy balance using the heat of reaction method at two reference temperatures (500°C and 25°C) and the heat of formation method.

Solution

Known quantities: Inlet and exit temperatures, percent conversion, inlet flow rate of carbon dioxide and hydrogen.

Find: The amount of heat added or removed.

Analysis: The schematic diagram is shown in Example Figure 9.7.1.

Basis: 1 mol of CO_2 and 4 mol of H_2. The process flow diagram is shown in Example Figure 9.7.1.

Material balance (Extent of reaction method)

$$CO_2 + 4H_2 \rightarrow 2H_2O + CH_4 \quad \xi$$

Component balance is calculated using the extent of reaction method as follows:

$$CO_2: \; n_{CO_2} = 1 - \xi$$

$$H_2: \; n_{H_2} = 4 - 4\xi$$

$$CH_4: \; n_{CH_4} = 0 + \xi$$

$$H_2O: \; n_{H_2O} = 0 + 2\xi$$

From the fractional conversion of CO_2,

$$f = \frac{n_{CO_2}^o - n_{CO_2}}{n_{CO_2}^o}$$

$$0.8 = \frac{1.0 - n_{CO_2}}{1.0}$$

$$n_{CO_2} = 1.0 - 0.8 = 0.2 \text{ mol}$$

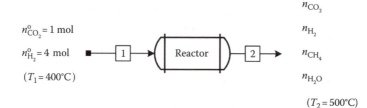

$n_{CO_2}^o = 1$ mol

$n_{H_2}^o = 4$ mol

$(T_1 = 400°C)$

Reactor

n_{CO_2}

n_{H_2}

n_{CH_4}

n_{H_2O}

$(T_2 = 500°C)$

EXAMPLE FIGURE 9.7.1
Process flow diagram of gas phase reaction.

Accordingly, the number of moles of carbon dioxide leaving the reactor is $n_{CO_2} = 0.2$ mol.

The extent of reaction, ξ, is calculated by substituting $n_{CO_2} = 0.2$ mol in the CO_2 mole balance equation: $0.2 = 1 - \xi$.

Solving for the extent of reaction, $\xi = 0.8$.

To calculate moles of exit components, substitute values of the extent of reaction in the mole component balance equations given earlier. The following results are obtained:

$$n_{H_2} = 0.8, \quad n_{CH_4} = 0.8, \quad n_{H_2O} = 1.6$$

Energy balance (Heat of reaction method)
The energy balance is performed at two reference temperatures; exit stream temperature (500°C) and at the standard heat of reaction temperature (25°C). First, prepare the following data table using the polynomial form of the specific heat capacity:

$$C_p(J/mol\,°C) = a + bT + cT^2$$

The standard heats of formation at 25°C can be obtained from Table A.2 (Appendix A.1). Components of heat capacity (Appendix A.2) and heats of formation (Table A.2):

Components	v_i	a	b	c	ΔH_f^o
CO_2	−1	36.11	0.04233	-2.887×10^{-5}	−393.5
H_2	−4	28.84	0.0000765	0.3288×10^{-5}	00.00
H_2O	2	33.46	0.006880	0.7604×10^{-5}	−241.83
CH_4	1	34.31	0.054690	0.3661×10^{-5}	−74.85
Δ		−50.24	0.025810	3.4600×10^{-5}	−165.00

Note: $\Delta = CH_4 + 2H_2O - 4H_2 - CO_2$

The ΔH_f^o in the last column is for the heats of formation. The Δ in the last row is the difference between the product and reactants based on the stoichiometric coefficient; for example, the Δa is calculated as

$$\Delta a = \sum v_i a = -1 \times 36.11 - 4 \times 28.84 + 2 \times 33.46 + 34.31 = -50.24$$

The same method is used for calculating the values of Δb, Δc, and ΔH_f^o.

Reference temperature = 500°C
In this approach, we perform the energy balance using the heat of reaction method at the reference temperature of 500°C: $T_{ref} = 500$°C. This requires knowledge of the heat of reaction at 500°C. The energy balance for this process includes the sensible heat to change the temperature of everything in the inlet stream from 500°C to 400°C and the heat of reaction at 500°C times the extent of reaction. Recall that the heat of reaction

was expressed in kilojoules per mole, where the mole basis was per molar extent of reaction:

$$Q = \xi \Delta H_{Rx}(T_{ref}) + H_{out} - H_{in}$$

For multiple input and exit components,

$$Q = \int_{500°C}^{500°C} \sum_{out} n_i C_{p_i} \, dT - \int_{500°C}^{400°C} \sum_{in} n_i^° C_{p_i} \, dT + \xi \, \Delta H_{Rx}\big|_{500°C}$$

Substituting the proper values,

$$Q = 0 + \int_{400°C}^{500°C} \sum_i n_i^° C_{p_i} \, dT + \xi \, \Delta H_{Rx}\big|_{500°C}$$

The heat of reaction at 500°C is then

$$\Delta H_{Rx}\big|_{500°C} = \Delta H_{Rx}^° + \int_{25°C}^{500°C} \Delta C_p \, dT = \Delta H_{Rx}^° + \int_{25}^{500} \left(\Delta a + \Delta bT + \Delta cT^2 \right) dT$$

The ΔC_p for the reaction is immediately obtained from the Δ row for use in integration from one temperature to another:

$$\Delta H_{Rx}\big|_{500°C} = -165 \text{ kJ/mol} + [(-50.24)(500 - 25) + \frac{1}{2}(0.02581)(500^2 - 25^2)$$

$$+ \frac{1}{3}(3.46 \times 10^{-5})(500^3 - 25^3)] \frac{J}{mol} \frac{kJ}{1000 \, J} = -184.6 \text{ kJ/mol}$$

The heat of reaction at 500°C is

$$\Delta H_{Rx}\big|_{500°C} = -184.6 \text{ kJ/mol}$$

Now the sensible heat term (first terms shown in the earlier equation) includes only the moles of those compounds in the inlet stream. This gives

$$H_{in} = \int_{500°C}^{400°C} \sum_i n_i^° C_{p_i} \, dT = (1) \int_{500}^{400} C_{p\,CO_2} \, dT$$

$$+ (4) \int_{500}^{400} C_{p\,H_2} \, dT = 1(-5) + 4(-2.95) \text{ kJ} = -16.80 \text{ kJ}$$

$$H_{\text{out}} = \int\limits_{500°C}^{500°C} \sum_i \dot{n}_i C_{p_i} dT = 0$$

So finally, substitute estimated inlet, exit enthalpies and heat of reaction in the general energy balance equation shown here:

$$Q = \xi \Delta H_{\text{Rx}}(T_{\text{ref}}) + H_{\text{out}} - H_{\text{in}}$$

$$Q = (0.8 \text{ mol})(-184.6 \text{ kJ/mol}) + 0 - (-16.80 \text{ kJ}) = -131 \text{ kJ}$$

Reference temperature $= 25°C$
The energy balance for this process at this reference temperature includes the sensible heat to change the temperature of everything in the inlet stream from 25°C to 400°C and also to change everything in the product stream from 25°C to 500°C. Again, the heat of reaction at 25°C must be multiplied by the extent of reaction. The heat of the process at 25°C is

$$Q = H_{\text{out}} - H_{\text{in}} + \xi \Delta H_{\text{Rx}}^o = \int\limits_{25°C}^{500°C} \sum_{\text{out}} n_i C_{p_i} dT - \int\limits_{25°C}^{400°C} \sum_{\text{in}} n_i^o C_{p_i} dT + \xi \Delta H_{\text{Rx}}^o$$

The enthalpy of inlet stream relative to reference temperature (25°C) is

$$H_{\text{in}} = \int\limits_{25°C}^{400°C} \sum_i n_i^o C_{p_i} dT = (1 \text{ mol}) \int\limits_{25°C}^{400°C} C_{pCO_2} dT$$

$$+ (4 \text{ mol}) \int\limits_{25°C}^{400°C} C_{pH_2} dT = 16.35 + 43.54 = 59.89 \text{ kJ}$$

The heat of reaction term at 25°C is found from the Δ term in the table for the heats of formation. Likewise, we have already found in method 1 that $\xi = 0.8$ mol. Thus, the reaction term is

$$\xi \Delta H_{\text{Rx}}^o = (0.8 \text{ mol})(-165 \text{ kJ/mol}) = -132 \text{ kJ}$$

We can now calculate the sensible heat term for heating everything in the outlet stream from 25°C to 500°C. The enthalpy change of the outlet stream with respect to reference temperature (25°C) is

$$H_{\text{out}} = \int\limits_{25°C}^{500°C} \sum_i n_i C_{p_i} dT = (0.2 \text{ mol}) \int\limits_{25}^{500} C_{pCO_2} dT + (0.8 \text{ mol}) \int\limits_{25}^{500} C_{pH_2} dT$$

$$+ (0.8 \text{ mol}) \int\limits_{25}^{500} C_{pCH_4} dT + (1.6 \text{ mol}) \int\limits_{25}^{500} C_{pH_2O} dT$$

Integration of the earlier equation after substitution of components heat capacity yields

$$H_{out} = \int_{25°C}^{500°C} \sum_i n_i C_{p_i}\, dT = (4.27 + 11.06 + 18.48 + 27.22)\, kJ = 61.03\, kJ$$

Finally, we obtain

$$Q = H_{out} - H_{in} + \xi \Delta H_{Rx}^o = \sum_{out} n_i \bar{h}_i - \sum_{in} n_i^o \bar{h}_i + \xi \Delta H_{Rx}^o$$

$$= 61.03 - 59.89 - 132 = -131\, kJ$$

The heat transferred from the system is 131 kJ.

Heat of formation method (Element balance)
In this case, no heat of reaction needs to be calculated at all. We write the energy balance as follows:

$$Q = \sum_{out} n_i \bar{h}_i - \sum_{in} n_i \bar{h}_i$$

Next, we compute the enthalpy of each component in each stream relative to their elements. The compound is formed from its elements at 25°C (this is the heat of formation), and then we raise the temperature of the compound up to the temperature of the stream.

The specific enthalpy of outlet compounds

$$CH_4:\ \bar{h}_{CH_4} = \Delta H_f^o(25) + \int_{25°C}^{500°C} C_{p,CH_4}\, dT = -74.85 + 23.10 = -51.75\, kJ/mol$$

$$H_2O:\ \bar{h}_{H_2O} = \Delta H_f^o(25) + \int_{25°C}^{500°C} C_{p,H_2O}\, dT = -241.83 + 17.01 = -224.8\, kJ/mol$$

$$CO_2:\ \bar{h}_{CO_2} = \Delta H_f^o(25) + \int_{25°C}^{500°C} C_{p,CO_2}\, dT = -393.5 + 21.34 = -372.2\, kJ/mol$$

$$H_2:\ \bar{h}_{H_2} = 0 + \int_{25°C}^{500°C} C_{p\,H_2}\, dT = 13.83\, kJ/mol$$

Thus, the sum of all of the outlet specific enthalpies is

$$H_{out} = \sum_{out} n_i h_i = (0.2)(-372.2) + (0.8)(13.83) + (0.8)(-51.75)$$
$$+ (1.6)(-224.8) = -464.5\, kJ$$

The specific enthalpies of inlet compounds

$$CO_2 : h^o_{CO_2} = \Delta H^o_f + \int_{25°C}^{400°C} C_{p,CO_2} dT = -393.5 + 16.35 = -377.2 \text{ kJ/mol}$$

$$H_2 : h^o_{H_2} = 0 + \int_{25°C}^{400°C} C_{p H_2} dT = 10.89 \text{ kJ/mol}$$

Thus, the sum of all of the inlet enthalpies is

$$H_{in} = \sum_{in} n_i h_i = (1)(-377.2) + (4)(10.89) = -333.6 \text{ kJ}$$

Finally, from the energy balance we obtain

$$Q = H_{out} - H_{in} = -464.5 \text{ kJ} + 333.6 \text{ kJ} = -131 \text{ kJ}$$

The heat calculated using the three methods are equal.

Example 9.8 Oxidation of Ammonia

Problem
Hundred moles per minute of ammonia (NH_3) and 200 mol/min of oxygen (O_2) at 25°C are fed into a continuous reactor in which ammonia is completely consumed. The product gas emerges at 300°C. Calculate the rate at which heat must be transferred to or from the reactor to maintain the reactor temperature at 300°C. The standard heat of reaction for the gas phase oxidation of ammonia is

$$NH_3 (g) + 1.25O_2 (g) \rightarrow NO (g) + 1.5H_2O \quad \Delta H^o_{Rx} = -225 \text{ kJ/mol}$$

Solution

Known quantities: Inlet molar flow rates, inlet temperature, exit temperature.

Find: Rate of heat transferred from or to the reactor.

Analysis: Use the first law of thermodynamics for an open system.

Basis: 100 mol/min of NH_3

Reference: 25°C and 1 atm
Using the first law of thermodynamics for an open system,

$$Q - W_s = \Delta \dot{H} + \Delta KE + \Delta PE$$

Since there is no shaft work in the process, no moving parts, no change in elevation between inlet and exit stream, and no change in velocity of inlet and exit streams, the general energy balance equation is reduced to

$$Q = \Delta \dot{H}$$

Using the extent of reaction or heat of reaction method,

$$Q = \Delta \dot{H} = \dot{\xi} \Delta H^{\circ}_{Rx} + \sum_{out} \dot{n}_i h_i - \sum_{in} \dot{n}_i h_i$$

Material balance (Extent of reaction method)

$$\dot{n}_{NH_3} = 100 - \xi$$

$$\dot{n}_{O_2} = 200 - 1.25\xi$$

$$\dot{n}_{NO} = 0.0 + \xi$$

$$\dot{n}_{H_2O} = 0.0 + 1.5\xi$$

Because of complete conversion of ammonia ($f=1$), no ammonia exits in the reactor exit stream; $\dot{n}_{NH_3} = 0$

$$0.0 = 100 - \xi \Rightarrow \xi = 100 \text{ mol}$$

Substituting the values of the extent of reaction in the material balance equation will give the following molar flow rate of exit stream components:

$\dot{n}_{NH_3} = 0.0$ mol/min, $\dot{n}_{O_2} = 75$ mol/min, $\dot{n}_{NO} = 100$ mol/min, $\dot{n}_{H_2O} = 150$ mol/min

Energy balance (Heat of reaction method)

Reference temperature: 25°C

$$Q = \Delta \dot{H} = \dot{\xi} \Delta H^{\circ}_{Rx} + \sum_{out} \dot{n}_i \bar{h}_i - \sum_{in} \dot{n}_i \bar{h}_i$$

Heat capacities of components involved in the process C_{p_i} (J/mol°C) are

$$C_{pO_2} = 29.1 + 0.01158T - 0.6076 \times 10^{-5} T^2$$

$$C_{pNO} = 29.5 + 0.008188T - 0.2925 \times 10^{-5} T^2$$

$$C_{pH_2O} = 33.46 + 0.00688T + 0.7604 \times 10^{-5} T^2$$

Substitute known values:

$$Q = \Delta \dot{H} = 100(-225 \text{ kJ/mol}) + \frac{\text{kJ}}{1000 \text{ J}}$$

$$\times \left[75 \int_{25°C}^{300°C} C_{pO_2} dT + 100 \int_{25}^{300} C_{pNO} dT + 150 \int_{25}^{300} C_{pH_2O} dT \right]_{out}$$

$$- \frac{\text{kJ}}{1000 \text{ J}} \left[100 \int_{25}^{25} C_{pNH_3} dT + 200 \int_{25}^{25} C_{pO_2} dT \right]_{in}$$

The change in sensible heat is in the units of J/mol, so it should be divided by 1000 to convert to the units of kJ/mol to be added to heat of reaction. Accordingly, the heat transfer rate is

$$Q = \Delta \dot{H} = 100 \text{ mol} \left[\frac{-225 \text{ kJ}}{\text{mol}} \right] + \left[75(8.47) + 100(8.45) + 150(9.57) \right] - 0$$

$$= -19{,}600 \text{ kJ/min}$$

The heat released from the reaction process is −19,600 kJ/min (−326 kW).

Example 9.9 Production of Formaldehyde

Problem
Formaldehyde is produced in a continuous reactor by oxidizing methane with pure oxygen (Example Figure 9.9.1). Feed streams of 50 mol/h of methane and 50 mol/h of pure oxygen are fed to a continuous reactor. The exit stream molar flow rate is 100 mol/h. The mole fractions of formaldehyde and carbon dioxide are 0.15 and 0.05, respectively. Calculate the rate of heat that must be added to or removed from the reactor to maintain the reactor temperature at 150°C.

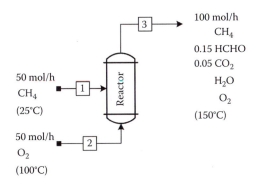

EXAMPLE FIGURE 9.9.1
Process flow sheet of the methane oxidation process.

Solution

Known quantities: Inlet and exit streams temperature and molar composition of formaldehyde and carbon dioxide.

Find: Heat transfer rate from or to the system.

Analysis: Use the extent of reaction method for both material and energy balance.

Basis: 100 mol/h of the exit gas stream

Material balance:

Since carbon dioxide appears in the exit gas stream, the following reactions take place in the reactor:

$$CH_4(g) + O_2 \rightarrow HCHO(g) + H_2O \quad \xi_1$$

$$CH_4(g) + 2O_2 \rightarrow CO_2 + 2H_2O \quad \xi_2$$

To calculate the extent of reaction of the first reaction, select a component that is available only in the first reaction, which is formaldehyde in this case:

$$n_{HCHO} = n^\circ_{CHHO} + \xi_1$$

$$0.15 \times 100 \, mol/h = 0 + \xi_1 \Rightarrow \xi_1 = 15 \, mol/h$$

The extent of the second reaction is obtained by selecting a component that is only available in the second reaction, which is in this case carbon dioxide:

$$n_{CO_2} = n^\circ_{CO_2} + \xi_2$$

Substitute inlet and exit molar flow rate of carbon dioxide:

$$0.05 \times 100 \, mol/h = 0 + \xi_2 \Rightarrow \xi_2 = 5 \, mol/h$$

Knowing the extent of the first and second reactions, the component molar flow rate can be found easily by performing the extent of reaction balance for each component:

$$n_{CH_4} = 50 - \xi_1 - \xi_2 = 50 - 15 - 5 = 30 \, mol/h$$

$$n_{H_2O} = 0 + \xi_1 + 2\xi_2 = 15 + 2 \times 5 = 25 \, mol/h$$

$$n_{O_2} = 50 - \xi_1 - 2\xi_2 = 50 - 15 - 2 \times 5 = 25 \, mol/h$$

Energy balance:
Reference state: 25°C

The standard heat of reaction for both reactions is calculated from standard heats of formation as follows:

The standard heat of the first reaction, ΔH°_{Rx1}, is given by

$$\Delta H^{\circ}_{Rx1} = \Delta H^{\circ}_{f,H_2O(v)} + \Delta H^{\circ}_{f,HCHO(g)} - \Delta H^{\circ}_{f,CH_4(g)}$$

$$\Delta H^{\circ}_{Rx1} = (-241.83) + (-115.9) - (-74.85) = -282.88 \text{ kJ/mol}$$

The standard heat of the second reaction, ΔH°_{Rx2}, is given by

$$\Delta H^{\circ}_{Rx2} = 2\Delta H^{\circ}_{f,H_2O(v)} + \Delta H^{\circ}_{f,CO_2(g)} - \Delta H^{\circ}_{f,CH_4(g)}$$

Substitute values of heats of formation from appendices:

$$\Delta H^{\circ}_{Rx2}(\text{kJ/mol}) = 2(-241.83) + (-393.5) - (-74.85) = -802.31 \text{ kJ/mol}$$

Overall energy balance of the reactor, no work, no moving parts, no change in elevation, hence, work, kinetic and potential energy, are neglected:

$$Q = \Delta \dot{H}$$

The rate of change in enthalpy, $\Delta \dot{H}$, is given by

$$\Delta \dot{H} = \sum_{\text{reactions}} \dot{\xi}_j \Delta H^{\circ}_{Rx,j} + \sum_{\text{out}} \dot{n}_i \bar{h}_i - \sum_{\text{in}} \dot{n}_i \bar{h}_i$$

Heat capacities of all components involved in the process, C_{pi} (J/mol°C), are as follows:

$$C_{pCH_4} = 34.31 + 0.05469T + 0.3661 \times 10^{-5}T^2$$

$$C_{pHCHO} = 34.28 + 0.04268T$$

$$C_{pCO_2} = 36.11 + 0.04233T - 2.887 \times 10^{-5}T^2$$

$$C_{pH_2O} = 33.46 + 0.00688T + 0.7604 \times 10^{-5}T^2$$

$$C_{pO_2} = 29.1 + 0.01158T - 0.6076 \times 10^{-5}T^2$$

In more detail, the energy balance equation is as follows:

$$Q = \dot{\xi}_1 \Delta H^{\circ}_{Rx1} + \dot{\xi}_2 \Delta H^{\circ}_{Rx2} + \left\{ \dot{n}_{CH_4} \bar{h}_{CH_4} + \dot{n}_{HCHO} \bar{h}_{HCHO} + \dot{n}_{CO_2} \bar{h}_{CO_2} \right.$$

$$\left. + \dot{n}_{H_2O} \bar{h}_{H_2O} + \dot{n}_{O_2} \bar{h}_{O_2} \right\}_{out} - \left\{ \dot{n}_{CH_4} \bar{h}_{CH_4} + \dot{n}_{O_2} \bar{h}_{O_2} \right\}_{in}$$

Substitute known quantities:

$$Q = \Delta \dot{H} = \dot{\xi}_1 \Delta H^{\circ}_{Rx1} + \dot{\xi}_2 \Delta H^{\circ}_{Rx2} + \left\{ 30 \int_{25^{\circ}C}^{150^{\circ}C} C_{P\,CH_4}\, dT \right.$$

$$+ 15 \int_{25}^{150} C_{P\,HCHO(g)}\, dT + 5 \int_{25}^{150} C_{PCO_2}\, dT + 25 \int_{25}^{150} C_{P\,H_2O}\, dT$$

$$\left. + 25 \int_{25}^{150} C_{P\,O_2}\, dT \right\} - \left\{ 50 \int_{25}^{25} C_{PCH_4}\, dT + 50 \int_{25}^{100} C_{P\,O_2}\, dT \right\}$$

Substituting the expressions of the heat capacities and integrating,

$$Q = \{15(-282.88) + 5(-802.31)\} + \{30(4.9) + 15(4.75) + 5(4.75)$$

$$+ 25(4.27) + 25(3.758)\} - \{50(0) + 50(2.235)\} = -7923\ \text{kJ/h}$$

The rate of heat transfer is −7923 kJ/h (−2.2 kW).

Example 9.10 Methane Combustion

Problem
Methane and oxygen are fed in stoichiometric proportions to a continuous reactor at 25°C and 1 atm. The reaction proceeds to completion and the effluent stream is found to be at 200°C. Calculate the heat transfer rate from the reactor. On the basis of 1 mol of methane set up an energy balance using the elements balance approach. The combustion reaction of methane is as follows:

$$CH_4 + 2O_2 \rightarrow CO_2 + 2H_2O$$

Solution

Known quantities: Inlet temperature and pressure, stoichiometric proportions of feed.

Find: Heat transfer rate from the reactor, Q.

Analysis: Start by calculating the enthalpy at the stream condition for each individual compound in each stream relatively to the elements at 25°C. We then plug these enthalpies into the normal energy balance expression.

Material balance:

Basis: 1 mol/s of methane and 2 mol/s of oxygen
Since the reaction goes to completion and the feed enters the reactor in stoichiometric proportions, the exit molar flow rate of methane and oxygen are zero. Based on the balance reaction stoichiometry for 1 mol of carbon dioxide reacted, 1 mol of carbon dioxide and 2 mol of water are

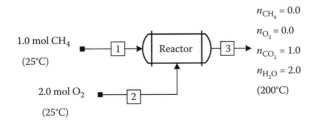

EXAMPLE FIGURE 9.10.1
Schematic of a methane combustion reactor.

produced. The schematic diagram of the combustion process is shown in Example Figure 9.10.1.

Energy balance:
The specific molar flow rate and molar enthalpies of inlet and exit components are arranged in the following table:

Species	n_{in}	h_{in}	n_{out}	h_{out}
CH_4	1	$\bar{h}_{CH_4}(T_1)$	0	—
O_2	2	$\bar{h}_{O_2}(T_2)$	0	—
CO_2	0	—	1	$\bar{h}_{CO_2}(T_3)$
H_2O	0	—	2	$\bar{h}_{H_2O}(T_3)$

From the following data

$$\Delta H^{\circ}_{f,CH_4} = -74.85 \text{ kJ/mol}$$

$$\Delta H^{\circ}_{f,CO_2} = -393.5 \text{ kJ/mol}$$

$$\Delta H^{\circ}_{f,H_2O} = -241.83 \text{ kJ/mol}$$

$$C_{pCO_2} (\text{J/mol}\,^{\circ}C) = 36.11 + 0.04233T - 2.887 \times 10^{-5}T^2$$

$$C_{pH_2O} (\text{J/mol}\,^{\circ}C) = 33.46 + 0.00688T + 0.7604 \times 10^{-5}T^2$$

The general energy balance equation is

$$Q = \sum_{out,\,200^{\circ}C} \dot{n}_i \bar{h}_i - \sum_{in,\,25^{\circ}C} \dot{n}_i^{\circ} \bar{h}_i = \left[\dot{n}_{CO_2} \bar{h}_{CO_2} + \dot{n}_{H_2O} \bar{h}_{H_2O} \right]_{out}$$

$$- \left[\dot{n}_{CH_4} \bar{h}_{CH_4} + \dot{n}_{O_2} \bar{h}_{O_2} \right]_{in}$$

Specific molar enthalpy of inlet streams at $T_1 = 25°C$ is

$$CH_4:\ \bar{h}_{CH_4}(T_1) = \Delta H^{\circ}_{f,CH_4} + \int_{25°C}^{25°C} C_{p\,CH_4}dT = -74.85\ kJ/mol + 0 = -74.85\ kJ/mol$$

Specific molar enthalpy of inlet oxygen at $T_2 = 25°C$ is

$$O_2:\ \bar{h}_{O_2}(T_2) = 0 + \int_{25°C}^{25°C} C_{p\,O_2}dT = 0 + 0 = 0$$

Specific molar enthalpies of the exit streams at $T_3 = 200°C$ are

$$CO_2:\ \bar{h}_{CO_2}(T_3) = \Delta H^{\circ}_{f,CO_2} + \frac{1\ kJ}{1000\ J}\int_{25°C}^{200°C} C_{p\,CO_2}dT = -393.5 + 7.08$$

$$= -386.42\ kJ/mol$$

$$H_2O:\ \bar{h}_{H_2O}(T_3) = \Delta H^{\circ}_{f,H_2O} + \frac{1\ kJ}{1000\ J}\int_{25°C}^{200°C} C_{p\,H_2O}dT = -241.83 + 6.01$$

$$= -235.82\ kJ/mol$$

Using the values calculated by the procedures shown earlier, we simply plug the values of specific enthalpies into the general energy balance equation using the element balance approach:

$$Q = \sum_{out} \dot{n}_i \bar{h}_i - \sum_{in} \dot{n}_i \bar{h}_i$$

Substitute the calculated values of specific enthalpies:

$$Q = \left[\dot{n}_{CO_2}\bar{h}_{CO_2} + \dot{n}_{H_2O}\bar{h}_{H_2O}\right]_{out} - \left[\dot{n}_{CH_4}\bar{h}_{CH_4} + \dot{n}_{O_2}\bar{h}_{O_2}\right]_{in}$$

$$Q = \left[1(-386.42) + 2(-235.82)\right] - \left[1(0) + 1(-74.85)\right] = -783.21\ kJ/s$$

9.4.1 Unknown Process Exit Temperature

Another set of problems involves the calculation of outlet temperature when the inlet conditions and heat input or output are specified. These types of problems require that the enthalpies be evaluated in terms of the unknown outlet temperature. The resulting enthalpy expressions are then substituted into the general energy balance equation and solved for the outlet temperature. The following example explains this case.

Example 9.11 Dehydrogenation of Ethanol

Problem
Dehydrogenation of ethanol to form acetaldehyde is carried out in an adiabatic reactor. Hundred moles per hour of ethanol at 400°C is fed to a continuous reactor. The reactor conversion is 30%. The gas average heat capacities of C_2H_5OH, CH_3CHO, and H_2 are 78, 96, and 29 (J/mol K), respectively. Calculate the reactor exit stream temperature. The following reaction takes place in gas phase:

$$C_2H_5OH \rightarrow CH_3CHO + H_2$$

Solution

Known quantities: Inlet ethanol molar flow rate, temperature, and percent conversion.

Find: The product stream temperature.

Analysis: The exit temperature is unknown, so a simultaneous material and energy balance is required. The schematic diagram of the dehydrogenation process is shown in Example Figure 9.11.1.

Basis: One hour of operation, accordingly all calculated flow rates are on per one hour basis.

Material balance
Using the extent of reaction method,

$$n_{C_2H_5OH} = 100 - \xi$$

$$n_{CH_3CHO} = 0 + \xi$$

$$n_{H_2} = 0 + \xi$$

A 30% conversion of methanol is achieved:

$$0.3 = \frac{100 - n_{C_2H_5OH}}{100} \Rightarrow n_{C_2H_5OH} = 70 \text{ mol}$$

Substitute known values ($n_{C_2H_5OH}$) in the ethanol mole balance equation:

$$70 = 100 - \xi \Rightarrow \xi = 30 \text{ mol}$$

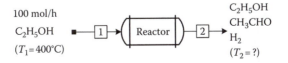

EXAMPLE FIGURE 9.11.1
Ethanol dehydration process, reactor system.

Substitute the extent of reaction in the material balance equations to get the following results:

$$n_{C_2H_5OH} = 70 \text{ mol}, \quad n_{CH_3CHO} = 30 \text{ mol}, \quad n_{H_2} = 30 \text{ mol}$$

Energy balance:
Reference temperature: 25°C

Using the heat of reaction approach, the general energy balance equation is

$$\Delta \dot{H} = \xi \Delta H^o_{Rx} + \sum_{out} n_i \bar{h}_i - \sum_{in} n_i \bar{h}_i$$

The standard heat of reaction is calculated from the heats of formation as follows:

$$\Delta H^o_{Rx} = \Delta H^o_{f, H_2} + \Delta H^o_{f, CH_3CHO} - \Delta H^o_{f, C_2H_5OH}$$

Substituting tabulated standard heat of formation values (from the appendix) yields

$$\Delta H^o_{Rx} = 0 + (-166.2) - (-235.31) = 69.11 \text{ kJ/mol}$$

Open system, adiabatic process, no change in potential and kinetic energy, accordingly the general energy balance is simplified to the following form: $Q = \Delta \dot{H} = 0$.

The change in enthalpy is a function of heat of reaction and sensible heat as follows:

$$0 = \xi \Delta H^o_{Rx} + \left\{ n_{C_2H_5OH} \bar{h}_{C_2H_5OH} + n_{CH_3CHO} \bar{h}_{CH_3CHO} + n_{H_2} \bar{h}_{H_2} \right\}_{out, T_2 = ?}$$

$$- \left\{ n_{C_2H_5OH} \bar{h}_{C_2H_5OH} \right\}_{in, 400 °C}$$

Substituting specific molar enthalpies in terms of heat capacities,

$$0 = \xi \Delta H^o_{Rx} + \left\{ n_{C_2H_5OH} \int_{25}^{T_2} C_{P C_2H_5OH} dT + n_{CH_3CHO} \int_{25}^{T_2} C_{P CH_3CHO} dT + n_{H_2} \int_{25}^{T_2} C_{PH_2} dT \right\}$$

$$- \left\{ n_{C_2H_5OH} \int_{25°C}^{400°C} C_{P C_2H_5OH} dT \right\}$$

Substitute the values of the molar flow rates of inlet and exit components and heat capacities of each component:

$$0 = 30 \times 69.11 \text{ kJ/mol} + \{70 \times 78(T_2 - 25) + 30 \times 96(T_2 - 25)$$

$$+ 30 \times 29(T_2 - 25)\} - \{100 \times 78(400 - 25)\}$$

Rearranging by taking $(T_2 - 25)$ as common factor,

$$0 = 30 \times 69.11 \frac{kJ}{mol} \frac{1000\ J}{kJ} + (T_2 - 25)\{70 \times 78 + 30 \times 96 + 30 \times 29\}$$

$$- \{100 \times 78(400 - 25)\}$$

Simplifying and rearranging leads to the following value of reactor exit temperature:

$$T_2 = \frac{851{,}700}{9210} + 25 = 117.5°C$$

Example 9.12 Methanol Dehydrogenation

Problem

Hundred moles per hour of methanol at 675°C and 1 bar is fed to an adiabatic reactor, where 25% of it is dehydrogenated to formaldehyde. Calculate the temperature of the gases leaving the reactor, assuming constant average heat capacities of 17, 12, and 7 cal/mol °C for methanol, formaldehyde, and hydrogen, respectively. The dehydrogenation of methanol proceeds according to the following reaction:

$$CH_3OH\ (g) \rightarrow HCHO\ (g) + H_2\ (g)$$

The heats of formation in kcal/mol are as follows:

$$\Delta H^{\circ}_{f,\ HCHO} = -27.7\ kcal/mol, \quad \Delta H^{\circ}_{f,\ CH_3OH} = -48.08\ kcal/mol$$

Solution

Known quantities: Inlet temperature, pressure, and molar flow rate are known.

Find: Exit stream temperature.

Analysis: The process flow sheet is shown in Example Figure 9.12.1. In the solution of this example, use the extent of reaction method.

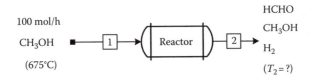

EXAMPLE FIGURE 9.12.1
Formaldehyde production in an adiabatic reactor.

Material balance

Basis: 100 mol/h of methanol
  The extent of reaction method is used to calculate exit number of moles:

$$\text{HCHO: } \dot{n}_{HCHO} = 0 + \xi$$

$$\text{CH}_3\text{OH: } \dot{n}_{CH3OH} = 100 - \xi$$

$$\text{H}_2\text{: } \dot{n}_{H_2} = 0 + \xi$$

The single-pass conversion of methanol

$$f = 0.25 = \frac{100 - \dot{n}_{CH3OH}}{100}, \quad \dot{n}_{CH3OH} = 75 \text{ mol/h}$$

Substitute $\dot{n}_{CH3OH} = 75$ mol/h in the methanol balance equation and calculate the extent of reaction: $\xi = 25$ mol/h.
  Substitute the extent of reaction in the formaldehyde and hydrogen mole balance equation to get the following results:

$$\dot{n}_{HCHO} = 25 \text{ mol/h}, \quad \dot{n}_{CH3OH} = 75 \text{ mol/h}, \quad \dot{n}_{H_2} = 25 \text{ mol/h}$$

Energy balance
The system is open so the first law of thermodynamics for open systems is used as follows:

$$Q - W_s = \Delta\dot{H} + \Delta KE + \Delta PE$$

Neglecting kinetic and potential energies, no shaft work is done, and the reactor is adiabatic. The energy balance equation is simplified to the following form:

$$0 - 0 = \Delta\dot{H} + 0 + 0$$

The enthalpy consists of enthalpy of reaction and the sensible heat:

$$0 = \Delta H = \xi\Delta H_{Rx}(T_{ref}) + \sum_{product} n_i \int_{T_{ref}}^{T_{out}} C_{pi}\, dT - \sum_{reactant} n_i \int_{T_{ref}}^{T_{in}} C_{pi}\, dT$$

Reference temperature $= 675°C$

$$\Delta\dot{H} = \xi\, \Delta H_{Rx}|_{675°C} + \sum_{product} n_i \int_{675°C}^{T} C_{pi}\, dT - \sum_{reactant} n_i \int_{675°C}^{675°C} C_{pi}\, dT$$

The heat capacity of this problem is constant and the effect of temperature on the heat capacity is negligible:

$$\Delta H_{Rx}\big|_{675°C} = \Delta H_{Rx}^{o} + \Delta C_{p}(675 - 25)$$

The standard heat of reaction at 25°C is

$$\Delta H_{Rx}^{o} = \Delta H_{f,\,HCHO}^{o} + \Delta H_{f,\,H_2}^{o} - \Delta H_{f,\,CH_3OH}^{o} = -27.7 + 0 - (-48.08)$$

$$= 20.38 \text{ kcal/mol}$$

The change in the heat capacity, ΔC_p, is given by

$$\Delta C_p = C_{p\,H_2} + C_{p\,HCHO} - C_{p\,CH_3OH}$$

Substituting known values to calculate the heat of reaction at 675°C,

$$\Delta H_{Rx}\big|_{675°C} = 20.38 \text{ kcal/mol} + (7 + 12 - 17)\left[\frac{1\,\text{kcal}}{1000\,\text{cal}}\right](675 - 25)$$

$$= 21.68 \text{ kcal/mol}$$

The rate of change in enthalpy is

$$\Delta \dot{H} = \xi\, \Delta H_{Rx}\big|_{675°C} + \left(n_{HCHO} C_{p HCHO} + n_{CH_3OH} C_{p CH_3OH} + n_{H_2} C_{p H_2}\right)(T_2 - 675) - 0$$

Substitute the extent of reaction, heat of reaction, and heat capacities to
calculate the exit temperature:

$$0 = 25 \times 21.68 \text{ kcal/mol} + (25 \times 12 + 75 \times 17 + 25 \times 7)\left(\frac{1\,\text{kcal}}{1000\,\text{cal}}\right)(T_2 - 675)$$

Rearranging and simplifying,

$$0 = 542 + 1.75(T_2 - 675)$$

Solving for T,

$$T = 365.3°C$$

9.5 Combustion Processes

Combustion or burning is an exothermic chemical reaction between a fuel
and an oxidant accompanied by the production of heat and conversion of

chemical species [4]. For example, the complete combustion of methane is given by the following reaction:

$$CH_4\,(g) + 2O_2\,(g) \rightarrow CO_2\,(g) + 2H_2O\,(g)$$

The result is carbon dioxide and water vapor, with a standard enthalpy of reaction at 25°C and 1 atm being −242 kJ/mol; complete combustion is almost impossible to achieve. As actual combustion reactions come to equilibrium, a wide variety of major and minor species will be present, such as carbon monoxide. Total inlet oxygen is the sum of the theoretical oxygen and excess oxygen. The equation for percent excess air is as follows:

$$\% \text{ excess air} = \frac{(\text{moles of air})_{\text{fed}} - (\text{moles of air})_{\text{theoretical}}}{(\text{moles of air})_{\text{theoretical}}}$$

Theoretical oxygen is the calculated amount of oxygen required to oxidize a compound to its final oxidation products. In the methane oxidation reaction, 2 mol of oxygen is required to oxidize 1 mol of methane. For 100 mol of CH_4 the theoretical oxygen is calculated as follows:

$$\text{Theoretical oxygen demand} = 100\text{ mol } CH_4 \times \frac{2\text{ mol }O_2}{1\text{ mol }CH_4} = 200 \text{ mol of }O_2$$

If the oxygen fed to the reactor is in excess amount than the theoretical oxygen, then the percent excess oxygen is

$$\% \text{ excess } O_2 = \frac{(\text{moles of }O_2)_{\text{fed}} - (\text{moles of }O_2)_{\text{theoretical}}}{(\text{moles of }O_2)_{\text{theoretical}}}$$

Example 9.13 Combustion of Carbon Monoxide

Problem

Carbon monoxide (CO) at 10°C is completely burned at 1 atm pressure with 50% excess air that is fed to a burner at a temperature of 540°C. The combustion products leave the burner chamber at a temperature of 425°C. Given 100 mol/h of carbon monoxide, calculate the heat evolved from the burner.

Solution

Known quantities: CO and air inlet temperature and pressure, complete combustion, 50% excess air.

Find: Heat evolved from the burner.

Analysis: Perform material balance and then energy balance.

Material balance

Basis: 100 mol/h of inlet CO

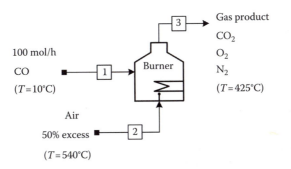

EXAMPLE FIGURE 9.13.1
Schematic of the combustion of carbon monoxide.

The following reaction takes place in the burner:

$$CO\,(g) + \frac{1}{2}O_2\,(g) \rightarrow CO_2\,(g)$$

The schematic diagram of the burning process is shown in Example Figure 9.13.1.

$$\text{Theoretical } O_2: 100 \text{ mol/h CO} \left(\frac{0.5 \text{ mol } O_2}{1 \text{ mol CO}} \right) = 50 \text{ mol/h } O_2$$

Total inlet oxygen is the sum of the theoretical oxygen and excess oxygen. For 50% excess air, the total inlet oxygen is the theoretical plus excess oxygen:

$$\dot{n}_{O_2}^o = 1.0\left(50 \text{ mol } O_2\right) + 0.5\left(500 \text{ mol } O_2\right) = 75 \text{ mol/h } O_2$$

Nitrogen is inert and the total inlet nitrogen is the number of moles of nitrogen in the air associated with the 75 mol of oxygen. That is,

$$\dot{n}_{N_2}^o = \frac{79}{21} \times \text{Total inlet oxygen} = \frac{79}{21} \times 75 \text{ mol} = 282 \text{ mol/h } N_2$$

Since the reaction goes to completion (i.e., complete composition), no carbon monoxide leaves the reactor, all is burned. Mole balance of carbon monoxide (CO) is calculated using the extent of reaction method:

$$n_{CO} = n_{CO}^o - \xi$$

$$0 = 100 - \xi \Rightarrow \xi = 100 \text{ mol}$$

Component mole balance

$$CO_2: \ n_{CO_2} = 0 + \xi$$

$$O_2: \ n_{O_2} = n_{O_2}^{o} - 0.5\xi$$

$$\dot{n}_{O_2} = 75 - 0.5 \times 100 = 25 \ \text{mol/h}$$

The exit number of moles of carbon dioxide is 100 mol; $n_{CO_2} = 100$ mol. Nitrogen is an inert gas and is not involved in the reaction. Accordingly, the inlet number of moles equals the exit number of moles:

$$\dot{n}_{N_2} = 282 \ \text{mol/h}$$

Energy balance
Reference temperature, $T_{ref} = 25°C$
   The standard heat of reaction is calculated as

$$\Delta H_{Rx}^{o} = \Delta H_{f, CO_2}^{o} - \Delta H_{f, CO}^{o} - \frac{1}{2}\Delta H_{f, O_2}^{o}$$

$$\Delta H_{Rx}^{o} = -393.5 \ \text{kJ/mol} - (-110.52 \ \text{kJ/mol}) - 0 = -282.98 \ \text{kJ/mol}$$

Heat capacities C_{pi} (J/mol °C) as a function of temperature of components involved in the process are

$$C_{pO_2} = 29.1 + 0.01158T - 0.6076 \times 10^{-5}T^2$$

$$C_{pN_2} = 29 + 0.002199T + 0.5723 \times 10^{-5}T^2$$

$$C_{pCO} = 28.95 + 0.00411T + 0.3548 \times 10^{-5}T^2$$

$$C_{pCO_2} = 36.11 + 0.04233T - 2.887 \times 10^{-5}T^2$$

Enthalpies of inlet components:
The enthalpies of inlet components are defined from the following expressions. The integration is from the reference temperature to the inlet temperature.
   Specific molar enthalpy of oxygen is obtained as follows:

$$\bar{h}_{in}(O_2) = \frac{1 \ \text{kJ}}{1000 \ \text{J}} \int_{25°C}^{540°C} C_{pO_2} \ dT = 16.38 \ \text{kJ/mol}$$

Specific inlet enthalpy of nitrogen is calculated as follows:

$$\bar{h}_{in}(N_2) = \frac{1\,kJ}{1000\,J} \int_{25°C}^{540°C} C_{P\,N2}\,dT = 15.49\,kJ/mol$$

Specific inlet enthalpy of carbon monoxide is obtained as follows:

$$\bar{h}_{in}(CO) = \frac{1\,kJ}{1000\,J} \int_{25°C}^{10°C} C_{P\,CO}\,dT = -0.4353\,kJ/mol$$

Enthalpies of exit components:
The enthalpies of exit components are defined from the following expressions. The integration is from the reference temperature to the exit temperature:

$$\text{Oxygen: } \bar{h}_{out}(O_2) = \frac{1\,kJ}{1000\,J} \int_{25°C}^{425°C} C_{P\,O2}\,dT = 12.54\,kJ/mol$$

$$\text{Nitrogen: } \bar{h}_{out}(N_2) = \frac{1\,kJ}{1000\,J} \int_{25°C}^{425°C} C_{P\,N2}\,dT = 11.92\,kJ/mol$$

$$\text{Carbon dioxide: } \bar{h}_{out}(CO_2) = \frac{1\,kJ}{1000\,J} \int_{25°C}^{425°C} C_{P\,CO2}\,dT = 17.58\,kJ/mol$$

Summary of the calculated enthalpies is shown in the following table:

Compound i	$\dot{n}_{i,\,in}$ (mol/h)	$\bar{h}_{i,\,in}$ (kJ/mol)	$\dot{n}_{i,\,out}$ (mol/h)	$\bar{h}_{i,\,out}$ (kJ/mol)
O_2	75	16.38	25	12.54
N_2	282	15.49	282	11.92
CO	100	−0.44	0	—
CO_2	0	—	100	17.58

The heat removed from the burner, Q, is given by

$$Q = \Delta H = \xi \Delta H_r^\circ + \sum_{out} n_i \bar{h}_i - \sum_{in} n_i \bar{h}_i$$

Substituting inlet and exit moles multiplied by specific enthalpies of inlet and exit streams, respectively, from the table in the earlier equation yields the heat released from the burner, Q:

$$Q = \xi \times \Delta H_{Rx}^\circ + \left\{ \dot{n}_{O_2}\bar{h}_{O_2} + \dot{n}_{N_2}\bar{h}_{N_2} + \dot{n}_{CO_2}\bar{h}_{CO_2} \right\}_{out} - \left\{ \dot{n}_{O_2}\bar{h}_{O_2} + \dot{n}_{N_2}\bar{h}_{N_2} + \dot{n}_{CO}\bar{h}_{CO} \right\}_{in}$$

Substitute component specific enthalpies:

$$Q = 100 \times (-282.98 \text{ kJ/mol}) + \{25(12.54) + 282(11.92) + 100(17.58)\}$$
$$- \{75(16.38) + 282(15.49) + 100(-0.44)\}$$

The heat lost from the reactor is

$$Q = -284{,}177 \text{ kJ/h} \left(-78.94 \text{ kW}\right)$$

The negative sign indicates that heat is released from the process.

9.6 Energy Balance in Bioprocesses

Energy contributions of bioprocesses contributions to sensible heat are insignificant compared with the total magnitude of ΔH_{Rx}^0 and can, therefore, be ignored without much loss of accuracy. This situation is typical of most reactions in bioprocessing where the actual temperature of reaction is not significantly different from 25°C [5–7].

Example 9.14 Fermentation and Citric Acid Production

Problem
An amount of 2500 kg of glucose and 860 kg of oxygen are consumed to produce 1500 kg citric acid, 500 kg biomass, and other products. Ammonia is used as a nitrogen source. Power input to the system by mechanical agitation of the broth is about 15 kW; approximately 100 kg water is evaporated during the culture period. Estimate the cooling requirements during 2 days of operation. The latent heat of evaporation of water at 30°C is 2430.7 kJ/kg. The heat of reaction at 30°C is −460 kJ/mol O_2 consumed. The batch reactor operates at 30°C. The reaction taking place in the fermenter is given by

$$\text{Glucose} + O_2 + NH_3 \rightarrow \text{Biomass} + CO_2 + H_2O + \text{citric acid}$$

Solution

Known quantities: Inlet flow and exit mass flow rate, shaft work, fermenter initial temperature.

Find: Heat transfer from the reactor.

Analysis: The reaction that takes place in the current fermentation process (Example Figure 9.14.1) follows the reaction shown above

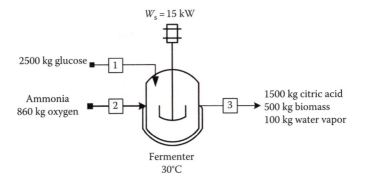

EXAMPLE FIGURE 9.14.1
Production of a citric acid fermenter.

The general energy balance equation is

$$Q - W_s = \xi \Delta H_{\text{reaction}} + m_v \Delta H_v$$

Heat released due to the reaction is

$$\xi \Delta H_{\text{reaction}} = 860 \text{ kgO}_2 \text{ consumed} \left| \frac{1000 \text{ g}}{1 \text{ kg}} \right| \frac{1 \text{ mol}}{32 \text{ g}} \left(\frac{-460 \text{ kJ}}{\text{mol}} \right) = -1.24 \times 10^7 \text{ kJ}$$

Heat needed for evaporation is

$$m_v \Delta H_v = (100 \text{ kg})(2430.7 \text{ kJ/kg}) = 2.43 \times 10^5 \text{ kJ}$$

The system shaft work is expressed as

$$W_s = \left(-15 \frac{\text{kJ}}{\text{s}} \right)(2 \text{ days}) \frac{3600 \text{ s}}{1 \text{ h}} \left| \frac{24 \text{ h}}{1 \text{ day}} \right. = -2.59 \times 10^6 \text{ kJ}$$

The general energy balance equation is

$$Q - W_s = \xi \Delta H_{\text{rxn}} + m_v \Delta H_v$$

Substitute known quantities to yield

$$Q - (-2.59 \times 10^6 \text{ kJ}) = -1.24 \times 10^7 \text{ kJ} + 2.43 \times 10^5 \text{ kJ}$$

$$Q = -1.475 \times 10^7 \text{ kJ}$$

The negative sign indicates that heat is removed from the system.

9.7 Energy Balance in Membrane Reactors

A membrane reactor is actually just a plug-flow reactor that contains an additional cylinder of some porous material within it. Its configuration is similar to that of a shell-and-tube heat exchanger, with a tube within the shell of the exchanger. This porous inner cylinder is the membrane that gives the membrane reactor its name.

Example 9.15 Membrane Reactor

Problem

A membrane reactor is used to produce formaldehyde by dehydrogenation of methanol. The conversion of methanol to formaldehyde takes place on the tube side of the membrane, which is an endothermic reaction:

$$CH_3OH \rightarrow HCHO + H_2 \quad \Delta H^\circ_{Rx} = 85.3 \text{ kJ/mol}$$

The following side reaction takes place as well on the shell side of the membrane. The reaction is exothermic and its heat is utilized to preserve the temperature of the gas stream at 150°C:

$$H_2 + \frac{1}{2}O_2 \rightarrow H_2O \quad \Delta H^\circ_{Rx} = -241.83 \text{ kJ/mol}$$

Methanol vapor stream at a flow rate of 100 mol/min and 250°C enters the tube side of the membrane. Assume that the entire methanol reacts to form formaldehyde and the whole produced hydrogen penetrates through the membrane tube walls to the shell side where sufficient amount of it is burned on the shell side. Sufficient amount of oxygen from air is supplied and reacts completely with the hydrogen. How much hydrogen must be burned to keep the reactor effluent temperature at 150°C? What is the molar flow rate of the stream leaving the reactor?

Data:
The average molar heat capacities of the various materials in this temperature range are: $CH_3OH = 0.0568$ kJ/(mol °C), $HCHO = 0.0380$ kJ/(mol °C), $H_2 = 0.0289$ kJ/(mol °C), $O_2 = 0.0309$ kJ/(mol °C), $N_2 = 0.0296$ kJ/(mol °C), and $H_2O = 0.0341$ kJ/(mol °C).

Solution

Known quantities: Inlet flow of methanol, complete conversion of methanol, and inlet oxygen.

Find: Amount of hydrogen burned to keep the membrane effluent streams at 150°C.

Analysis: First we need to determine how much heat is required for the methanol to formaldehyde reaction to be kept at the temperature

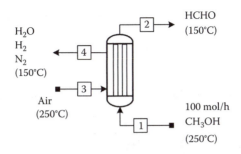

EXAMPLE FIGURE 9.15.1
Schematic of a membrane reactor for formaldehyde production.

of 150°C. The schematic diagram of the membrane reactor is shown in Example Figure 9.15.1. The hydrogen reaction takes place on the shell side with sufficient amount of oxygen associated with inlet air. All inlet oxygen is consumed.

Material balance (tube side):
The exit stream from the tube side contains formaldehyde, and hydrogen penetrates membrane walls to the shell side; since complete reaction is achieved, 100 mol/min of formaldehyde leave the tube side, and 100 mol/min of hydrogen penetrate the membrane walls to the shell side.

Energy balance (tube side):
Tube side energy balance

$$Q_{tub} = \xi_1 \Delta H_{Rx,1}^o + \sum_{out} n_i \bar{h}_i - \sum_{in} n_i \bar{h}_i$$

The inlet stream to the tube side contains only methanol, where it reacts completely:

$$\sum_{in} n_i \bar{h}_i = n_{CH_3OH}^o C_{pCH3OH} \left(T_1 - T_{ref} \right)$$

$$\sum_{in} n_i \bar{h}_i = 100 \frac{mol}{min} 0.0568 \frac{kJ}{mol \, °C} (250 - 25)°C = 12.78 \, kJ/min$$

The heat of reaction term for methanol dehydrogenation can be computed from heats of formation of the two compounds. Thus,

$$\Delta H_{Rx}^o = -115.90 - (-201.2) = 85.3 \, kJ/mol$$

$$\sum_{out} n_i \bar{h}_i = \left(\dot{n}_{HCHO} C_{p,HCHO} + \dot{n}_{H_2} C_{p,H_2} \right) \left(T_2 - T_{ref} \right)$$

Moles of formaldehyde and hydrogen produced are equal (100 mol/min):

$$\sum_{\text{out}} n_i \bar{h}_i = 100 \frac{\text{mol}}{\text{min}} (0.038 + 0.0289) \frac{\text{kJ}}{\text{mol} \, ^{\circ}\text{C}} (150^{\circ}\text{C} - 25^{\circ}\text{C}) = 836.25 \, \text{kJ/min}$$

Thus, the heat required to maintain the desired reactor temperature is

$$Q_{\text{tub}} = \dot{\xi} \Delta H_{\text{Rx}}^{\circ} + \sum_{\text{out}} \dot{n}_i \bar{h}_i - \sum_{\text{in}} \dot{n}_i \bar{h}_i = (100 \times 85.3) \frac{\text{kJ}}{\text{min}}$$

$$+ \, 836.25 \, \text{kJ/min} - 12.78 \, \text{kJ/min} = 8088.25 \, \text{kJ/min}$$

That is, 8088 kJ/min of heat must be supplied by the hydrogen combustion reaction taking place on the shell side in order for the temperature at the completion of the reaction to be 150°C. Since the hydrogen combustion is also occurring at 150°C we need its heat of reaction at 150°C. This would correspond to the following enthalpy calculation pathway.

Material balance (shell side)
The following reaction takes place:

$$H_2 + \frac{1}{2} O_2 \rightarrow H_2O \quad \xi_2$$

The specific molar flow rates of components leaving the shell side are

Unreacted hydrogen: $n_{\text{H}_2} = 100 - \xi_2$
Water vapor generated: $n_{\text{H}_2\text{O}} = \xi_2$
Nitrogen (inert): $n_{\text{N}_2} = n_{\text{N}_2}^{\circ} = \frac{0.79}{0.21} 0.5 \xi_2$
Oxygen (complete conversion): $n_{\text{O}_2} = 0$

Energy balance (shell side)
On the shell side, hydrogen is burned with sufficient amount of oxygen from the air fed to the shell side. The exit stream of the shell side is water vapor, unreacted hydrogen, and nitrogen; accordingly, the following overall energy balance for the shell side can be written:

$$Q_{\text{shell}} = \xi_2 \Delta H_{\text{Rx},2}^{\circ} + \sum_{\text{out}} n_i \bar{h}_i - \sum_{\text{in}} n_i \bar{h}_i$$

Inlet stream to the shell

$$\sum_{\text{in}} n_i \bar{h}_i = \left\{ n_{\text{H}_2}^{\circ} C_{\text{P}, \text{H}_2} + n_{\text{O}_2}^{\circ} C_{\text{P}, \text{O}_2} + n_{\text{N}_2}^{\circ} C_{\text{P}, \text{N}_2} \right\} (T_3 - T_{\text{ref}})$$

$$n_{\text{H}_2}^{\circ} = 100 \, \text{mol/min}$$

$$n_{N_2}^o = \frac{0.79}{0.21} n_{O_2}^o$$

$$\sum_{in} n_i \bar{h}_i = \left\{ 100(0.0289) + 0.5 n_{H_2}^o (0.0309) + \frac{0.79}{0.21} 0.5 n_{H_2}^o (0.0296) \right\}$$

$$\times (250°C - 25°C)$$

Simplifying yields

$$\sum_{in} n_i \bar{h}_i = 650.25 + 15.98 n_{H_2}^o$$

The shell side outlet stream enthalpy:

$$\sum_{out} n_i \bar{h}_i = \left\{ n_{H_2} C_{P,H_2} + n_{H_2O} C_{P,H_2O} + n_{N_2} C_{P,N_2} \right\} (T_4 - T_{ref})$$

Substituting heat capacity values and molar flow rates of the shell side exit stream gives

$$\sum_{out} n_i \bar{h}_i = \left\{ (100 - \xi_2)(0.0289) + \xi_2(0.0341) + \frac{0.79}{0.21} 0.5 \xi_2 (0.0296) \right\} (150 - 25)°C$$

Thus we would get

$$\sum_{out} n_i \bar{h}_i = \{ 2.89 - 0.06 \xi_2 \} (150 - 25)°C$$

Simplifying further,

$$\sum_{out} n_i \bar{h}_i = 361.25 - 7.5 \xi_2$$

The heat of reaction term is just the heat of formation of water as a vapor; this is –241.83 kJ/mol. So for the hydrogen combustion reaction we get

$$Q_{shell} = \xi_2 \Delta H_{Rx,2}^o + \sum_{out} n_i \bar{h}_i - \sum_{in} n_i \bar{h}_i$$

$$Q_{shell} = \xi_2(-241.83) + 361.25 - 7.5 \xi_2 - 650.25 + 15.98 \xi_2$$

Simplifying,

$$Q_{shell} = -233.35\xi_2 - 289$$

Since heat lost from the shell side is gained by components on the tube side to maintain 150°C in the reactor,

$$Q_{shell} = -Q_{tube}$$

Substituting heat lost from the shell side and heat gained by the tube side yields

$$-233.35\xi_2 - 289 = -8088.25 \text{ kJ/min}$$

Solving for ξ_2,

$$\xi_2 = 33.42 \text{ mol/min}$$

Thus the combustion rate of hydrogen must be 33.4 mol/min.

9.8 Summary

The main difference between using a component balance and an element balance is that we must calculate the heat of reaction when using a component balance, but we simply use the heats of formation when writing down a balance based on the elements as the reference.

General procedure for energy balance with reaction

1. Draw the process flow diagram.
2. Complete the material balance calculations for the reactor (using either extent of reaction or atomic species balances).
3. Prepare the inlet and outlet enthalpy table, inserting known molar amounts (or flow rates) for each stream component (and phase).
4. Choose your reference state for specific enthalpy calculations.
5. Calculate each unknown stream component enthalpy, h_i.
6. Calculate $\Delta \dot{H}$ for the reactor.
7. Using the general energy balance equation, solve for the unknown quantity.

Steps 4–6 depend on the energy balance method used (i.e., heat of reaction or heat of formation).

Homework Problems

9.1 Determine the heat of reaction for the liquid phase of lactic acid ($C_3H_6O_3$) with ethanol (C_2H_5OH) to form ethyl lactate ($C_5H_{10}O_3$) and liquid water at 25°C. The heat of combustion of ethyl acetate is -2685 (kJ/mol). (-32.7 kJ/mol)

The liquid-phase reaction of lactic acid with ethanol is

$$C_3H_6O_3 + C_2H_5OH \rightarrow C_5H_{10}O_3 + H_2O$$

The combustion reaction of ethyl lactate is

$$C_5H_{10}O_3 + 6O_2 \rightarrow 5CO_2 + 5H_2O_{(l)}$$

The following table shows the standard heats of formation data:

Compound	ΔH_f^o (kJ/mol)
$C_3H_6O_3$	-687.0
C_2H_5OH	-277.6
$C_5H_{10}O_3$	—
H_2O (liquid)	-285.8
CO_2 (gas)	-393.5
O_2 (gas)	0

9.2 Superheated steam at 40 bar and 350°C is produced from liquid water at 40 bar and 50°C in a methane-fired boiler. To ensure complete combustion of the methane, 10% excess air is provided. Both methane and combustion air enter the boiler at 25°C. Determine the outlet temperature of the flue gas from the boiler, if 19.85 kg/min of superheated steam is produced from the combustion of 1.4 kg/min of methane. Assume the boiler is perfectly insulated. (458°C)

9.3 Ammonia is synthesized through the reaction of nitrogen with hydrogen. The reactor feed temperature is 400°C. The fresh feed consists of 1 mol% argon and stoichiometric amount of nitrogen and hydrogen. The fractional conversion of N_2 to NH_3 in the reactor is 0.15. The converter is operated adiabatically and the heat of reaction at 400°C is -53 kJ/mol. Given 100 mol of feed stream, estimate the temperature of the effluent gases from the converter. The average heat capacities at the pressure of the reactor for ammonia, hydrogen, nitrogen, and argon are 49.4, 29.5, 31.0, and 20.8 (J/mol °C), respectively. (467.8°C)

9.4 Toluene reacts to form benzene and *o*-xylene according to the following reaction:

$$2C_7H_8 \rightarrow C_6H_6 + C_8H_{10}$$

Toluene may also react with hydrogen in the reactor to form benzene and methane:

$$C_7H_8 + H_2 \rightarrow C_6H_6 + CH_4$$

In this process, toluene reacts with a fractional conversion of 0.80, resulting in benzene and xylene yields of 0.505 and 0.495, respectively. Here yields are defined as moles of component produced/moles of toluene reacted. If the fresh toluene stream, fresh hydrogen stream, and product stream are all at 400°C and 15 bar absolute, determine the heat requirements for the reactor to maintain reaction temperature at 400°C and a pressure of 15 bar absolute in the vapor phase. The fresh feed stream contains 225 mol/h of toluene. Average heat capacities: benzene = 82.44 J/(mol K), toluene = 103.7 J/(mol K), o-xylene = 353.6 J/(mol K), methane = 35.69 J/(mol K), and hydrogen = 28.82 J/(mol K). (7725.4 kJ/h)

9.5 Methane at 25°C is burned in a boiler furnace with 10.0% excess air. The air enters the burner at a temperature of 100°C. Ninety percent of the methane fed is consumed; the product gas is analyzed and found to contain 10.0 mol CO_2 per 1 mol of CO. The exhaust gases exit the furnace at 400°C. Calculate the rate of heat transferred from the furnace, given that a molar flow rate of 100 mol/s CH_4 is fed to the furnace. (−58,626 kJ/s)

9.6 A certain bacterium is grown in a continuous culture at 30°C. Glucose is used as carbon source and ammonia is the nitrogen source. A mixture of glycerol and ethanol is produced. The reactant contains 36 kg/h glucose and 0.4 kg/h ammonia. The product stream contains 2.81 kg/h cells, 7.94 kg/h glycerol, 11.9 kg/h ethanol, and 0.15 kg/h water. Estimate the cooling requirement. (1.392×10^4 kJ)

Heat of combustion:

$$Glucose = -1.558 \times 10^4 \text{ kJ/kg}$$

$$NH_3 = -2.251 \times 10^4 \text{ kJ/kg}$$

$$Glycerol = -1.799 \times 10^4 \text{ kJ/kg}$$

$$Ethanol = -2.971 \times 10^4 \text{ kJ/kg}$$

$$Cell = -2.120 \times 10^4 \text{ kJ/kg}$$

The reaction:

$$Glucose + NH_3 \rightarrow Biomass + glycerol + ethanol + CO_2 + H_2O$$

9.7 One mole of C_8H_{14} and 14.0 mol of oxygen are placed into a lab scale batch reactor. The reactor then is placed into a water bath at 25°C; the water bath contains 20.0 kg of water. The contents of the reactor are burned. The C_8H_{14} combusts completely to give H_2O and CO_2. At the end of the combustion, the water bath and the reactor and its contents are at 90°C. Under these conditions, the water produced during the combustion can be assumed to be completely in its liquid form. The top surface of the batch reactor is insulated such that all heat lost from the reactor goes to the water bath. What is the standard state heat of combustion of the C_8H_{14}? (−5500 kJ/mol)

The following reaction took place:

$$C_8H_{14} + \frac{23}{2}O_2 \rightarrow 8CO_2 + 7H_2O$$

9.8 Natural gas stream contains 90 mol% methane and the balance H_2S. The stream flowing at a molar flow rate of 100 mol/min and 25°C enters a combustor chamber. Water in the shell side of the combustor is supplied to maintain the combustor exit stream temperature at 25°C. If 50% excess air is used for the combustion, what is the molar flow rate of the gases leaving the combustion process and what is the heat released from the combustor? (−77,395 kJ/min)

The stream is burned completely where the following two reactions take place:

$$CH_4 + 2O_2 \rightarrow CO_2 + 2H_2O$$

$$H_2S + 1.5O_2 \rightarrow H_2O + SO_2$$

9.9 Propane (C_3H_8) enters a combustion chamber at 200°C at a rate of 100 mol/h. The gas is mixed and burned with 50% excess air that enters the combustion chamber at 200°C. An analysis of the combustion gases reveals that 90% of the propane carbon burns to CO_2, with the remaining 10% forming CO, if the exit temperature of the combustion gases is 200°C. The average specific heats of propane, oxygen, carbon dioxide, carbon monoxide, and water vapor are 73.5, 29.4, 37.4, 28.6, and 34.7 J/mol K, respectively. The standard heats of formation of propane, carbon dioxide, carbon monoxide, and water vapor are −103.85, −393.51, −110.52, and −241.86 kJ/mol, respectively. Determine the rate of heat transfer from the combustion chamber. (−54 kW)

9.10 Liquid octane (C_8H_{18}) enters the combustion chamber of a gas turbine steadily at 1 atm and 25°C, and it is burned with air that enters the combustion chamber at the same state. Determine the adiabatic flame temperature for a complete combustion at 400% theoretical air given 1 kmol octane. Assume constant heat capacities 45, 35, 30, and 30 kJ/kmol K, for CO_2, H_2O, O_2, and N_2, respectively. (977 K)

References

1. Felder, R.M. and R.W. Rousseau (1999) *Elementary Principles of Chemical Processes*, 3rd edn., John Wiley, New York.
2. Himmelblau, D.M. (1996) *Basic Principles and Calculations in Chemical Engineering*, 6th edn., Prentice-Hall, Upper Saddle River, NJ.
3. Reklaitis, G.V. (1983) *Introduction to Material and Energy Balances*, John Wiley & Sons, New York.
4. Whirwell, J.C. and R.K. Toner (1969) *Conservation of Mass and Energy*, Blaisdell, Waltham, MA.
5. Atkinson, B. and F. Mavituna (1991) *Biochemical Engineering and Biotechnology Handbook*, 2nd edn., Macmillan, Basingstoke, U.K.
6. Doran, P. (1995) *Bioprocess Engineering Principles*, Academic Press, San Diego, CA.
7. Bailey, J.E. and D.F. Ollis (1986) *Biochemical Engineering Fundamentals*, 2nd edn., McGraw-Hill, Boston, MA.

10

Simultaneous Material and Energy Balances

Material and energy balances are very important in the chemical industries. Material quantities can be described by material balances. Similarly, energy quantities can be described by energy balances. If there is no accumulation, what goes into a process must come out. Material and energy balances are fundamental to the control of processing, particularly in the control of yields of the products. After completing this chapter, the following learning objectives should be accomplished.

Learning Objectives

1. Understand the basic definitions needed in solving material balance problems, for example, conversion, yield, extent of reaction, and standard heat of reaction (Section 10.1).
2. Write an energy balance for a reacting system (Section 10.2).

10.1 Material Balances

There are frequently used definitions that should be known when solving material and energy balance problems involving chemical reactions. As the definitions include conversion, yield, selectivity, and extent of reaction, these are briefly explained in subsequent sections. The general material balance equation takes the form [1,2]

$$\text{Accumulation} = (\text{in} - \text{out}) + (\text{generation} - \text{consumption})$$

10.1.1 Conversion

Generally, syntheses of chemical products do not involve a single reaction but rather multiple reactions. The purpose, in this case, is to maximize the production of the desirable product and minimize the production of unwanted by-products. Conversion is the ratio of the moles that react to the moles that are fed to a reactor. Relative to species (*i*), the fractional conversion can be calculated using the following equation: Fractional conversion of component *i*

$$f_i = \frac{\{\text{moles of component } i\}_{\text{in}} - \{\text{moles of component } i\}_{\text{out}}}{\{\text{moles of component } i\}_{\text{in}}} \tag{10.1}$$

$$f_i = \frac{n_{i0} - n_i}{n_{i0}}$$

10.1.2 Yield

The yield of a reaction is the ratio of the desired product formed (in moles) to the total amount that could have been produced if conversion of the limiting reactant was complete (i.e., 100%) and no side reactions occurred [3].

$$\text{Yield} = \frac{\text{moles of desired product formed}}{\text{moles formed if there were no side reactions and limiting reactant reacted completely}} \tag{10.2}$$

10.1.3 Selectivity

The selectivity of a reaction is the ratio of the desired product formed (in moles) to the undesired product formed (in moles):

$$\text{Selectivity} = \frac{\text{moles of desired product formed}}{\text{moles of undesired product formed}} \tag{10.3}$$

10.1.4 Extent of Reaction (ξ)

The concept of extent of reaction can also be applied to multiple reactions, with each reaction having its own extent. The extent of reaction is the amount in moles (or molar flow rate) that is converted in a given reaction (we used ξ in Chapter 5). If a set of reactions take place in a batch or continuous steady-state reactor, we can write

$$n_i = n_{i0} + \sum_j v_{ij} \xi_j \tag{10.4}$$

where

v_{ij} is the stoichiometric coefficient of substance *i* in reaction *j*
ξ_j is the extent of reaction for reaction *j*
n_{i0} is the inlet molar flow rate of component *i*

For a single reaction, the earlier equation reduces to the following equation:

$$\dot{n}_i = \dot{n}_{i0} + v_i \dot{\xi} \tag{10.5}$$

10.2 Energy Balances

The general energy balance equation for an open system at steady state is as follows [4]:

$$\dot{Q} - \dot{W}_s = \Delta \dot{H} + \Delta KE + \Delta PE \tag{10.6}$$

$$\Delta KE = \frac{1}{2} \dot{m} \left(v_2^2 - v_1^2 \right) \tag{10.7}$$

$$\Delta PE = \dot{m} g (z_2 - z_1) \tag{10.8}$$

Methods differ in reference state (and thus in the calculation of $\Delta \dot{H}$).

10.2.1 Heat of Reaction Method

In this method, the reference state is such that the reactants and products are at 25°C and 1 atm:

$$\Delta \dot{H} = \sum_{\text{reactions}} \dot{\xi}_j \Delta H^0_{\text{Rx},j} + \sum_{\text{out}} \dot{n}_i \bar{h}_i - \sum_{\text{in}} \dot{n}_i \bar{h}_i \tag{10.9}$$

$$\bar{h}_i = \int_{T_{\text{ref}}}^{T} C_{p_i} dT \tag{10.10}$$

$$\Delta H^0_{\text{Rx},j} = \sum_i^n v_i \Delta H^0_{f,i} \tag{10.11}$$

10.2.2 Heat of Formation Method

In this method, the reference state is the elemental species that constitutes the reactants and products in the states they occur in nature at 25°C and 1 atm:

$$\Delta \dot{H} = \sum_{\text{out}} \dot{n}_i \bar{h}_i - \sum_{\text{in}} \dot{n}_i \bar{h}_i \tag{10.12}$$

The enthalpy in this case includes the sensible heat and the enthalpy of formation:

$$\Delta \dot{H}_i \Big|_{\text{at } T} = \Delta H_f^0 + \int_{T_{\text{ref}}}^{T} C_{p_i} dT \tag{10.13}$$

10.2.3 Concept of Atomic Balances

Consider the reaction of hydrogen with oxygen to form water:

$$H_2 + O_2 \rightarrow H_2O$$

We may attempt to do our calculations with this reaction, but there is something seriously wrong with this equation. It is not balanced; as written, it implies that an atom of oxygen is somehow "lost" in the reaction, but this is in general impossible. Therefore, we must compensate by writing

$$H_2 + \frac{1}{2}O_2 \rightarrow H_2O$$

The number of atoms of any given element does not change in any reaction (assuming that it is not a nuclear reaction).

10.2.4 Mathematical Formulation of the Atomic Balance

Now recall the general balance equation:

$$(\text{In} - \text{out}) + (\text{generation} - \text{consumption}) = \text{accumulation}$$

Moles of atoms of any element are conserved; therefore, generation $= 0$. So we have the following balance on a given element A:

$$\sum \dot{n}_{A,\text{in}} - \sum \dot{n}_{A,\text{out}} = 0 \tag{10.14}$$

When analyzing a reacting system you must choose either an atomic balance or a molecular species balance but not both. An atomic balance often yields simpler algebra, but also will not directly tell you the extent of reaction, and will not tell you whether the system specifications are actually impossible to achieve for a given set of equilibrium reactions.

10.2.5 Degrees of Freedom Analysis for the Atomic Balance

As before, to do a degrees of freedom analysis, it is necessary to count the number of unknowns and the number of equations one can write, and then subtract them. However, there are a couple of important things to be aware of with these balances: when doing atomic balances, the extent of reaction does not count as an unknown, while with a molecular species balance it does. This is the primary advantage of this method. The extent of reaction does not matter since atoms of elements are conserved regardless of how far the reaction has proceeded. When doing an atomic balance, only reactive species are included, and not inert.

Example 10.1 Natural Gas Burner

Problems

Suppose you have a gas mixture that contains nitrous oxide, oxygen, and methane. The natural gas is burned. The following chemical reaction occurs in it. How many atomic balance equations can you write?

$$CH_4 + 2O_2 \rightarrow 2H_2O + CO_2$$

Solution

Known quantities: Balanced reaction.

Find: The number of atomic balance equations that could be written.

Analysis: There would be four equations that you can write: three atomic balances (C, H, and O) and a molecular balance on nitrous oxide (inert and not involved in reaction). You would not include the moles of nitrous oxide in the atomic balance on oxygen.

Example 10.2 Equilibrium Reactions

Problem

An amount of 10 kg of compound A is added to 100 kg of 16 wt% aqueous solution of B, which has a density of 0.90 kg/L. A has a molecular weight of 25 kg/kmol and B has a molecular weight of 47 g/mol. If the equilibrium constant, K, for this reaction is 200 at 300 K, how much of compound C could you obtain from this reaction? Adding 10 kg of A to the solution causes the volume to increase by 9 L. The following reaction occurs:

$$A + B \rightleftarrows C + D$$

Solution

Known quantities: Mass of components A and B.

Find: Amount of compound C that could be obtained from this reaction.

Analysis: The process flowchart is shown in Example Figure 10.2.1.

Since all of the species are dissolved in water, we should write the equilibrium constant in terms of molarities (mol/L):

$$K = 200 = \frac{C_C C_D}{C_A C_B}$$

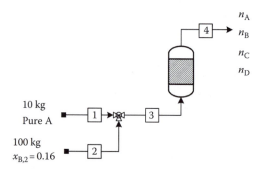

EXAMPLE FIGURE 10.2.1
Schematic diagram of the equilibrium reaction process.

The number of moles of A we have initially is

$$n_{A0} = 10 \text{ kg A} \times \frac{1 \text{ kmol}}{25 \text{ kg}} = 0.4 \text{ kmol}$$

The number of moles of B we have initially is

$$n_{B0} = 100 \text{ kg solution} \times \frac{0.16 \text{ kg B}}{\text{kg solution}} \frac{1 \text{ kmol}}{47 \text{ kg}} = 0.34 \text{ kmol}$$

Now, the volume contributed by 100 kg of 16% B solution is

$$V_B = \frac{m}{\rho} = \frac{100 \text{ kg}}{0.90 \dfrac{\text{kg}}{\text{L}}} = 111 \text{ L}$$

After adding 9 L of A to the volume, the new volume is

$$V_{\text{final}} = 111 + 9 \text{ L} = 120 \text{ L}$$

There is no C or D in the solution initially:

$$C_{C0} = C_{D0} = 0$$

Plugging all the known values into the equilibrium equation for liquids, the following equation is obtained:

$$200 = \frac{\left(\dfrac{\xi}{120}\right)\left(\dfrac{\xi}{120}\right)}{\left(\dfrac{0.40 - \xi}{120}\right)\left(\dfrac{0.34 - \xi}{120}\right)}$$

Simplifying,

$$200 = \frac{\xi^2}{(0.40 - \xi)(0.34 - \xi)}$$

Rearranging,

$$200(0.40 - \xi)(0.34 - \xi) = \xi^2$$

$$27.2 - 148\xi + 200\xi^2 = \xi^2$$

$$27.2 - 148\xi + 199\xi^2 = 0$$

This equation can be solved using trial and error or any available software package:

$$\xi = 0.411 \text{ kmol}$$

$$n_C = \xi = 0.411 \text{ kmol}$$

A total of 411 mol of component C can be produced by this reaction.

10.2.6 Implementing Recycle on the Separation Process

Recycle may improve reaction conversion enough to eliminate the need for a second reactor to achieve an economical conversion. Recycle reduces the amount of waste that a company generates. Not only it is the most environmentally sound way to go about but it also saves the company money in disposal costs. Using recycle, it is possible to recover expensive catalysts and reagents. Catalysts are not cheap, and if we do not try to recycle them into the reactor, they may be lost in the product stream. This not only gives us a contaminated product but also wastes a lot of catalyst.

Example 10.3 Separation of Binary Liquid Mixture

Problem
Fresh feed stream (100 kg/h) contains equal mass fractions of A and B joining a recycle stream and fed to a separator. The top product stream of the separator contains 60% of A and 50% of the B that is fed to the separator and not that of the fresh feed stream. The recycle system is set up in which half of the separator bottom product stream is recycled and recombined with the fresh feed. Calculate the compositions of A in all streams.

Solution

Known quantities: Fresh feed stream flow rate and composition.

Find: Exit stream molar flow rates.

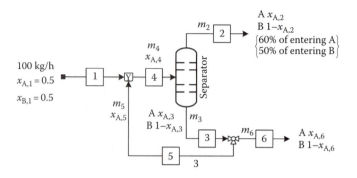

EXAMPLE FIGURE 10.3.1
Schematic diagram of the binary separation process.

Analysis: The process flow sheet is shown in Example Figure 10.3.1. The number of degrees of freedom (NDF) of each process:

	Mixing	Separator	Splitter	Overall
Number of unknowns	4	6	6	4
Number of independent equations	2	2	1	2
Number of auxiliary relations	0	2	1	0
NDF	2	2	4	2

Since none of the units got zero degree of freedom, by contrast, the whole process is solvable because the number of independent equations plus auxiliary relations $(7+3)$ equals the total number of unknowns (10 unknowns) in the process:

System: Overall

Overall mass balance: $100 \dfrac{\text{kg}}{\text{h}} = \dot{m}_2 + \dot{m}_6$

Overall mass balance on A: $50 \dfrac{\text{kg}}{\text{h}} = \dot{m}_2 \times x_{A,2} + \dot{m}_6 \times x_{A,6}$

We have four unknowns and two equations at this point. First, combine this information with the splitting ratio and constant composition at the splitter.

Splitting ratio: $\dot{m}_6 = \dfrac{\dot{m}_3}{2}$

Constant composition: $x_{A,6} = x_{A,3}$

Plugging these into the overall balances, we have $100 = \dot{m}_2 + \dfrac{\dot{m}_3}{2}$

Component balance (A): $50 = x_{A,2}\dot{m}_2 + x_{A,3}\dfrac{\dot{m}_3}{2}$

Relations: Translating words in the relations into algebraic equations.

If 60% of A entering the separator goes into stream 2, then 40% of A entering the separator goes into stream 3, and then

$$x_{A,2}\dot{m}_2 = 0.6x_{A,4}\dot{m}_4$$

$$x_{A,3}\dot{m}_3 = 0.4x_{A,4}\dot{m}_4$$

If 50% of B entering the separator goes into stream 2, then 50% of B entering the separator goes into stream 3, and then

$$(1 - x_{A,2})\dot{m}_2 = 0.5(1 - x_{A,4})\dot{m}_4$$

$$(1 - x_{A,3})\dot{m}_3 = 0.5(1 - x_{A,4})\dot{m}_4$$

Plugging in all of these into the existing balances, we finally obtain two equations in two unknowns:

System: Overall

Component balances:

$$A: \quad 50 = 0.6x_{A,4}\dot{m}_4 + \frac{0.4}{2}x_{A,4}\dot{m}_4 \Rightarrow x_{A,4}\dot{m}_4 = 62.5$$

$$B: \quad 50 = 0.5(1 - x_{A,4})\dot{m}_4 + \frac{0.5}{2}(1 - x_{A,4})\dot{m}_4 \Rightarrow 50 = 0.75\dot{m}_4 - 0.75x_{A,4}\dot{m}_4$$

Substitute $x_{A,4}\dot{m}_4 = 62.5$ in B component balance equation:

$$50 = 0.75\dot{m}_4 - 0.75(62.5)$$

Solving these equations gives

$$\dot{m}_4 = 129.17\frac{kg}{h}, \quad x_{A,4} = 0.484$$

System: Mixer
Total balance

$$100 + \dot{m}_5 = \dot{m}_4 = 129.17 \Rightarrow \dot{m}_5 = 29.17 \text{ kg/h}$$

System: Splitter
The recycle stream ($\dot{m}_5$) is half of the separator bottom product stream ($\dot{m}_3$): $\dot{m}_3 = 2\dot{m}_5 = 2 \times 29.17 = 58.34$ kg/h

System: Overall

$$100 = \dot{m}_2 + 29.165 \Rightarrow \dot{m}_2 = 70.835 \text{ kg/h}$$

Relations:

$$x_{A,2}\dot{m}_2 = 0.6x_{A,4}\dot{m}_4$$

$$x_{A,2} \times 70.835 = 0.6 \times 0.484 \times 129.17 \Rightarrow x_{A,2} = 0.530$$

$$x_{A,3}\dot{m}_3 = 0.4x_{A,4}\dot{m}_4$$

$$x_{A,3} \times 58.34 = 0.4 \times 0.484 \times 129.17 \Rightarrow x_{A,3} = 0.429$$

Example 10.4 Methane Oxidization

Problem

Methane and oxygen at 25°C are fed to a continuous reactor in stoichiometric amounts according to the following reaction to produce formaldehyde:

$$CH_4(g) + O_2(g) \rightarrow HCHO(g) + H_2O(g)$$

In a side reaction, methane is oxidized to carbon dioxide and water:

$$CH_4(g) + 2O_2(g) \rightarrow CO_2(g) + 2H_2O(g)$$

The product gases emerge at 400°C, and the number of moles of CO_2 in the effluent gases is 0.15, and there is no remaining O_2 found in the effluent gases stream. Determine the composition of effluent gas per mole of CH_4 fed to the reactor. Determine the amount of heat removed from the reactor per mole of CH_4 fed to the reactor.

Solution

Known quantities: Inlet component molar flow rate and compositions.

Find: The amount of heat removed from the reactor per mole of CH_4 fed to the reactor.

Analysis: The process flow sheet is shown in Example Figure 10.4.1.

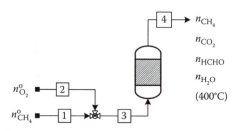

EXAMPLE FIGURE 10.4.1
Schematic diagram of the methane oxidation process.

Material balance
Material balance using the extent of reaction method is given by

$$n_{CH_4} = 1 - \xi_1 - \xi_2$$

There is no remaining O_2:

$$n_{O_2} = 0 = 1 - \xi_1 - 2\xi_2$$

$$n_{HCHO} = \xi_1$$

The number of moles of CO_2 in the effluent stream is 0.15 mol:

$$n_{CO_2} = \xi_2 = 0.15$$

$$n_{H_2O} = \xi_1 + 2\xi_2$$

Hence, $1 - \dot{\xi}_1 - 2\dot{\xi}_2 = 0$ and $n_{CO_2} = \dot{\xi}_2 = 0.15$
Substituting $\dot{\xi}_2 = 0.15$ in the earlier equation and solving for $\dot{\xi}_1$, we get $\dot{\xi}_1 = 0.7$
Substitute values of $\dot{\xi}_2 = 0.15$ and $\dot{\xi}_1 = 0.7$ in the mole balance equations:

$$\dot{n}_{CH_4} = 0.15 \, mol/s, \quad \dot{n}_{HCHO} = 0.7 \, mol/s, \quad \dot{n}_{CO_2} = 0.15 \, mol/s,$$

$$\dot{n}_{H_2O} = 1 \, mol/s$$

The amount of heat removed from the reactor per mole of CH_4 fed to the reactor. The heat capacities are taken from tabulated values as a function of temperature.

Energy balance
Reference temperature $= 25°C$
The standard heats of reaction for both reactions at 25°C are calculated from the standard heats of formation.
The heats of formation are

$$\Delta H^0_{f,CH_4} = -74.85 \, kJ/mol$$

$$\Delta H^0_{f,O_2} = 0$$

$$\Delta H_{f,HCHO} = -115.9 \, kJ/mol$$

$$\Delta H^0_{f,H_2O} = -241.83 \, kJ/mol$$

$$\Delta H^0_{f,CO_2} = -393.5 \, kJ/mol$$

The standard heats of reaction are

$$\Delta H^0_{Rx,1} = \Delta H^0_{f,H_2O}(g) + \Delta H^0_{f,HCHO}(g) - \Delta H^0_{f,CH_4}(g) - \Delta H^0_{f,O_2}(g)$$

$$\Delta H^0_{Rx,1} = -115.9 - 241.83 - (-74.85) - (0) = -282.88 \frac{kJ}{mol}$$

$$\Delta H^0_{Rx,2} = 2\Delta H^0_{f,H_2O}(g) + \Delta H^0_{f,CO_2}(g) - \Delta H^0_{f,CH_4}(g) - 2\Delta H^0_{f,O_2}(g)$$

$$\Delta H^0_{Rx,2} = 2(-241.83) + (-393.5) - (-74.85) - 2(0) = -802.31 \frac{kJ}{mol}$$

The sensible heat is calculated using specific heat:

$$C_p\left(\frac{J}{mol\ {}^\circ C}\right) = a + bT + cT^2 + dT^3$$

The specific heat can be found from Table A.3:

HCHO: $a = 34.28$, $b = 4.268 \times 10^{-2}$, $c = 0.0$, $d = -8694 \times 10^{-9}$

CH$_4$: $a = 34.31$, $b = 5.469 \times 10^{-2}$, $c = 0.3661 \times 10^{-5}$, $d = -11 \times 10^{-9}$

CO$_2$: $a = 36.11$, $b = 4.233 \times 10^{-2}$, $c = -2.887 \times 10^{-5}$, $d = 7.464 \times 10^{-9}$

H$_2$O: $a = 33.46$, $b = 0.688 \times 10^{-2}$, $c = 0.7604 \times 10^{-5}$, $d = -3.593 \times 10^{-9}$

Calculation of enthalpies of outlet and inlet streams relative to the reference temperature (i.e., 25°C) is shown in the following.
Enthalpy of outlet stream components:

$$\bar{h}_{HCHO} = \int_{25}^{400} C_p\, dT = \int (34.28 + 4.268 \times 10^{-2}T + 0T^2 - 8694 \times 10^{-9}T^3)\, dT$$

$$\bar{h}_{HCHO} = \int_{25}^{400} C_p\, dT = \left[34.28T + 4.268 \times 10^{-2}\frac{T^2}{2} - 3(8694 \times 10^{-9})\frac{T^4}{4} \right]_{25}^{400}$$

$$\bar{h}_{HCHO} = \frac{kJ}{1000\ J}\left[34.28(400-25) + 4.268 \times 10^{-2}\frac{(400^2 - 25^2)}{2} \right.$$
$$\left. -8694 \times 10^{-9} \times \frac{(400^4 - 25^4)}{4} \right]$$

Note that $(400^2 - 25^2) \neq (400 - 25)^2$

The results are

$$\bar{h}_{HCHO} = 16.2 \frac{kJ}{mol}, \quad \bar{h}_{CH_4} = 17.23 \frac{kJ}{mol}, \quad \bar{h}_{CO_2} = 16.35 \frac{kJ}{mol},$$

$$\bar{h}_{H_2O} = 13.23 \frac{kJ}{mol}$$

The general energy balance equation is

$$Q = \sum_{out} \dot{n}_i \hat{H}_i - \sum_{in} \dot{n}_i \hat{H}_i + \xi_1 \Delta H_{rxn1}^0 + \xi_2 \Delta H_{rxn2}^0$$

Substituting known quantities,

$$\dot{Q} = \left\{ \dot{n}_{CH_4} \left(\bar{h}_{CH_4} \right) + \dot{n}_{CO_2} \left(\bar{h}_{CO_2} \right) + \dot{n}_{HCHO} \left(\bar{h}_{HCHO} \right) + \dot{n}_{H_2O} \left(\bar{h}_{H_2O} \right) \right\}_{400°C}$$

$$- \left\{ \dot{n}_{CH_4} \left(\bar{h}_{CH_4} \right) + \dot{n}_{O_2} \left(\bar{h}_{O_2} \right) \right\}_{25°C} + \dot{\xi}_1 \Delta H_{Rx,1}^0 + \dot{\xi}_2 \Delta H_{Rx,2}^0$$

The enthalpy of inlet stream components is zero because inlet temperature is at reference temperature; therefore, the enthalpy of inlet components relative to reference temperature of 25°C is zero:

$$\dot{Q} = \{0.15(17.23) + 0.15(16.35) + 0.7(16.2) + 1(13.23)\}_{out} - \{0 + 0\}_{in}$$

$$+ 0.7(-282.88) + 0.15(-802.31)$$

$$= -288.76 \text{ kJ/s}$$

Example 10.5 Adiabatic Saturation Temperature

Problem

Air at a temperature of 50°C and 10% relative humidity is to be humidified adiabatically (constant wet-bulb temperature line) to 40% relative humidity. Use the psychrometric chart to estimate the adiabatic saturation temperature of the air, the rate at which water must be added to humidify 15 kg/min of the entering air and the final temperature of air.

Solution

Known quantities: Inlet air temperature and relative humidity.

Find: The adiabatic saturation temperature of the air.

Analysis: Refer to a psychrometric chart:

$$\text{at } T_{db} = 50°C \text{ and } h_r = 10\% \Rightarrow T_{as} = T_{wb} = 23.5°C$$

$$h_a = 0.0077 \frac{kg\ H_2O}{kg\ DA}$$

The state of the exit air lies on the line $T_{wb} = 23.5°C$. From the intersection of this line and the 40% relative humidity curve, the absolute humidity of exit air is determined to be

$$h_a = 0.014 \text{ kg H}_2\text{O/kg dry air}$$

The rate at which water must be added to humidify 15 kg/min of the entering air is

$$\frac{15 \text{ kg air}}{\text{min}} \left| \frac{1 \text{ kg DA}}{1.0077 \text{ kg air}} \right| \frac{0.014 - 0.0077}{\text{kg DA}} \right| = 0.0938 \frac{\text{kg H}_2\text{O}}{\text{min}}$$

From the intersection of the 40% relative humidity curve and the $T_{wb} = 23.6°C$ line, the dry-bulb temperature of exit gas is found to be around 35°C.

Example 10.6 Partial Condensation of Cyclopentane

Problem

A stream of pure cyclopentane vapor is flowing at a rate of 1550 L/s, at a temperature of 150°C, and at a pressure of 1 atm it enters a cooler. Seventy-five percent of the feed is condensed and exits the cooler at 1 atm. What is the temperature of the exiting streams from the cooler?

Solution

Known quantities: Inlet stream volumetric flow, inlet temperature, and pressure.

Find: Temperature of the exiting streams from the cooler.

Analysis: The process flow diagram is shown in Example Figure 10.6.1.
    The exit stream temperature must be 49.3°C for cyclopentane, because the boiling point of cyclopentane at 1 atm pressure is 49.3°C. Taking the reference temperature at 49.3°C and cyclopentane in liquid state

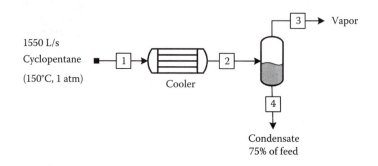

EXAMPLE FIGURE 10.6.1
Partial condensation of cyclopentane.

(i.e., $T_{ref} = 49.3°C$, liquid state), the number of moles of inlet cyclopentane in vapor phase, using the ideal gas law, is

$$n = \frac{PV}{RT} = \frac{1 \text{ atm} \times 1550 \text{ L/s}}{0.08206 \dfrac{\text{L} \cdot \text{atm}}{\text{mol} \cdot \text{K}} \times (150 + 273.15)\text{K}} = 44.64 \frac{\text{mol}}{\text{s}}$$

The inlet specific molar enthalpy of vapor cyclopentane at 150°C relative to reference temperature (49.3°C, liquid) is shown in the following equation:

$$\bar{h}_{in} = \Delta\bar{h}_{vap} + \int_{49.3°C}^{150°C} C_{p,v} \, dT$$

Heat capacity of cyclopentane vapor (C_{pv}) as a function of temperature is shown here:

$$C_{pv}\left(\frac{\text{J}}{\text{mol K}}\right) = 73.39 + 0.3928T - 2.554 \times 10^{-4}T^2$$

The change in specific molar enthalpy of cyclopentane vapor is

$$\Delta\bar{h}_v = \int_{49.3°C}^{150°C} C_p \, dT = \left[aT + \frac{b}{2}T^2 + \frac{c}{3}T^3\right]_{49.3°C}^{150} = 11,060 \frac{\text{J}}{\text{mol}}$$

The heat of vaporization of cyclopentane at 1 atm and its boiling point 49.3°C is

$$\Delta\bar{h}_{vap} = 27.30 \frac{\text{kJ}}{\text{mol}}$$

Substituting heat of vaporization and sensible heat,

$$\bar{h}_{in} = \Delta\bar{h}_v + \int_{49.3}^{150} C_{pv} \, dT = 27.3 + 11.06 = 38.36 \text{ kJ/mol}$$

The resultant enthalpies can be summarized in the following table.
Enthalpies Relative to Reference Conditions ($T_{ref} = 49.3°C$, liquid state, 1 atm)

Substance	$\dot{n}_{in}$ (mol/s)	$\bar{h}_{in}$ (kJ/mol)	$\dot{n}_{out}$ (mol/s)	$\bar{h}_{out}$ (kJ/mol)
Cyclopentane (v)	44.64	38.36	11.14	27.3
Cyclopentane (l)	—	—	33.5	0

Total energy balance: $Q = \sum_{out} n_i \bar{h}_i - \sum_{in} n_i \bar{h}_i$

$$Q = \left(11.14\ \frac{\text{mol}}{\text{s}}\right)\left(27.30\ \frac{\text{kJ}}{\text{mol}}\right) - 44.64\ \frac{\text{mol}}{\text{s}}(38.36)\ \frac{\text{kJ}}{\text{mol}} = -1408\ \frac{\text{kJ}}{\text{s}}$$

Example 10.7 Methanol Combustion Process

Problem

Methanol (CH_3OH) is fed at a rate of 297 mol/h and burned with excess air. The product gas is analyzed, and the following dry-basis mole percentages are determined: $CH_3OH = 0.45\%$, $CO_2 = 9.03\%$, and $CO = 1.81\%$. Calculate the fractional conversion of methanol, the percentage excess air fed, and the mole fraction of water in the product gas. The equations of the chemical reactions taking place in the heater are

$$CH_3OH + \frac{3}{2}O_2 \rightarrow CO_2 + 2H_2O$$

$$CH_3OH + O_2 \rightarrow CO + 3H_2O$$

Solution

Known quantities: Methanol volumetric flow rate, dry gas composition.

Find: The fractional conversion of methanol, the percentage excess air fed, and the mole fraction of water in the product gas.

Analysis: The labeled flowchart is shown in Example Figure 10.7.1.
 Degrees of Freedom Analysis (Atomic Balance)

Number of unknowns	4
Atomic balance equations (C, H, O)	3
Number of relations	1
NDF	0

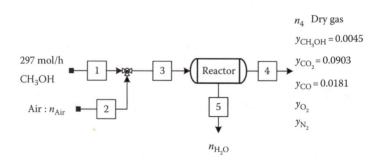

EXAMPLE FIGURE 10.7.1
Schematic of the methanol combustion process.

Atomic balance

Basis: 297 mol/h of inlet methanol

C balance:

$$n_{\text{dry gas}} = \frac{297 \text{ mol/h}}{0.1129} = 2631 \text{ mol/h}$$

$$297 \frac{\text{mol}}{\text{h}} = (0.0045 + 0.0903 + 0.0181)n_{\text{dry gas}}$$

H balance:

$$297 \text{ mol} \frac{CH_3OH}{h} \times \frac{4 \text{ mol H}}{\text{mol } CH_3OH}$$

$$= 2n_{H_2O} + 0.0045 \text{ mol } CH_3OH \frac{4 \text{ mol H}}{\text{mol } CH_3OH} n_{\text{dry gas}}$$

$$n_{H_2O} = 570.3 \text{ mol/s}$$

Total moles out $= 570.3 + 2631 = 3201$ mol

$$\text{Fraction of water in the product stream} = \frac{570.3}{3201} = 0.178$$

$$\text{Fractional conversion of methanol} = x = \frac{297 - 0.0045 \times 2631}{297} = 0.96$$

Calculation of the amount of excess air:
   O balance:

$$2(0.21 \times n_{\text{Air}}) + 297 = (0.0045 + 2 \times 0.0903 + 0.0181 + 2 \times y_{O_2}) \times 2631 + 569.2$$

N balance:

$$2(0.79 \times n_{\text{Air}}) = 2 \times (1 - 0.0045 - 0.0903 - 0.0181 - y_{O_2}) \times 2631$$

Solving for air, oxygen, and nitrogen,

$$n_{\text{Air}} = 2733.41 \text{ mol/s}, \quad y_{N_2} = 0.822, \quad y_{O_2} = 0.065$$

To calculate the percent excess air, first calculate the theoretical oxygen using the complete combustion reaction:

$$\text{Theoretical oxygen} = \frac{297 \text{ mol } CH_3OH}{h} \left| \frac{\frac{3}{2} \text{mol } O_2 \text{ conusmed}}{1 \text{ mol } CH_3OH} = 445.5 \frac{\text{mol}}{h} \right.$$

$$\text{Theoretical air} = \frac{\dot{n}_{O_2}}{0.21} = \frac{445.5\ \text{mol}\ O_2/h}{0.21} = 2121.43\ \text{mol/h}$$

$$\% \text{Excess air} = \frac{\text{Total inlet air} - \text{Theoretical air}}{\text{Theoretical air}} = \frac{2733.41 - 2121.43}{2121.43}$$

$$\times 100\% = 28.85$$

Example 10.8 Methanol Synthesis

Problem

A fresh feed stream is flowing at 100 mol/h, containing 31% carbon monoxide (CO), 66 mol% hydrogen (H_2), and 3 mol% nitrogen. The fresh feed stream joins a recycle stream, and the combined stream is fed to a catalytic reactor for methanol synthesis. This stream is mixed with a recycle stream in a ratio of 4 mol recycle to 1 mol of fresh feed to enter the reactor, and the stream entering the reactor contains 11 mol% N_2. The reactor effluent goes to a condenser, from which two streams emerge: a liquid stream containing pure liquid CH_3OH and a gas stream containing all the CO, H_2, and N_2. The gas stream from the condenser is split into a purge stream, and the remainder is recycled to mix with the fresh feed to enter the reactor. Calculate the production rate of methanol (mol/h), the molar flow rate and composition of the purge gas, and the overall conversion of CO.

Solution

Known quantities: Fresh feed stream flow rate and composition.

Find: The production rate of methanol (mol/h), the molar flow rate and composition of the purge gas, and the overall conversion of CO.

Analysis: The process flowchart is shown in Example Figure 10.8.1.

Basis: 100 mol/h of fresh feed

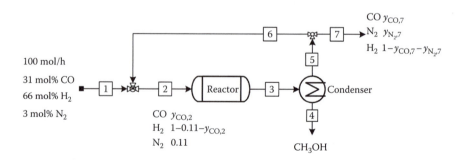

EXAMPLE FIGURE 10.8.1
Schematic diagram of the methanol production process.

Degrees of Freedom Analysis

Degrees of Freedom Analysis	Overall Process	Mixing Point
Number of unknowns	$5\,(\dot{n}_4, \dot{n}_7, y_{CO,7}, y_{N_2,7}, \xi)$	$5\,(\dot{n}_2, y_{CO,2}, \dot{n}_6, y_{N_2,6}, y_{CO,6})$
Number of equations	4	4
Number of relations	—	1
NDF	1	0

System: Mixing point balances

From the relation recycle stream -4 times of the fresh feed stream, $n_6 = 400$ mol/h

$$n_2 = 500 \text{ mol/h}$$

N_2 balance:

$$3 + y_{N_2,6} \times 400 = 0.11 \times 500$$

$$y_{N_2,6} = 0.13$$

System: Overall balance

N_2 balance:

$$3 = y_{N_2,7} \times n_7$$

Since the composition of recycle stream and purge stream are the same,

$$y_{N_2,7} = y_{N_2,6} = 0.13$$

The flow rate of purge is

$$n_7 = 23 \text{ mol/h}$$

Atomic C balance:

$$31 \text{ mol CO} \times 1 = n_4 \text{ mol CH}_3\text{OH} + y_{CO,7} \times 23$$

Simplify

$$n_4 = 31 - 23 \times y_{CO,7}$$

Atomic H balance:

$$2 \times 66 \text{ mol H}_2 = \left(n_4 \text{ mol CH}_3\text{OH} \frac{4 \text{ mol H}}{\text{mol CH}_3\text{OH}} \right)$$

$$+ \left((1 - 0.13 - y_{CO,7}) \frac{2 \text{ mol H}}{\text{mol H}_2\text{O}} \times 23 \right)$$

Substituting $n_4 = 31 - 23 \times y_{CO,7}$ in the earlier equation yields

$$132 = (31 - 23y_{CO,7}) \times 4 + 40.02 - 46y_{CO,7}$$

$$-32.02 = -138 \times y_{CO,7}$$

$$y_{CO,7} = 0.23$$

$$y_{H_2,7} = 0.87 - 0.23 = 0.64 \, \frac{\text{mol H}_2}{\text{mol}}$$

$$n_4 = 25.71 \text{ mol CH}_3\text{OH}$$

$$\text{Overall conversion} = \frac{31 \text{ mol} - 0.23 \times 23}{31} = 0.83$$

Overall, 83% conversion of CO is achieved.

Example 10.9 Heating of Propane Gas

Problem

Propane gas at 40°C and 250 kPa enters a continuous adiabatic heat exchanger where no heat is lost from the outside of the unit while heat is transferred between streams. The stream exits at 240°C. The flow rate of propane is 100 mol/min, and superheated steam at 5 bar absolute pressure and 300°C enters the heat exchanger with a flow rate of 6 kg/min. The steam exits the heat exchanger at 1 bar absolute pressure. Calculate the temperature of the exit steam.

Solution

Known quantities: Propane gas temperature, pressure flow rate.

Find: Temperature of the exit steam.

Analysis: The process flow sheet is shown in Example Figure 10.9.1.
For a heat capacity of propane,

$$a = 68.032, \quad b = 0.2259, \quad c = -1.311 \times 10^{-4}, \quad d = 3.171 \times 10^{-8}$$

Substituting values of a, b, c, and d and integrating,

$$\Delta \bar{h}_{C_3H_8} = \int_{40°C}^{240°C} C_P \left(\frac{J}{\text{mol °C}} \right) dT = \int_{40°C}^{240°C} (a + bT + cT^2 + dT^3) \, dT$$

$$\Delta \bar{h}_{C_3H_8} = 0.68[T_2 - T_1] + \frac{0.2259}{2} \left[T_2^{\,2} - T_1^{\,2} \right] + \frac{-1.311 \times 10^{-4}}{3} \left[T_2^{\,3} - T_1^{\,3} \right]$$

$$+ \frac{3.171 \times 10^{-8}}{4} \left[T_2^{\,4} - T_1^{\,4} \right]$$

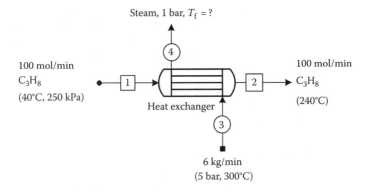

EXAMPLE FIGURE 10.9.1
Heating of propane.

Substituting the given values of inlet and exit temperatures,

$$= 0.68[240 - 40] + \frac{0.2259}{2}[240^2 - 40^2] + \frac{-1.311 \times 10^{-4}}{3}[240^3 - 40^3]$$

$$+ \frac{3.171 \times 10^{-8}}{4}[240^4 - 40^4]$$

The change in specific molar enthalpy $\bar{h}(\text{J/mol})$ is

$$\Delta \bar{h}_{C_3H_8} = 13{,}606.4 + 6325.2 - 13 + 26.3 = 19{,}360 \frac{\text{J}}{\text{mol}}$$

The amount of heat transferred from the steam to heat propane is

$$\dot{n}\Delta\bar{h} = 100 \frac{\text{mol}}{\text{min}} \times 19.36 \frac{\text{kJ}}{\text{mol}} = 1936 \frac{\text{kJ}}{\text{min}}$$

The change in mass specific enthalpy of steam, $h(\text{J/g})$, is

$$\dot{m}\Delta h = 6 \frac{\text{kg}}{\text{min}}[h_{\text{out}} - 3065] \frac{\text{kJ}}{\text{kg}}$$

$$\dot{Q} = 0 = 1936 \frac{\text{kJ}}{\text{min}} + 6 \frac{\text{kg}}{\text{min}}[h_{\text{out}} - 3065] \frac{\text{kJ}}{\text{kg}}$$

$$\frac{1936}{6} \frac{\text{kJ}}{\text{kg}} + 3065 \frac{\text{kJ}}{\text{kg}} = h_4$$

Using the superheated steam table calculate the temperature at 1 bar absolute pressure and enthalpy, $h_4 = 2742.3$ kJ/kg. Since the value of the calculated enthalpy at 1 bar does not exist in the steam table, interpolation is required to get the value of the exit temperature:

$$\frac{2776 - 2676}{150 - 100} = \frac{2776 - 2742.3}{150 - T_f} \Rightarrow T_f = 133°C$$

Example 10.10 Heating of Liquid Methanol

Problem

Liquid methanol at 25°C is heated and vaporized for use in a chemical reaction. How much heat is required to heat and vaporize 2 kmol/h of methanol to 600°C.

Solution

Known quantities: Inlet liquid methanol temperature and flow rate.

Find: Heat required to heat and vaporize 2 kmol of methanol to 600°C.

Analysis: The inlet and exit conditions of the heater are shown in Example Figure 10.10.1.

Consider $T_{ref} = 25°C$ and methanol in liquid phase as reference conditions. The normal boiling point of methanol is 64.7°C. Accordingly, the change of methanol enthalpy is the sum of change in the enthalpy of the liquid methanol, heat of vaporization, and the change in enthalpy of vapor methanol from its boiling point to its final temperature.

The enthalpy change for methanol liquid is

$$\Delta \bar{h}_{CH_3OH} = \int_{25°C}^{64.7°C} C_{P CH_3OH,l} \, dT = 0.07586[64.7 - 25]$$

$$+ \frac{16.83 \times 10^{-5}}{2}[64.7^2 - 25^2] = 3.312 \text{ kJ/mol}$$

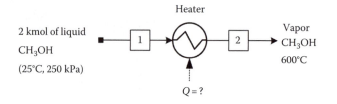

Heater

2 kmol of liquid

CH$_3$OH

(25°C, 250 kPa)

Vapor
CH$_3$OH
600°C

$Q = ?$

EXAMPLE FIGURE 10.10.1
Schematic of heating liquid methanol.

The enthalpy change for methanol vapor is

$$\Delta \bar{h}_{CH_3OH} = \int_{64.7°C}^{600°C} C_{p_v} \, dT = 0.68[600 - 64.7] + \frac{0.2259}{2}[600^2 - 64.7^2]$$

$$+ \frac{-1.31 \times 10^{-4}}{3}[600^3 - 64.7^3] - \frac{3.71 \times 10^{-8}}{4}[600^4 - 64.7^4]$$

$$= 68.21 \, kJ/mol$$

The methanol latent heat of vaporization is

$$\Delta \bar{h}_{vapor} = 36.14 \, \frac{kJ}{mol}$$

The change in enthalpy of the methanol heated from 25°C to 600°C is

$$Q = 2000 \, \frac{mol}{h} \times [3.312 + 68.21 + 36.14] \frac{kJ}{mol} = 2,15,324 \, \frac{kJ}{h}$$

Example 10.11 Turbine Power Plant

Problem

A hydrocarbon fuel whose composition is unknown but may be represented by the expression C_xH_y is burned with excess air. An analysis of the product gas gives the following results in mol% on a moisture-free basis: 9.51% CO_2, 1.0% CO, 5.3% O_2, and 84.2% N_2. Given 100 mol/h of flue gas dry gas, determine the molar ratio of hydrogen to carbon in the fuel, r, where $r = y/x$, and the percentage of excess air used in the combustion.

Solution

Known quantities: Dry gas mole fraction.

Find: The ratio of hydrogen to carbon in the used fluid.

Analysis: The number of independent chemical reactions in the process is two since the flue gases contain CO, which means that there is a side reaction and the combustion is not complete.

Basis: 100 mol of flue gases (dry basis).

$$C_xH_y + \frac{(2x + y/2)}{2} O_2 \rightarrow xCO_2 + \frac{y}{2} H_2O$$

$$C_xH_y + \frac{(x + y/2)}{2} O_2 \rightarrow xCO + \frac{y}{2} H_2O$$

The flowchart of this process is shown in Example Figure 10.11.1.

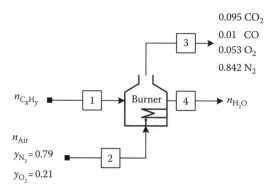

EXAMPLE FIGURE 10.11.1
Schematic diagram of burning hydrocarbon.

Degrees of Freedom Analysis

Degrees of Freedom Analysis	Overall Process (Atomic Balance)
Number of unknowns	$5\ (n_{C_xH_y}, n_{H_2O}, x, y, n_2)$
Number of independent equations	4
Number of relations	—
NDF	1

Atomic balance:
  C atomic balance: $x\, n_{C_xH_y} = 9.5 + 1.0$
  H atomic balance: $y\, n_{C_xH_y} = 2 \times n_{H_2O}$
  N atomic balance: $2(0.79 n_{Air}) = 2(84.2\ \text{mol})$

$$n_{air} = 106.6\ \text{mol}$$

O atomic balance: $2 \times (106.6 \times 0.21) = (2 \times 9.5) + 1.0 + (2 \times 5.3) + n_{H_2O}$

$$n_{H_2O} = 14.17$$

The molar ratio of hydrogen to carbon in the fuel $r = y/x$, and the percentage of excess air used in the combustion can thus be calculated from the following equations:
  C atomic balance: $x \times n_{C_xH_y} = 9.5 + 1.0$
  H atomic balance: $y \times n_{C_xH_y} = 2 \times 14.17$
  Divide H atomic balance equation by C atomic balance:

$$\frac{y \times \cancel{C_xH_y}}{x \times \cancel{C_xH_y}} = \frac{2 \times 14.17}{9.5 + 1.0}$$

$$r = \frac{y}{x} = \frac{28.34}{10.5}$$

For complete combustion, the following reaction is taking place:

$$C_xH_y + \left(\frac{y}{4} + x\right)O_2 \rightarrow xCO_2 + \frac{y}{2}H_2O$$

For 106.6 mol air fed, 10.5 mol of CO_2 produced, $x = 10.5$, $y = 28.34$.
The percentage of excess air used in the combustion:
Moles of C_xH_y can be found from the equation of C atomic balance:

$$10.5 \times n_{C_xH_y} = 9.5 + 1.0$$

$$n_{C_xH_y} = \frac{10.5}{10.5} = 1 \text{ mol}$$

Per one mole of C_xH_y fed to the burner,

$$\text{Theoretical } O_2 = 1 \text{ mol } C_xH_y \frac{\left(\dfrac{y}{4} + x\right)O_2}{\text{mol } C_xH_y} = \frac{28.34}{4} + 10.5 = 17.59 \text{ mol } O_2$$

$$\text{Theortical } N_2 = 17.59 \text{ mol } O_2 \times \frac{0.79 \text{ mol } N_2}{0.21 \text{ mol } O_2} = 66.15 \text{ mol } N_2$$

The associated theoretical nitrogen is thus calculated.

$$\text{Theoretical air} = \text{moles of theoretical oxygen} + \text{nitrogen}$$
$$= 17.59 + 66.15 = 83.74 \text{ mol}$$

Excess air: $106.6 - 83.74 = 27.86$ mol

$$\% \text{ excess air} = \frac{22.8}{83.7} \times 100\% = 27\% \text{ excess air}$$

Example 10.12 Ethanol Production

Problem
Fresh feed containing 20.0% C_2H_4 and 80.0% H_2O is fed to a continuous reactor. The reaction products are fed to a condenser that has two product streams: a vapor stream that contains C_2H_4, C_2H_5OH, and water vapor; and a liquid stream that contains the remaining ethanol, ethylene oxide, and water. The vapor stream from the condenser is recycled and mixed with the fresh feed to be fed to the reactor. The overall process yield is 80% of ethanol produced. In the reactor, ethanol (C_2H_5OH) is produced by steam (H_2O) reformation with ethylene (C_2H_4). An undesirable side reaction, ethylene oxide is formed:

$$C_2H_4 + H_2O \rightarrow C_2H_5OH$$

$$2C_2H_5OH \rightarrow (C_2H_5)_2O + H_2O$$

If the ratio of water in the recycle stream to water in the product stream is 1:10, and the ratio of the ethylene in the recycle to ethylene in the fresh feed is 4:1. The mole fraction of ethanol in the reactor exit stream is 0.157.

Determine the recycle flow rate and the single-pass conversion of ethylene in the reactor.

Solution

Known quantities: Fresh feed containing 20.0% C_2H_4 and 80.0% H_2O is fed to a continuous reactor.

Find: The recycle flow rate and the single-pass conversion of ethylene in the reactor.

Analysis: The flowchart of the process is shown in Example Figure 10.12.1.
Degrees of Freedom Analysis

Degrees of Freedom Analysis	Overall Process
Number of unknowns	$3\ (n_{3,H2O}, n_{3,C2H4OH}, n_{3,(C2H5)2O})$
Number of reactions	1
Number of atomic balances	3
Number of relations	1 (overall conversion)
NDF	0

System: Overall process

$$\text{Overall yield}: 0.8 = \frac{n_{C_2H_4OH}}{20}$$

$$n_{3,C2H4OH} = 16 \text{ mol}$$

C atomic balance: $20 \times 2 = 4n_{3,(C2H4)2O} + 16 \times 2$

$$n_{3,(C2H4)2O} = 2 \text{ mol}$$

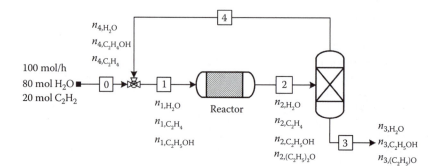

EXAMPLE FIGURE 10.12.1
Schematic of the ethanol production process.

O atomic balance: $80 = 2 + 16 + n_{3H_2O}$

$$n_{3,H_2O} = 62 \text{ mol}$$

Relation: The ratio of water vapor in the recycle stream (4) to liquid water in the product stream (3) is 1/10:

$$\frac{n_{4,H_2O}}{n_{3,H_2O}} = \frac{1}{10}$$

$$n_{4,H_2O} = \frac{1}{10} n_{3,H_2O} = \frac{1}{10} \times 62$$

$$n_{4,H_2O} = 6.2 \text{ mol}$$

Mixing point balance: water in the recycle stream + water in the fresh stream

$$n_{1,H_2O} = 80 + 6.2 = 86.2 \text{ mol}$$

Water balance around the condenser:

$$n_{2,H_2O} = 62 + 6.2 = 68.2 \text{ mol}$$

$$\frac{\text{Ethylene in recycle}}{\text{Ethylene in fresh feed}} = \frac{n_{4,C_2H_4}}{20} = \frac{4}{1}$$

$$n_{4,C_2H_4} = 80 \text{ mol}$$

Also 80 mol C_2H_4 enters the condenser. The exit of the reactor contains the following:

$$80 \text{ mol } C_2H_4$$

$$2 \text{ mol}(C_2H_5)_2O$$

$$68.2 \text{ mol } H_2O$$

$$y_{C_2H_5OH} = 0.157$$

$$f_{C_2H_4} = \frac{n_{1,C_2H_4} - n_{2,C_2H_4}}{n_{1,C_2H_4}}$$

$$\% \text{ Conversion} = \frac{100 - 80}{100} = 20\%$$

Total moles of reactor effluent stream

$$80 + 2 + 68.2 = (1 - 0.157)n_2$$

Solving for n_2,

$$n_2 = \frac{150.2}{0.843} = 178.10 \text{ mol}$$

Total moles of ethanol present in the reactor effluent

$$n_{2,\text{CH}_3\text{OH}} = 0.157 \times 178.10 = 27.96 \text{ mol}$$

The flow rate of recycle stream is $27.96 - 16 = 11.96$ mol of ethanol:

$$\text{Recycle} = 80 \text{ mol C}_2\text{H}_4 + 6.2 \text{ mol H}_2\text{O} + 11.9 \text{ mol ethanol} = 98.1 \text{ mol}$$

Example 10.13 Methanol Combustion

Problem

Methanol (CH_3OH) at 240 mol/min and oxygen (O_2) at a rate of 240 mol/min are fed to an isothermal reactor operating at 25°C. The reactor operates at steady state. Two reactions take place:

$$CH_3OH + O_2 \rightarrow HCOOH + H_2O$$

$$CH_3OH + \frac{3}{2}O_2 \rightarrow CO_2 + 2H_2O$$

The flow rate out of the reactor is 520 mol/min. No oxygen was found in the reactor product stream. Determine the heat that must be withdrawn to keep the reactor at constant temperature. Determine the fractional conversion of methanol. Determine the selectivity for the conversion of methanol to formic acid.

Solution

Known quantities: Inlet and product stream flow rates.

Find: The fractional conversion of methanol, selectivity for the methanol.

Analysis: The process flowchart is shown in Example Figure 10.13.1.

Material balance

   Using the extent of reaction method,

$$CH_3OH: \quad n_{\text{CH}_3\text{OH}} = 240 - \xi_1 - \xi_2$$

$$HCOOH: \quad n_{\text{HCOOH}} = 0 + \xi_1$$

$$H_2O: \quad n_{\text{H}_2\text{O}} = 0 + \xi_1 + 2\xi_2$$

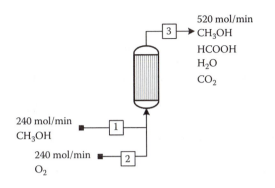

EXAMPLE FIGURE 10.13.1
Process flow sheet of the methanol combustion process.

$$CO_2: \quad n_{CO_2} = 0 + \xi_2$$

$$O_2: \quad n_{O_2} = 240 - \xi_1 - \frac{3}{2}\xi_2$$

$$n_3 = 480 + \frac{1}{2}\xi_2$$

The total molar flow rate out of the reactor

$$\dot{n}_3 = 480 + \frac{1}{2}\dot{\xi}_2 = 520$$

Solving : $\dot{\xi}_2 = 80$ gmol/min

O_2 mole balance: (note that no oxygen is found in the reactor effluent stream)

$$\dot{n}_{3,O_2} = 240 - \dot{\xi}_1 - \frac{3}{2}\dot{\xi}_2 = 240 - \dot{\xi}_1 - 120 = 0$$

Solving for the extent of reaction 1,

$$\dot{\xi}_1 = 120 \text{ gmol/min}$$

Energy balance
    Since the reactor is isothermal (25°C), the heat released is the heat of reaction:

$$Q = \Delta H_{Rx}^0$$

The standard heat of reaction

$$Q = \Delta H_{Rx}^0 = \xi_1 \Delta H_{Rx,1}^0 + \xi_2 \Delta H_{Rx,2}^0$$

The standard heat of reaction from the heat of formations

$$\Delta H_{Rx,1}^0 = \Delta H_{f,H_2O}^0 + \Delta H_{f,HCOOH}^0 - \Delta H_{f,O_2}^0 - \Delta H_{f,CH_3OH}^0$$

Substitute standard heat of formation found in Table A.2:

$$\Delta H_{Rx,1}^0 = -241.83 + (-115.9) - 0 - (-238.6) = -119.13 \text{ kJ/mol}$$

The standard heat of reaction for the second reaction

$$\Delta H_{Rx,2}^0 = 2\Delta H_{f,H_2O}^0 + \Delta H_{f,CO_2}^0 - \frac{3}{2} \Delta H_{f,O_2}^0 - \Delta H_{f,CH_3OH}^0$$

Substitute the values of standard heat of formation for the second reaction:

$$\Delta H_{Rx,2}^0 = 2(-241.83) + (-393.5) - \frac{3}{2}(0) - (-238.6) = -638.56 \text{ kJ/mol}$$

The heat evolved from the reactor, Q, is given by

$$Q = \xi_1 \Delta H_{Rx,1}^0 + \xi_2 \Delta H_{Rx,2}^0$$

Substitute the values of heat of reaction and extent of reaction:

$$Q = 120(-119.13) + 80(-638.56) = -65,380.4 \text{ kJ/min}$$

Fractional conversion of methanol: $\dfrac{\dot{\xi}_1 + \dot{\xi}_2}{\dot{n}_{1,CH_3OH}} = \dfrac{120 + 80}{240} = 0.833$

Fractional selectivity to formic acid: $\dfrac{\dot{\xi}_1}{\dot{\xi}_1 + \dot{\xi}_2} = \dfrac{120}{120 + 80} = 0.60$

Example 10.14 Ethane Combustion

Problem

Ethane (C_2H_6) at a molar flow rate of 750 mol/h is mixed with 20% excess air and fed to a burner where the mixture is completely combusted iso-thermally at 25°C, using cold water. What is the air flow rate to the burner (mol/h) and what is the amount of heat released to the cold water?

Solution

Known quantities: Ethane molar flow rate and percent excess air.

Find: The air flow rate to the burner (mol/h) and the amount of heat released to the cooling water.

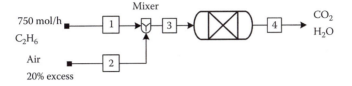

EXAMPLE FIGURE 10.14.1
Ethane combustion process flow sheet.

Analysis: The process flowchart is shown in Example Figure 10.14.1.

Balanced reaction is $C_2H_6 + \dfrac{7}{2}O_2 \rightarrow 2CO_2 + 3H_2O$

Air flow at 20% above the stoichiometric requirement is calculated as

$$\text{Theoretical air} = \left(\frac{750 \text{ mol ethane}}{h}\right)\left(\frac{3.5 \text{ mol } O_2}{\text{mol ethane}}\right)$$

$$\times \left(\frac{1 \text{ mol air}}{0.21 \text{ mol } O_2}\right) = 12,500 \text{ mol/h}$$

Excess air:

Excess air $= 0.2$(Theoretical air) $= 0.2(12,500 \text{ mol/h}) = 2500 \text{ mol/h}$

Total air $=$ Theoretical $+$ excess $= 12,500 + 2500 = 15,000 \text{ mol/h}$

The heat released to the coolant is given by

$$Q = \dot{\xi} \times \Delta \hat{H}^0_{rxn}$$

$$\Delta H^0_{Rx} = 3\Delta H^0_{f,H_2O} + 2\Delta H^0_{f,CO_2} - \Delta H^0_{f,C_2H_6} - \frac{7}{2}\Delta H^0_{f,O_2}$$

$$\Delta H^0_{Rx} = 3(-241.83) + 2(-393.5) - (-84.67) - 0 = -1427.8 \text{ kJ/mol}$$

$$Q = 750 \text{ mol/h} (-1427.8 \text{ kJ/mol}) = -1.24 \times 10^6 \text{ kJ/h}$$

Example 10.15 Chemical Reactors with Recycle

Problem
180 kmol/h of propylene (C_3H_6) is mixed with 240 kmol/h of a mixture containing 50% CO and 50% H_2 and with a recycle stream containing only unreacted propylene and then fed to a reactor. A single-pass conversion of propylene of 30% is achieved. Butanol (C_4H_8O), used to make laundry detergents, is made by the reaction of propylene (C_3H_6) with CO and H_2:

$$C_3H_6 + CO + H_2 \rightarrow C_4H_8O$$

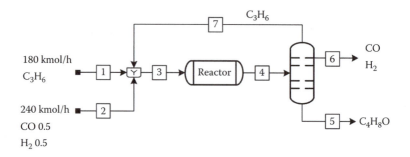

EXAMPLE FIGURE 10.15.1
Process flow sheet for butanol production.

The desired product butanol (B) is removed in one stream, unreacted CO and H_2 are removed in a second stream, and unreacted C_3H_6 is recovered and recycled. Calculate production rate of butanol (kmol/h) and the flow rate of the recycle stream (kmol/h).

Solution

Known quantities: Inlet stream molar flow rate and compositions.

Find: Production rate of butanol (kmol/h) and the flow rate of the recycle stream (kmol/h).

Analysis: The process flow sheet is shown in Example Figure 10.15.1. First choose the entire process as the system. Since no propylene leaves the process, the overall conversion is 100% (fractional conversion = 1.0).

Basis: 180 kmol/h of propylene feed.
   Therefore,

$$1 = \frac{\dot{\xi}}{180} \quad \text{or} \quad \dot{\xi} = 180 \text{ kgmol/h}$$

The balance on butanol is simply $\dot{n}_{5,C_4H_8O} = \dot{\xi} = 180$ kg mol/h.
   Butanol (C_4H_8O) production rate is 180 kmol/h.
   Now choose the reactor as the system. The fractional single-pass conversion = 0.3. The extent of reaction is the same as that for overall process because the reactor is the only unit where reaction takes place in the process. Therefore,

$$0.3 = \frac{\dot{\xi}}{\dot{n}_{3,C_3H_6}} = \frac{180}{\dot{n}_{3,C_3H_6}}$$

Solving for the molar flow rate of propylene fed to the reactor,

$$\dot{n}_{3,C_3H_6} = 600 \text{ kmol/h}$$

From a balance around the mixer, we can find that the recycle rate must be

$$600 - 180 = 420 \text{ kgmol/h}$$

Homework Problems

10.1 An adiabatic pot is used to cool and condense 10 kg of hot ethanol (150°C, 1.2 atm) by mixing it with cold ethanol (5°C, 1.2 atm). If the final ethanol product is to be at 25°C and 1.2 atm, the heat capacity of liquid and vapor ethanol is 112 and 65.6 J/mol °C. Boiling point of ethanol is 78.3°C. The latent heat of vaporization of ethanol is 36,600 J/mol. How much cold ethanol (kg) must be added? (220 kg)

10.2 A quantity of 100 mol/h acetylene (C_2H_2) is mixed with 2000 mol/h air (79 mol% N_2, 21 mol% O_2) and the mixture (at 298 K and 1 atm) fed to a reactor, where complete combustion takes place. The reactor is equipped with cooling tubes. The combustion mixture leaving the reactor is at 1000 K and 1 atm. Draw and label the process flowchart. How much heat (kJ/h) was removed in the reactor? (−82,370 kJ/h)

Suppose the coolant supply was suddenly shut off. What reactor outlet temperature would be reached?

$$C_{PCO_2} = 37 \text{ J/mol °C}, \quad C_{PH_2O} = 33.6 \text{ J/mol °C},$$

$$C_{PO_2} = 29.3 \text{ J/mol °C}, \quad C_{PN_2} = 29.3 \text{ J/mol °C}$$

Assume these values to be constant and independent of temperature.

10.3 Ethanol (C_2H_5OH) is dehydrogenated in a catalytic reactor to acetaldehyde (CH_3CHO), with hydrogen (H_2) as a by-product. In an existing process, 100 mol/min liquid ethanol at 25°C and 1 atm pressure is first heated to 300°C in a heat exchanger, and then fed to the reactor. One hundred percent of the ethanol is converted to products, and the product stream leaves the reactor at 300°C and 1 atm (760 mmHg). The product stream leaving the reactor is cooled to −15°C, and sent to a flash drum, where vapor and liquid streams are separated (Problem Figure 10.3.1).

How much heat must be supplied to the first heat exchanger? (6400 kJ/min)

How much heat must be supplied to or removed from (state which one) the reactor in order to maintain a constant temperature of 300°C? (7100 kJ/min)

What are the flow rates of the vapor and liquid streams leaving the flash drum? (126 mol/min, 74 mol/min)

Species	T_b °C	$\Delta H_{vap}(T_b)$ kJ/gmol	C_p(liquid) kJ/gmol °C	C_p(gas) kJ/gmol °C	$\Delta H_f^0(25\,°C)$ kJ/gmol
H_2	−252.76	0.904	—	0.029	0 (g)
CH_3CHO	20.2	25.1	0.089	0.055	−166.2 (g)
C_2H_5OH	78.5	38.58	0.158	0.077	−277.63 (L)
					−235.31 (g)

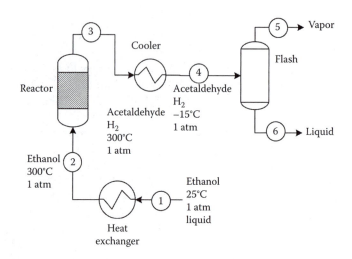

PROBLEM FIGURE 10.3.1
Ethanol (C_2H_5OH) dehydrogenation process.

Antoine equation:

$$\text{Ethanol:} \quad \log_{10}P_{sat}\left(mmHg\right) = 8.04494 - \frac{1554.3}{222.65 + T(^\circ C)}$$

$$\text{Acetaldehyde:} \quad \log_{10}P_{sat}\left(mmHg\right) = 6.81089 - \frac{992}{230 + T \ (^\circ C)}$$

10.4 A gas contains the following compounds: CH_4, C_2H_6, O_2, CO_2, CO, H_2O, and H_2. What is the maximum number of independent chemical reactions that can be written involving these compounds? (Four independent equations)

10.5 Cumene (C_9H_{12}) is synthesized from propylene (C_3H_6) and benzene (C_6H_6). Unfortunately, a side reaction also occurs, in which diisopropylbenzene ($C_{12}H_{18}$) is generated by reaction of propylene with cumene. The two balanced reactions are

$$C_3H_6 + C_6H_6 \rightarrow C_9H_{12}$$

$$C_3H_6 + C_9H_{12} \rightarrow C_{12}H_{18}$$

A block flow diagram for the cumene manufacturing process is shown in Problem Figure 10.5.1. A quantity of 100 kmol/h of a gas containing 95 mol% Propylene and 5 mol% Inert is mixed with 80 kmol/h benzene plus a recycle stream. The mixer outlet is fed to a reactor. The fractional conversions achieved in the reactor based on the reactor feed stream are: the fractional conversion of benzene is 0.9 and the

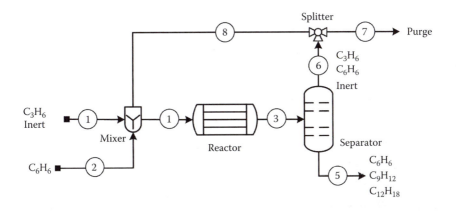

PROBLEM FIGURE 10.5.1
Cumene (C_9H_{12}) synthesis process.

fractional conversion of propylene is 0.7. The reactor outlet is sent to a separator, where all of the propylene, all of the Inert, and 10% of the benzene are recovered in stream 6, and the remaining benzene and all of the cumene and di-isopropyl benzene are recovered as bottom product. Stream 6 is sent to a splitter. Eighty-five percent of the splitter feed is recycled to the mixer and the remainder is purged. Calculate the following:

(a) Flow rates of propylene and benzene (kmol/h) in stream 3. (80.7 kmol/h)

(b) Mol% Inert in purge stream. (46%)

(c) Selectivity for converting benzene to cumene achieved by the overall process. (0.77)

10.6 Ammonia is synthesized through the reaction of nitrogen with hydrogen as follows:

$$N_2 + 3H_2 \rightarrow 2NH_3$$

Problem Figure 10.6.1 shows a process flow sheet for the ammonia synthesis process. In this process, the fresh feed consists of argon (1 mol%) and stoichiometric amounts of N_2 and H_2. The reactor feed has a molar flow rate of 100 mol/min and a composition of 15 mol% argon, 21.25 mol% N_2, and 63.75 mol% H_2. The reactor feed temperature is 400°C. The fractional conversion of N_2 to NH_3 in the reactor is 0.15 mol N_2 reacted/mol N_2 feed to the reactor. The hot reactor effluent gas is used to heat the recycle gas from the separator in a combined reactor effluent/recycle heat exchanger. After passing through this heat exchanger, the reactor effluent gas passes through a condenser where the NH_3 product is condensed. The liquid NH_3 is separated from the noncondensable recycle gases. A purge stream is taken

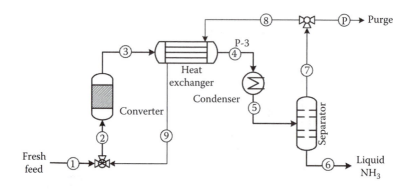

PROBLEM FIGURE 10.6.1
Ammonia synthesis process.

off the separator off-gas to maintain the level of argon at 15 mol% feed to the reactor. In this particular process, the converter is operated adiabatically and the heat of reaction at 400°C was found to be −53.109 kJ/mol at the pressure of the reactor. The following table gives the heat capacities at the pressure of the reactor. Note that the given heat capacities are assumed to be constant over the temperature range found in the reactor:

Compound	C_p (J/mol °C)
NH_3	49.4
H_2	29.5
N_2	31.0
Argon	20.8

Determine the flow rates in moles per minute and compositions in mole percent of

(a) Fresh feed stream. (12.5 mol/min)

(b) Separator purge gas stream. (0.74 mol/min)

(c) Recycle gas stream. (87.5 mol/min)

(d) Estimate the temperature of the effluent gases from the converter. (460.776°C)

(e) If the recycle gas stream enters the heat exchanger at 50°C and leaves the exchanger at 400°C, determine the outlet temperature of the reactor effluent stream from the heat exchanger. Assume no condensation of ammonia in the heat exchanger. (147.271°C)

10.7 Toluene (225 kmol/h) is fed to a reactor to produce benzene. In this process (Problem Figure 10.7.1), toluene reacts with a fractional conversion of 0.80, resulting in benzene and xylene yields of 0.505 and 0.495, respectively:

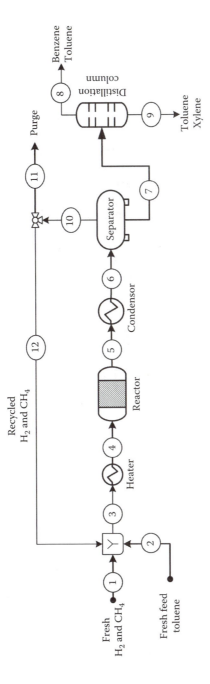

PROBLEM FIGURE 10.7.1

Process flow sheet of benzene production process.

$$2C_7H_8 \rightarrow C_6H_6 + C_8H_{10}$$

Toluene also may dealkylate in the reactor to form benzene and methane:

$$C_7H_8 + H_2 \rightarrow C_6H_6 + CH_4$$

Yields are defined as moles of product/moles of toluene reacted. In this process, the reactor effluent is condensed and separated. The separator liquid is sent to a benzene distillation column where benzene with a purity of 99.5 mol% (balance toluene) is removed as the overhead product. The bottom products contain benzene, unreacted toluene, and xylene. A purge stream containing 90% hydrogen and 10% methane is taken off the separator gas stream. The rest of this separator gas stream is combined with a fresh hydrogen stream containing hydrogen and methane with concentrations of 95.0 and 5.0 mol%, respectively. These combined hydrogen streams are added to the fresh toluene feed and sent to the reactor heater. If the liquid toluene fresh feed stream, fresh hydrogen stream, and recycled streams are all at 25°C and 15 bar absolute, determine the heat requirements for the reactor heater to provide the reactor with a combined feed at 400°C and 15 bar absolute in the vapor phase.

10.8 The dehydrogenation of propane is carried out in a continuous reactor. Pure propane is fed to the reactor at 1300°C and at a rate of 100 mol/s. Heat is supplied at a rate of 1.34 kW. If the product temperature is 1000°C, calculate the extent of reaction. ($\xi = 26$)

10.9 A fresh feed stream contains 5% inert (propane) and 95% propylene. The fresh feed of propylene and inert (propane) is mixed with 210 mol/h carbon dioxide and same amount of hydrogen. At your reactor conditions, propane (I) is an inert, and it is too expensive to separate propane from propylene, so you decide to install a purge stream. The single-pass conversion of propylene in the reactor is 0.3. The production rate of butanol is 180 kmol/h. An overall conversion of 0.90 can be achieved. The purge stream is necessary to avoid inert accumulation in the process. Calculate the flow rate of the contaminated propylene stream to the process (Problem Figure 10.9.1). (210.5 kmol/h)

10.10 Steam flowing at a mass flow rate of 1500 kg/h, a pressure of 20 bar, and 350°C is fed to a turbine that operates adiabatically and at steady state. The steam leaves the turbine at 1.0 bar and 150°C and is cooled in a heat exchanger to a saturated liquid. Draw and label the process flow diagram.

How much work (kJ/h) is extracted in the turbine? (541,650 kW)

How much heat (kJ/h) is removed in the heat exchanger? (-3.54×10^6 kJ/h)

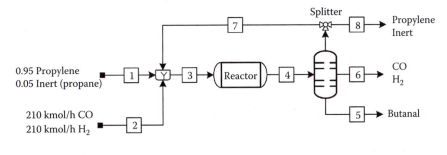

PROBLEM FIGURE 10.9.1
Butanal production process.

References

1. Reklaitis, G.V. (1983) *Introduction to Material and Energy Balances*, John Wiley & Sons, New York.
2. Himmelblau, D.M. (1996) *Basic Principles and Calculations in Chemical Engineering*, 6th edn., Prentice-Hall, Upper Saddle River, NJ.
3. Whirwell, J.C. and R.K. Toner (1969) *Conservation of Mass and Energy*, Blaisdell, Waltham, MA.
4. Felder, R.M. and R.W. Rousseau (1999) *Elementary Principles of Chemical Processes*, 3rd edn., John Wiley & Sons, New York.

11

Unsteady-State Material and Energy Balances

This chapter focuses briefly on unsteady-state processes in which the value of the state, dependent variable, as a function of time is of interest. The term "unsteady state" refers to processes in which quantities or operating conditions within the system change with time. The word transient state applies to such processes. A wide variety of important industrial problems fall into this category, such as start-up/shut-down of process equipment, batch processing, the change from one set of operating conditions to another, and the perturbations that develop as process conditions fluctuate. The following items outline the principal learning objectives of this chapter.

Learning Objectives

1. Develop unsteady-state material balance equations and solve simultaneous first-order ordinary differential material balance equations using MATLAB®/Simulink® software package (Section 11.1).

2. Develop unsteady-state energy balance equations and explain the rational changes in concentration or temperature versus time (Section 11.2).

11.1 Unsteady-State Material Balance

Unsteady or transient state refers to processes in which quantities or operating conditions within the system change with time [1–3]. For such processes, the accumulation term in the mass balance equation cannot be neglected and must be accounted for (Figure 11.1).

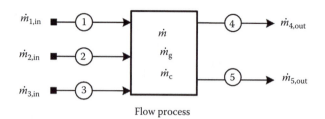

Flow process

FIGURE 11.1
Open system with multiple input and output streams.

The general material balance equation takes the following form:

$$\frac{dm}{dt} = \dot{m}_{in} - \dot{m}_{out} + \dot{m}_g - \dot{m}_c \tag{11.1}$$

where
m is the mass accumulated in the system [mass]
$\dot{m}_{in}$ is the inlet mass flow rate [mass/time]
$\dot{m}_{out}$ is the outlet mass flow rate [mass/time]
$\dot{m}_g$ is the generated mass flow rate [mass/time]
$\dot{m}_c$ is the consumed mass flow rate [mass/time]

$$\dot{m}_{in} = \sum_{in} \dot{m}_{i,in}$$

$$\dot{m}_{out} = \sum_{out} \dot{m}_{i,out}$$

Example 11.1 Filling Controlled Level Storage Tank

Problem
A storage tank that is 2.0 m in diameter is filled at a rate of 2.0 m³/min. When the height of the liquid is 2 m in the tank, a control valve installed on the exit stream at the bottom of the tank opens up, and the fluid flows at a rate proportional to the head of the fluid, that is, 0.4h m³/min, where h is the height of fluid in meters. Plot the height of the liquid as a function of time. What is the steady-state height of the fluid in the tank?

Solution

Known quantities: Inlet and exit tank volumetric flow rate.

Find: Plot the height of the tank as a function of time.

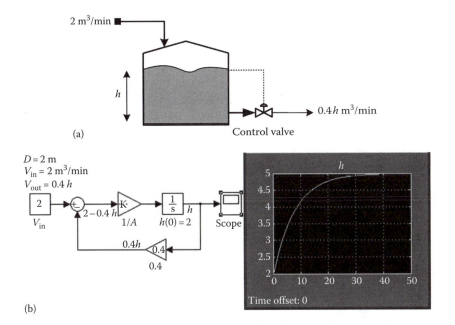

EXAMPLE FIGURE 11.1.1
(a) Schematic of a storage tank. (b) Solution using Simulink.

Analysis: The tank flowchart is shown in Example Figure 11.1a.1.
Unsteady-state mass balance:
Neither generation nor consumption occurs in the process:

$$\frac{dm}{dt} = \dot{m}_{in} - \dot{m}_{out}$$

$$m = \rho V = \rho(Ah), \quad A = \pi D^2/4$$

where
 A is the cross-sectional area of the tank
 V is the volume of the system
 $\dot{V}$ is the volumetric flow rate

$$\dot{m} = \rho \times \dot{V}$$

$\dot{V}_{in}$ and $\dot{V}_{out}$ are the volumetric flow rates of inlet and exit streams, respectively. Simplifying the material balance equations in terms of one variable (h) with time gives

$$\frac{d(\rho Ah)}{dt} = \rho \dot{V}_{in} - \rho \dot{V}_{out}$$

Assume density is constant; the equation is then simplified to

$$A\frac{dh}{dt} = \dot{V}_{in} - \dot{V}_{out}$$

where
$\dot{V}_{in}$ = constant = 2 m³/min
$\dot{V}_{out}$ is a function of the height of fluid in the tank, which is $0.4h$

The set of generated material balance equations is solved using MATLAB/Simulink as shown in Example Figure 11.1b.1. It can be seen that the height of the tank increases sharply with time at the very beginning of the process and slowly with time later on until it reaches the steady state, which is at approximately 5 m. The steady-state height is calculated by setting the differential term to zero. The height is found to be approximately 5 m.

Example 11.2 Dilution of a Salt Solution

Problem
A tank holds 100 L of a salt–water solution in which 5.0 kg of salt is dissolved. Water runs into the tank at a rate of 5 L/min, and salt solution overflows at the same rate. Plot the concentration of the salt versus time. How much salt is in the tank at the end of 10 min?

Solution

Known quantities: Volume, mass, and inlet flow rate are known.

Find: The amount of salt in the tank at the end of 10 min.

Analysis: Assume that the solution in the tank is well mixed and the density of the salt solution is essentially the same as that of pure water. The process flow sheet is shown in Example Figure 11.2a.1.
  The general unsteady-state mass balance is

$$\frac{dm}{dt} = \dot{m}_{in} - \dot{m}_{out} + \dot{m}_g - \dot{m}_c$$

Since there are no reactions, generation and consumption terms are dropped and the general material balance equation is reduced to

$$\frac{dm}{dt} = \dot{m}_{in} - \dot{m}_{out}$$

The inlet mass flow rate as a function of salt concentration is

$$\dot{m}_{in} = \dot{V}_{in}C_{in}[=]\frac{L}{min}\left|\frac{kg}{L}\right| = \frac{kg}{min}$$

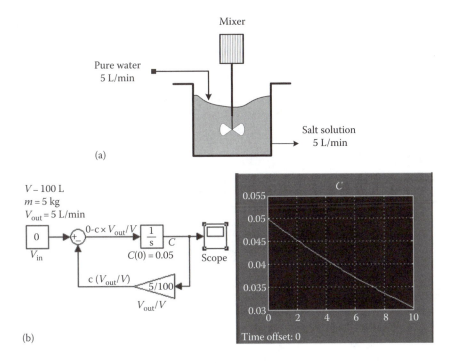

EXAMPLE FIGURE 11.2.1
(a) Dilution of salt–water solution. (b) Solution of a salt concentration tank.

The outlet mass flow rate is

$$\dot{m}_{out} = \dot{V}_{out} C_{out} [=] \frac{L}{min} \left| \frac{kg}{L} \right| = \frac{kg}{min}$$

Accumulated mass (note that, in the accumulated mass, V is the volume of the fluid in the tank) is

$$\frac{d(m)}{dt} = \frac{d(VC)}{dt} [=] \frac{L \frac{kg}{L}}{min} = \frac{kg}{min}$$

Substitution of these terms in the simplified material balance equation yields the following equation:

$$\frac{d(VC)}{dt} = \left(\dot{V}_{in} C_{in} \right) - \left(\dot{V}_{out} C_{out} \right)$$

Assuming the tank is well mixed, the outlet salt concentration equals the concentration in the tank:

$$\frac{d(VC)}{dt} = \left(\dot{V}_{in} C_{in} \right) - \left(\dot{V}_{out} C \right)$$

The inlet is pure water, which means inlet salt concentration is zero $\Rightarrow C_{in} = 0$.

Inlet and outlet volumetric flow rates are equal $\Rightarrow \dot{V}_{in} = \dot{V}_{out}$.

Volume of fluid in the tank is constant $\Rightarrow V = $ constant:

$$V \frac{d(C)}{dt} = 0 - \dot{V}_{out} C$$

Solving the resulting differential equation requires an initial condition, which is the concentration of salt in the tank at time zero:

$$C|_{at\ t=0} = \frac{5\ kg}{100\ L} = 0.05\ kg/L$$

Integrating the developed ordinary differential equation and solving for concentration at time equal to 10 min will give the analytical solution as follows:

$$\frac{dC}{C} = -\frac{\dot{V}_{out}}{V} dt$$

Integrating,

$$\ln \frac{C}{C_o} = -\frac{\dot{V}_{out}}{V} t$$

Substituting known quantities,

$$\ln \frac{C}{0.05} = -\frac{5\ L/min}{1\ L} \times 10\ min$$

Rearranging,

$$C = 0.05 \exp\left(-\frac{5\ L/min}{100\ L} \times 10\ min\right) = 0.03\ kg/L$$

Example Figure 11.2b.1 reveals that the salt concentration in the tank decreases with time. At the end of 10 min the salt concentration in the tank is 0.03 kg/L.

Example 11.3 Dilution of Salt Solution

Problem

The average ocean water of salinity 35 ppt flows into a 100 L tank containing 1.5 kg salt at a rate of 5 L/min. The salt solution overflows out of the tank at 5 L/min. How much salt remains in the tank at the end of 15 min?

Solution

Known quantities: Water concentration, exit flow rate are known.

Find: The amount of salt in the tank at the end of 15 min.

Analysis: The dilution process flow sheet is shown in Example Figure 11.3a.1. Assume the fluid in the tank is well mixed and the density of salt solution is constant and equal to that of water. If we have 1 g of salt and 1000 g of water, the salinity is 1 ppt.

The general unsteady-state material balance is

$$\frac{dm}{dt} = \dot{m}_{in} - \dot{m}_{out} + \dot{m}_g - \dot{m}_c$$

Since there are no reactions, generation and consumption terms are dropped and the general material balance equation is reduced to the following form:

$$\frac{dm}{dt} = \dot{m}_{in} - \dot{m}_{out}$$

Replacing mass flow rates in terms of concentration, as done in Example 11.2, yields

$$\frac{d(VC)}{dt} = \left(\dot{V}_{in} C_{in} \right) - \left(\dot{V}_{out} C_{out} \right)$$

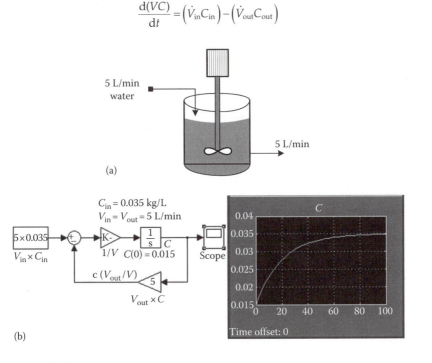

(a)

(b)

EXAMPLE FIGURE 11.3.1

(a) Schematic of a salt dilution tank. (b) Simulink solution of Example 11.3.

Assuming the solution in the tank is well mixed, the outlet salt concentration equals the concentration in the tank; this assumption reduces the earlier equation to the following form:

$$\frac{d(VC)}{dt} = \left(\dot{V}_{in}C_{in}\right) - \left(\dot{V}_{out}C\right)$$

Inlet and outlet volumetric flow rates are equal $\Rightarrow \dot{V}_{in} = \dot{V}_{out}$.
Volume of fluid in the tank is constant $\Rightarrow V = $ constant:

$$V\frac{dC}{dt} = \dot{V}_{in}C_{in} - \dot{V}_{out}C$$

Solving the resultant differential equation requires an initial condition, which is the concentration of salt in the tank at time zero:

$$C\big|_{at\,t=0} = \frac{1.5\,\text{kg}}{100\,\text{L}} = 0.015\,\text{kg/L}$$

The ocean salt concentration in kilograms per liter is

$$C_{in} = 35\,\text{ppt}\left|\frac{\frac{1\,\text{g salt}}{1000\,\text{g water}}}{1\,\text{ppt}}\right|\frac{1000\,\text{g water}}{\text{L}}\left|\frac{1\,\text{kg}}{1000\,\text{g}}\right. = 0.035\,\text{kg/L}$$

Results are obtained using Simulink as shown in Example Figure 11.3b.1. The concentration of salt in the tank after 15 min (90 s) is approximately 0.035 kg/L. Since the volume of the tank is 100 L, accordingly, the amount of salt is 3.5 kg.

Example 11.4 Sewage Treatment

Problem
In a sewage treatment plant, a large concrete tank initially contains 440,000 L liquid and 10,000 kg fine suspended solids. To flush this material out of the tank, water is pumped into the vessel at a rate of 40,000 L/h, and liquid containing solids leave at the same rate. Estimate the concentration of suspended solids in the tank at the end of 4 h.

Solution

Known quantities: Tank volume, initial fine solid concentration, flow rates.

Find: The concentration of suspended solids in the tank at the end of 4 h.

Analysis: The process flow diagram is shown in Example Figure 11.4a.1. The general unsteady-state material balance equation is

$$\frac{dm}{dt} = \dot{m}_{in} - \dot{m}_{out} + \dot{m}_{g} - \dot{m}_{c}$$

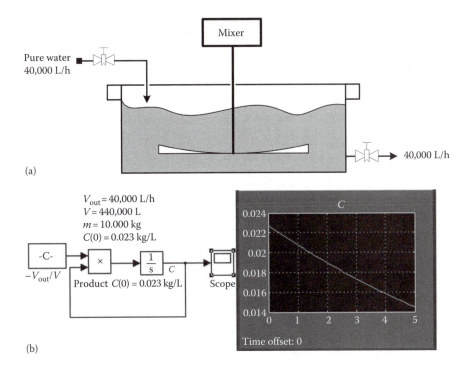

EXAMPLE FIGURE 11.4.1
(a) Sewage treatment tank. (b) Simulink solution of a sewage treatment tank.

Since there are no reactions, generation and consumption terms are dropped, and the general material balance equation is reduced to the following form:

$$\frac{dm}{dt} = \dot{m}_{in} - \dot{m}_{out}$$

Replacing mass flow rates in terms of concentration, as done in Example 11.3, yields

$$\frac{d(VC)}{dt} = \left(\dot{V}_{in}C_{in}\right) - \left(\dot{V}_{out}C_{out}\right)$$

Assuming the solution in the tank is well mixed, the outlet suspended solid concentration equals the concentration in the tank; this assumption reduces the preceding equation to the following form:

$$\frac{d(VC)}{dt} = \left(\dot{V}_{in}C_{in}\right) - \left(\dot{V}_{out}C\right)$$

Inlet and outlet volumetric flow rates are equal $\Rightarrow \dot{V}_{in} = \dot{V}_{out} = 40,000$ L/h. Volume of fluid in the tank is constant $\Rightarrow V = $ constant $= 440,000$ L.

Pure water is pumped into the vessel at a rate of 40,000 L/h; concentration of solids in the inlet pure water is zero, $C_{in} = 0$, accordingly, solid component balance is

$$V\frac{d(C)}{dt} = 0 - \left(\dot{V}_{out}C\right)$$

Solving the resultant differential equation requires an initial condition, which is the concentration of solids in the sewage tank at time zero:

$$C|_{at\,t=0} = \frac{10,000\,kg}{440,000\,L} = 0.023\,kg/L$$

The analytical solution shows the following results:

$$\frac{d(C)}{C} = -\frac{\dot{V}_{out}}{V}, \quad \ln\frac{C_f}{C_o} = -\frac{\dot{V}_{out}}{V}t \Rightarrow C_f = C_o\exp\left(-\frac{\dot{V}_{out}}{V}t\right)$$

Substitute known quantities:

$$C_f = 0.023\,\frac{kg}{L}\exp\left(-\frac{40,000\,L/h}{440,000\,L}\,4h\right) = 0.016\,kg/L$$

Solving the equation numerically using Simulink the plot of concentration of suspended solids versus time is shown in Example Figure 11.4b.1. The concentration of suspended solids in the tank after 4 h of operation is approximately 0.016 kg/L, which is the same as the values of the concentration obtained analytically.

Example 11.5 Diffusion of a Solid into a Liquid

Problem
A compound dissolves in water at a rate proportional to the product of the amount of undissolved solid and the difference between the concentration in a saturated solution and the actual solution, that is, $C_{sat} - C(t)$. The dissolution rate is 0.257 h^{-1}. A saturated solution of this compound contains 0.4 g solid/g water. In a test run starting with 20 kg of undissolved compound in 100 kg of water, how many kilograms of compound will remain undissolved after 10 h? Assume that the system is isothermal.

Solution

Known quantities: The dissolution rate is 0.257 h^{-1}. A saturated solution of this compound contains 0.2 g solid/g water.

Find: The kilograms of solid compound that will remain undissolved after 10 h.

Analysis: Example Figure 11.5a.1 is a schematic of the process of diffusion of solids in water. Let us assign m for the mass of the undissolved

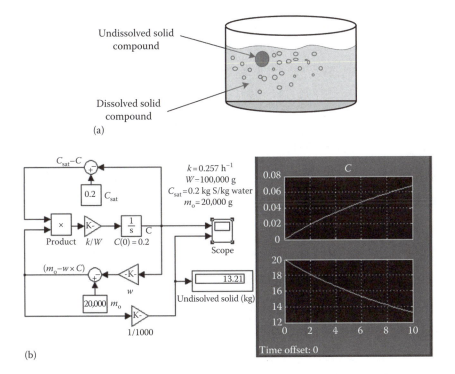

EXAMPLE FIGURE 11.5.1
(a) Diffusion of solid into water. (b) Simulink solution of solid dissolved into a water tank.

compound at any time, m_0 is the initial mass of the undissolved compound at time zero, and C is the concentration of the dissolved compound in water.

General material balance on the undissolved compound is

$$\frac{dm}{dt} = \dot{m}_{in} - \dot{m}_{out} + \dot{m}_g - \dot{m}_c$$

There is no inlet or outlet mass flow rate to the tank: $\dot{m}_{in} = \dot{m}_{out} = 0$.

There is no generation of the undissolved solid: $\dot{m}_g = 0$.

The rate of consumption: $m_c = k_m(C_{sat} - C)$

Rearranging the general material balance equation leads to the following equation:

$$\frac{dm}{dt} = -k_m(C_{sat} - C)$$

Mass of solids in the tank as a function of concentration is

$$C = \frac{m_0 - m}{W} \Rightarrow m = m_0 - C \times W$$

$$C = \frac{m_0}{W} - \frac{1}{W}m$$

Differentiation of this relation leads to

$$\frac{dC}{dt} = -\frac{1}{W}\frac{dm}{dt}$$

Rearranging so as to replace mass by solid concentration yields

$$\frac{dm}{dt} = -W\frac{dC}{dt}$$

$$\frac{dC}{dt} = \frac{k_m}{W}(C_{sat} - C)$$

Substituting concentration instead of mass of undissolved compound,

$$\frac{dC}{dt} = \frac{k_m(m_0 - C \cdot W)}{W}(C_{sat} - C)$$

With the following data, solve the earlier equations using Simulink:

$$k = 0.257\,h^{-1}$$

$$W = 100\,\text{kg}\left|\frac{1000\,\text{g}}{1\,\text{kg}}\right| = 100,000\,\text{g}$$

$$C_{sat} = 0.2\frac{\text{g solid}}{\text{g water}}$$

Initial undissolved solid: $m_0 = 20\,\text{kg}\left|\frac{1000\,\text{g}}{1\,\text{kg}}\right| = 20,000\,\text{g}$

The concentration of the dissolved solid in liquid water is shown in Example Figure 11.5b.1. The diagram reveals the solid concentration is increasing with time. The amount of undissolved solid after 10 h is around 13.21 kg (13,210 g).

11.2 Unsteady-State Energy Balance

Consider the mixing tank shown in Figure 11.2 where heat is added or removed from the tank through jacketed inlet and exit streams [4,5]. The general form of the energy balance under unsteady-state condition takes the form

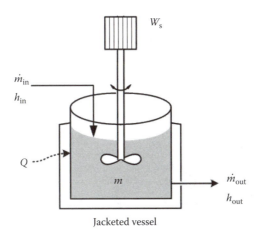

Jacketed vessel

FIGURE 11.2
Energy balance on unsteady state process.

$$\frac{d(U)}{dt} = \sum \dot{m}_{in} h_{in} - \sum \dot{m}_{out} h_{out} + \dot{Q} - \dot{W}_s \qquad (11.2)$$

where
    U is the internal energy; $U = mC_vT$
    h_{in} is the specific enthalpy of the inlet stream
    h_{out} is the specific enthalpy of the exit stream
    Q is the heat added to the system (+). If the heat is lost or transferred from the system to the surroundings, then the sign is negative.
    W_s is the work done by the system on the surroundings (+). If the work is done on the system, then the sign is negative (−).

For liquids and solids $C_v \approx C_p$ because $\dfrac{d(PV)}{dt} \approx 0$:

$$H = U + PV$$

$$\frac{dH}{dt} = \frac{dU}{dt} + \frac{d(PV)}{dt}$$

This leads to $\dfrac{dH}{dt} = \dfrac{dU}{dt}$

Example 11.6 Heating of a Closed System

Problem
A kettle used to boil water containing 1.00 L of water at 20°C is placed on an electric heater ($Q = 2200$ J/s). Find out the time at which water begins to boil.

Solution

Known quantities: Kettle volume, initial water temperature, heat supplied.

Find: The time at which water begins to boil.

Analysis: The tank flowchart is shown in Example Figure 11.1a.1. The normal boiling water temperature is 100°C. The kettle heater diagram is shown in Example Figure 11.6a.1.

The general energy balance equation is

$$\frac{d(U)}{dt} = \Sigma \dot{m}_{in} h_{in} - \Sigma \dot{m}_{out} h_{out} + \dot{Q} - \dot{W}_s$$

The kettle is batch; inlet and outlet mass flow rates are zero; $\dot{m}_{in} = \dot{m}_{out} = 0$.
No shaft work: $W_s = 0$
For liquids: $C_p \approx C_v$
Specific heat of water: $C_p = 4.18$ J/g°C

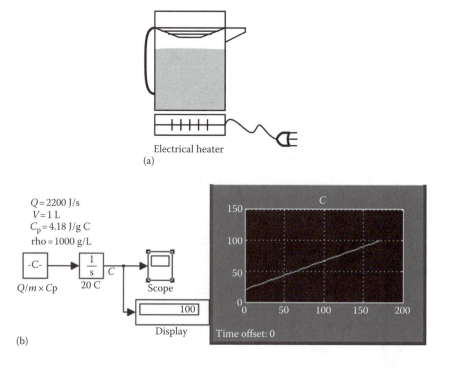

Electrical heater
(a)

(b)

EXAMPLE FIGURE 11.6.1

(a) Schematic of an electrical heater kettle. (b) Simulink solution of the electrical heater.

The general energy balance equation is simplified to the following equation:

$$\frac{d(mC_pT)}{dt} = \dot{Q}$$

The mass of liquid water inside the kettle is

$$m = \rho V$$

Substituting m in the simplified energy balance equation gives

$$\frac{d\left(mC_pT\right)}{dt} = \dot{Q}$$

Rearranging,

$$\frac{dT}{dt} = \frac{\dot{Q}}{\rho V C_p}$$

Solve this ordinary differential equation using the following data:
   Heat added to the kettle: $\dot{Q} = 2200$ J/s
   Specific heat of water: $C_p = 4.18$ J/g°C
   Volume of the kettle: $V = 1.0$ L
   Density of the water in the kettle: $\rho = 1000$ g/L
   The analytical solution is obtained by integrating the first-order energy balance equation:

$$T_2 - 20 = \frac{\dot{Q}}{\rho V C_p} t$$

Since water boils at 100°C, accordingly $T_2 = 100$. Substitute known values and solve for time:

$$200 - 20 = \frac{2200 \, \text{J/s}}{\dfrac{1000 \, \text{g}}{\text{L}} \times 1 \, \text{L} \, 4.18 \, \text{J/g C)}} t$$

$$\frac{80 \times 1000 \times 4.18}{2200} s = t$$

$t = $ time $= 152$ s (2.53 min)
   Results obtained are as shown in Example Figure 11.6b.1. The time required to heat the water to its boiling temperature in the mentioned kettle is 171.5 s (2.86 min).

Example 11.7 Heating of a Stirred Tank

Problem

Oil at 20°C is being heated in a stirred tank. Oil enters the tank at a rate of 500 kg/h at 20°C and leaves at temperature T. The tank holds 2300 kg of oil, which is initially at 20°C. The heat is provided by steam condensing at 130°C in coils submerged in the tank. The rate of heat transfer is given by

$$Q = h(T_{steam} - T_{oil})$$

The heat capacity of the oil is given by C_p = 2.1 J/(g °C) and the heat transfer coefficient is h = 115 J/s °C. The shaft work of the stirrer is 560 W. Once the process is started, how long does it take before the oil leaving the tank is at 30°C?

Solution

Known quantities: Inlet and exit tank volumetric flow rate.

Find: Plot the height of the tank as a function of time.

Analysis: The tank flowchart is shown in Example Figure 11.1a.1. The general energy balance equation is

$$\frac{d(u)}{dt} = \sum \dot{m}_{in} h_{in} - \sum \dot{m}_{out} h_{out} + \dot{Q} - \dot{W}_s$$

Reference: Inlet oil temperature (i.e., $T_{ref} = T_0 = 20°C$).

System: Oil in the tank (Example Figure 11.7a.1).
   For liquids, C_p and C_v are approximately equal.
   Specific heat of oil: C_p = 2.1 J/g °C.
   The shaft work of the stirrer is 560 W.
   The work done on the system W_s = −560 W = −560 J/s.
   The heat added to the oil from the steam is $Q = h (T_{steam} - T)$.
   The heat transfer coefficient is h = 115 J/(s°C).

$$\text{Mass of oil in the tank } m = 2300 \text{ kg} \left| \frac{1000 \text{ g}}{1 \text{ kg}} \right. = 2.3 \times 10^6 \text{ g}$$

$$\text{Inlet and exit mass flow rates: } \dot{m} = \frac{500 \text{ kg}}{h} \left| \frac{1000 \text{ g}}{1 \text{ kg}} \right| \frac{h}{3600 \text{ s}} = 139 \text{ g/s}$$

The general energy balance equation is simplified to the following equation:

$$\frac{d(mC_p T)}{dt} = \sum \dot{m}_{in} \hat{H}_{in} - \sum \dot{m}_{out} \hat{H}_{out} + \dot{Q} - \dot{W}_s$$

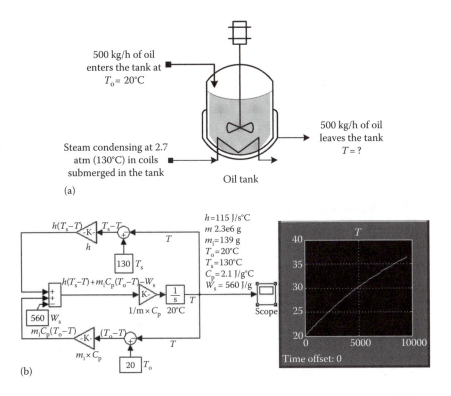

EXAMPLE FIGURE 11.7.1
(a) Schematic of an oil heating tank. (b) Simulink solution of a heating tank with flow.

Rearranging the equation in order to collect and separate variables leads to

$$mC_p \frac{d(T)}{dt} = 0 - \dot{m}C_p(T - T_0) + h(T_{\text{steam}} - T) - \dot{W}_s$$

$$\frac{dT}{dt} = \frac{[0 - \dot{m}C_p(T - T_0) + h(T_{\text{steam}} - T) - \dot{W}_s]}{mC_p}$$

This differential equation is solved using Simulink (Example Figure 11.7b.1). The result reveals that the temperature of oil reaches 35°C in approximately 123 min.

Example 11.8 Quenching of an Iron Bar

Problem

The volume of a cubic iron bar is 60 cm³ at a temperature of 95°C, the heat transfer area of the block is 112 cm². The iron bar is dropped into a barrel of water at 25°C. Density of the iron bar is 11.34 g/cm³. The barrel is large enough so that the water temperature rise is negligible as the bar cools down. The rate at which heat is transferred from the bar to the water is given by the expression

$$Q(J/min) = UA(T - T_w)$$

where U is the heat transfer coefficient, which is 0.050 J/(min cm^2 °C). The heat capacity of the iron bar is 0.460 J/(g °C). Plot the temperature of the bar as a function of time, and calculate the time for the bar to cool to 30°C.

Solution

Known quantities: Inlet and exit tank volumetric flow rate.

Find: Plot the height of the tank as a function of time.

Analysis: The tank flowchart is shown in Example Figure 11.1a.1. Assume the heat conduction in iron is rapid enough for the temperature of the bar to be uniform throughout. This latter concept is an important approximation called lumped capacitance, and it allows us to considerably simplify the problem because we do not have to worry about heat transfer within the solid bar itself. Assume also temperature of water to remain constant.

System: Iron block.
  Reference temperature: 25°C
  Schematic of the process is shown in Example Figure 11.8a.1.
  The general energy balance equation is

$$\frac{d(mC_pT)}{dt} = \sum \dot{m}_{in}\hat{H}_{in} - \sum \dot{m}_{out}\hat{H}_{out} + \dot{Q} - \dot{W}_s$$

Note that for solids and liquids C_p and C_v are equal.

No inlet or exit flowing streams: $\sum \dot{m}_{in}h_{in} = \sum \dot{m}_{out}h_{out} = 0$.

The barrel is large enough so that the water temperature rise is negligible as the bar cools; this means the water temperature remains constant, T_w = constant.
  No stirrer or shaft work: $W_s = 0$.
  The heat transfer from the iron block: $Q(J/s) = -UA(T - T_w)$

Heat transfer coefficient, $U = \dfrac{0.050\ J}{min \cdot cm^2 \cdot °C}\left|\dfrac{1\ min}{60\ s}\right. = 8.33 \times 10^{-4}\ \dfrac{J}{s \cdot cm^2 \cdot °C}$.

The heat capacity of the bar: $C_{pb} = 0.460$ J/(g °C).
  Heat conduction in iron is rapid enough for the temperature of the bar to be uniform throughout.
  Density of the iron bar: $\rho_b = 11.34$ g/cm^3
  Mass of the block: $m = \rho_b V_b$
  Heat transfer area of the block: $A = 112$ cm^2

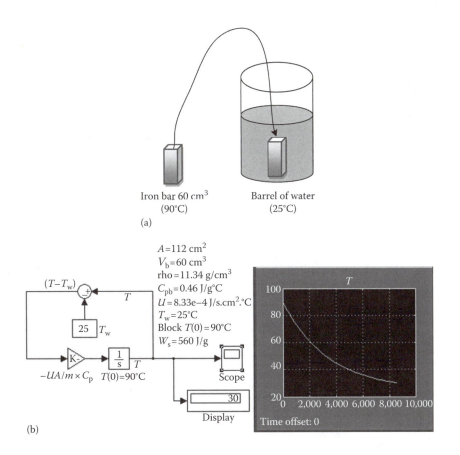

EXAMPLE FIGURE 11.8.1
(a) Schematic of cooling of iron bar block in a large tank of water. (b) Simulink solution of a quenching tank.

We now rearrange the equation to obtain it in terms of the two variables T and t (T dependent variable, t independent variable):

$$\frac{dT}{dt} = \frac{\dot{Q}}{mC_p}$$

Substituting the heat transferred from the block (i.e., $Q = -UA(T-T_w)$),

$$\frac{dT}{dt} = \frac{-UA(T-T_w)}{mC_p}$$

Solution of this differential equation using Simulink gives the time required to cool the block to 30°C, which is around 2.39 h as shown in Example Figure 11.8b.1.

Example 11.9 Heating of a Solution

Problem

An electric heating coil is immersed in a stirred tank. The shaft work of the stirrer is 560 W. A solvent at 15°C with a heat capacity of 2.1 J/g °C is fed into the tank at a rate of 15 kg/h. Heated solvent is discharged at the same flow rate. The tank is filled initially with 125 kg cold solvent at 10°C. The rate of heating the electric coil is 2000 W. Calculate the time required for the temperature of the solvent to reach 60°C.

Solution

Known quantities: Inlet and exit flow rate, initial tank temperature.

Find: The time required for the temperature of the solvent to reach 60°C.

Analysis: The tank flowchart is shown in Example Figure 11.1a.1. Assume the tank reference temperature is 15°C. The general energy balance equation is simplified to the following equation:

$$\frac{d(mC_pT)}{dt} = \sum \dot{m}_{in}h_{in} - \sum \dot{m}_{out}h_{out} + \dot{Q} - \dot{W}_s$$

Rearranging the equation in order to collect and separate variables gives

$$mC_p\frac{d(T)}{dt} = 0 - \dot{m}C_p(T - T_0) + \dot{Q} - \dot{W}_s$$

$$\frac{dT}{dt} = \frac{\left[0 - \dot{m}C_p(T - T_0) + \dot{Q} - \dot{W}_s\right]}{mC_p}$$

The schematic diagram is shown in Example Figure 11.9a.1.
Solving the earlier equation necessitates the units to be consistent.

Mass of oil: $m = 125 \text{ kg}\left|\frac{1000 \text{ g}}{1 \text{ kg}}\right. = 1.25 \times 10^5 \text{ g}.$

Inlet and exit oil flow rates: $\dot{m} = \frac{15 \text{ kg}}{h}\left|\frac{1000 \text{ g}}{1 \text{ kg}}\right|\frac{h}{3600 \text{ s}} = 4.17 \text{ g/s}.$

Heat capacity: $C_p = 2.1 \text{ J/g °C}.$
Heat added to the oil from the cooling coil: $Q = 2000 \text{ J/s}.$
Work applied on the system from the stirrer: $W_s = -560 \text{ J/s}.$
The solution of the set of differential and algebraic equations using Simulink is shown in Example Figure 11.9b.1. The time required for temperature to reach 60°C is around 5521 s (1.53 h).

Example 11.10 Heating a Glycol Solution

Problem

An adiabatic stirred tank is used to heat 100 kg of a 45 wt% glycol solution in water (mass heat capacity 3.54 J/g °C). An electrical coil delivers

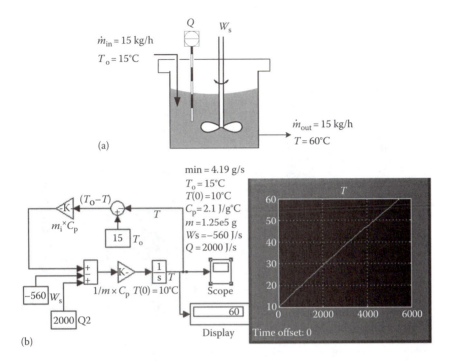

EXAMPLE FIGURE 11.9.1

(a) Schematic of a heating tank. (b) Simulink solution of a heating tank.

2.5 kJ/s of power to the tank; 88% of the energy delivered by the coil goes into heating the vessel contents. The shaft work of the stirrer is 500 W. The glycerol solution is initially at 15°C. How long will the solution take to reach 90°C?

Solution

Known quantities: Mass of glycol in the tank, heat capacity, heat supplied from coil.

Find: The time taken for the glycol solution to reach 90°C.

Analysis: Assume a reference temperature of 15°C. The schematic diagram of the adiabatic stirred tank heater is shown in Figure 11.10a.

The general energy balance equation is simplified to the following equation:

$$\frac{d(u)}{dt} = \sum \dot{m}_{in} h_{in} - \sum \dot{m}_{out} h_{out} + \dot{Q} - \dot{W}_s$$

$$\frac{d(mC_v T)}{dt} = \sum \dot{m}_{in} h_{in} - \sum \dot{m}_{out} h_{out} + \dot{Q} - \dot{W}_s$$

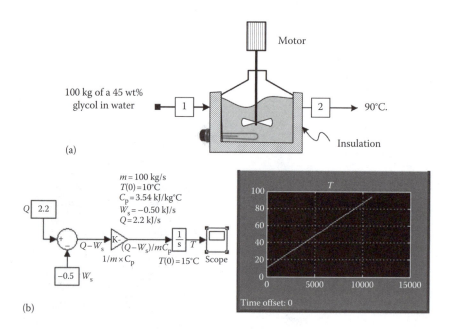

FIGURE 11.10
(a) Schematic of a glycol heating tank. (b) Simulink solution for heating glycol solution.

Note that for solids and liquids C_p and C are equal:

$$\frac{d(mC_pT)}{dt} = \sum \dot{m}_{in}h_{in} - \sum \dot{m}_{out}h_{out} + \dot{Q} - \dot{W}_s$$

The system is adiabatic, that is, no heat is transferred to or from the sur-roundings, but still there is heat added by the electrical coil:

$$\dot{Q}_{net} = \dot{Q}_e + \dot{Q}_{sur}$$

Since the system is adiabatic, $\dot{Q}_{sur} = 0$. Heat added to the glycol solution is 88% of that of the heating coil:

$$\dot{Q}_e = 0.88 \times 2.5 \text{ kJ/s} = 2.2 \text{ kJ/s}$$

The mass of the glycol solution in the heated tank: $m = 100$ kg.
  The heat capacity of the glycol solution: $C_p = 3.54$ J/g °C.
  Rearranging the equation in order to collect and separate variables gives

$$\frac{dT}{dt} = \frac{[\dot{Q} - \dot{W}_s]}{mC_p}$$

Results of this equation solved using Simulink are shown in Example Figure 11.10b.1.

Use the following data:

$$\dot{Q} = \frac{2.2\ kJ}{s}\left|\frac{1000\ J}{kJ}\right. = 2200\ J/s$$

$$W_s = -500\ J/s, \quad m = 100\ kg$$

$$C_p = 3.54\ J/g\ °C$$

As shown in Example Figure 11.10b.1, the solution obtained by Simulink shows that it will take around 10,000 s (2.78 h) for the glycol solution to reach the temperature of 90°C.

Homework Problems

11.1 A storage tank that is 2.0 m in diameter is filled at a rate of 2.0 m³/ min. The exit fluid flow rate is proportional to the head of the fluid (0.5h m³/min), where h is the height of fluid in meters. Plot the height of the liquid as a function of time. What is the steady-state height of the fluid in the tank? (4 m)

11.2 A boiler used to boil water, containing 100 L of water at a temperature of 25°C, is placed on an electric heater (Q = 3000 J/s). Find the time at which water begins to boil. (100 min)

11.3 A tank containing 1000 kg water at 25°C is heated using saturated steam at 130°C. The rate of heat transfer from the steam is given by the following equation:

$$\dot{Q} = UA(T_s - T)$$

$\dot{Q}$ is the rate of heat transfer to the system. U is the overall heat transfer coefficient, A is the surface area for heat transfer, and T is the temperature. The heat transfer area provided by the coil is 0.3 m², and the heat transfer coefficient is 220 (kcal)/m² h°C. The condensate leaves the coil as saturated steam.

a. The tank has a surface area of 0.9 m² exposed to the ambient air. The tank exchanges heat through this exposed surface at a rate given by an equation similar to that given earlier. For heat transfer to or from the surrounding air, the heat transfer coefficient is 25 (kcal)/m² h°C. If the air temperature is 20°C, calculate the time required to heat the water to 80°C. (5.9 h)

b. How much time can be saved if the tank is insulated? (4.2 h)

11.4 A stirred tank is used to heat 100 kg of a solvent (mass heat capacity 2.5 J/g°C). An electrical coil delivers 2.0 kJ/s of power to the tank; the shaft work of the stirrer is 560 W. The solvent is initially at 25°C. The heat lost from the walls of the tank is 200 J/s.

    a. Write a differential equation for the energy balance.
    b. Solve the equation using the available software package.
    c. How long will the solution take to reach 70°C? (1.33 h)

11.5 The following series reaction takes place in a constant volume batch reactor:

$$A \xrightarrow{\ k_1\ } B \xrightarrow{\ k_2\ } C$$

Each reaction is of first order and is irreversible. If the initial concentration of A is 1 mol/L and if only A is present initially, find an expression for the concentrations of A, B, and C as a function of time ($k_1 = 0.1/s$, $k_2 = 0.2/s$).

References

1. Himmelblau, D.M. (1996) *Basic Principles and Calculations in Chemical Engineering*, 6th edn., Prentice-Hall, Upper Saddle River, NJ.
2. Doran, P. (1995) *Bioprocess Engineering Principles*, Academic Press, San Diego, CA.
3. Reklaitis, G.V. (1983) *Introduction to Material and Energy Balances*, John Wiley & Sons, New York.
4. Felder, R.M. and R.W. Rousseau (1999) *Elementary Principles of Chemical Processes*, 3rd edn., John Wiley, New York.
5. Atkinson, B. and F. Mavituna (1991) *Biochemical Engineering and Biotechnology Handbook*, 2nd edn., Macmillan, Basingstoke, UK.

Appendix

Physical properties (Appendix A.1), heat capacities (Appendix A.2), and saturated and superheated steam table (Appendix A.3) are adapted with permission from the following references:

1. Himmelblau, D.M. (1996) *Basic Principles and Calculations in Chemical Engineering*, 6th edn., Prentice-Hall, Upper Saddle River, NJ (Appendices A.1 and A.2).
2. Haywood, R.W. (1968) *Thermodynamic Tables in SI (Metric) Units*, Cambridge University Press, London, U.K. (Appendix A.3).

Appendix A.1

Table A.1 contains physical properties of various organic and inorganic substances such as molecular weight, critical temperature, critical pressure, specific gravity, melting temperature, heat of fusion, boiling temperature, and heat of vaporization. Table A.2 contains standard heats of formation and standard heats of combustion.

TABLE A.1

Physical Properties of Various Organic and Inorganic Substances

No.	Compound	Formula	Molecular Weight	T_c K	P_c atm	Sp. Gr.	T_m K	$\Delta \bar{h}_m$ KJ/mol	T_b K	$\Delta \bar{h}_v$ KJ/mol
1	Acetaldehyde	CH_3CHO	44.05	461.0		0.78	149.5		293.2	
2	Acetic acid	CH_3COOH	60.05	594.8	57.1	1.05	289.9	12.09	390.4	24.4
3	Acetone	C_3H_6O	58.08	508.0	47.0	0.79	178.2	5.69	329.2	30.2
4	Acetylene	C_2H_2	26.04	309.5	61.6		191.7	3.7	191.7	17.5
5	Air					1.00				
6	Ammonia	NH_3	17.03	405.5	111.3	0.81	195.40	5.653	239.73	23.35
7	Ammonium carbonate	$(NH_4)_2CO_3 \cdot H_2O$	114.11							
8	Ammonium chloride	NH_4Cl	53.50			2.53				
9	Ammonium nitrate	NH_4NO_3	80.05			1.73	442.8	5.4		
10	Ammonium sulfate	$(NH_4)_2SO_4$	132.14			1.77	786			
11	Aniline	C_6H_7N	93.12	699	52.4	1.02	266.9		457.4	
12	Benzaldehyde	C_6H_5CHO	106.12			1.05	247.16		452.16	38.40
13	Benzene	C_6H_6	78.11	562.6	48.6	0.88	278.693	9.837	353.26	30.76
14	Benzoic acid	$C_7H_6O_2$	122.12			1.32	395.4		523.0	
15	Benzyl alcohol	C_7H_8O	108.13			1.05	257.8		478.4	
16	Boron oxide	B_2O_3	69.64			1.85	723	22.0		
17	Bromine	Br_2	159.83	584	102	3.12	265.8	10.8	331.78	31.0
18	1,2-Butadiene	C_4H_6	54.09	446		0.65	136.7		283.3	
19	1,3-Butadiene	C_4H_6	54.09	425	42.7	0.62	164.1		268.6	
20	Butane	$n\text{-}C_4H_{10}$	58.12	419.6	39.7	0.58	134.83	4.661	272.66	22.31
21	iso-Butane	$iso\text{-}C_4H_{10}$	58.12			0.56	113.56	4.540	261.43	21.29
22	1-Butene	C_4H_8	56.10	419.6	39.7	0.60	87.81	3.848	266.91	21.92
23	Butyl phthalate	$C_8H_{22}O_4$	278.34			1.05			613	

24	n-Butyric acid	n-C4H8O2	88.10			0.96	267		437.1	
25	iso-Butyric acid	iso-C4H8O2	88.10			0.95	226		427.7	
26	Calcium arsenate	Ca3(AsO4)2	398.06				1723			
27	Calcium carbide	Ca2C2	64.10			2.22	2573			
28	Calcium carbonate	CaCO3	100.09			2.93				
29	Calcium chloride	CaCl2	110.99			2.15	1055	28.4		
		CaCl2·H2O	129.01							
		CaCl2·2H2O	147.03							
		CaCl2·6H2O	219.09			1.78	303.4	37.3		
30	Calcium cyanamide	CaCN2	80.11			2.29				
31	Calcium cyanide	Ca(CN)2	92.12							
32	Calcium hydroxide	Ca(OH)2	74.10			2.24				
33	Calcium oxide	CaO	56.08			2.62	2873	50	3123	
34	Calcium phosphate	Ca3(PO4)2	310.19			3.14	1943			
35	Calcium silicate	CaSiO3	117.17			2.92	1803	48.62		
36	Calcium sulfate	CaSO4·2H2O	172.18			2.32				
37	Carbon	C	12.010			2.26	3873	46.0	4473	
38	Carbon dioxide	CO2	44.01	304.2	72.9		217.0	8.32		
39	Carbon disulfide	CS2	76.14			1.26	161.1	4.39	319.41	26.8
40	Carbon monoxide	CO	28.01	133.0	34.5		68.10	0.837	81.66	6.04
41	Carbon tetrachloride	CCl4	153.84	556.4	45.0	1.60	250.3	2.5	349.9	30.0
42	Chlorine	Cl2	70.91	417.0	76.1		172.16	6.406	239.10	20.41
43	Chlorobenzene	C6H5Cl	112.56	632.4	44.6	1.11	228		405.26	36.5
44	Chloroform	CHCl3	119.39	536.0	54.0	1.49	209.5		334.2	
45	Chromium	Cr	52.01			7.1				
46	Copper	Cu	63.54			8.92	1356.2	13.0	2855	305
47	Cumene	C9H12	120.19			0.86	177.125	7.1	425.56	37.5

(Continued)

TABLE A.1 (*Continued*)

Physical Properties of Various Organic and Inorganic Substances

No.	Compound	Formula	Molecular Weight	T_c K	P_c atm	Sp. Gr.	T_m K	$\Delta \bar{h}_m$ KJ/mol	T_b K	$\Delta \bar{h}_v$ KJ/mol
48	Cupric sulfate	$CuSO_4$	159.61			3.61				
49	Cyclohexane	C_6H_{12}	84.16	553.7	40.4	0.78	279.83	2.677	353.90	30.1
50	Cyclopentane	C_5H_{10}	70.13	511.8	44.55	0.75	179.71	0.6088	322.42	27.30
51	Decane	$C_{10}H_{22}$	142.28	619.0	20.8	0.73	243.3		447.0	
52	Dibutyl phthalate	$C_8H_{22}O_4$	278.34			1.05			613	
53	Diethyl ether	$(C_2H_5)_2O$	74.12	467	35.6	0.71	156.86	7.301	307.76	26.05
54	Ethane	C_2H_6	30.07	305.4	48.2		89.89	2.860	184.53	14.72
55	Ethanol	C_2H_5OH	46.07	516.3	63.0	0.79	158.6	5.021	351.7	38.6
56	Ethyl acetate	$C_4H_8O_2$	88.10	523.1	37.8	0.90	189.4		350.2	
57	Ethyl benzene	C_8H_{10}	106.16	619.7	37.0	0.87	178.185	9.163	409.35	36.0
58	Ethyl bromide	C_2H_5Br	108.98	504	61.5	1.46	154.1		311.4	
59	Ethyl chloride	C_2H_5Cl	64.52	460.4	52.0	0.90	134.83	4.452	285.43	25
60	3-Ethyl hexane	C_8C_{18}	114.22	567.0	26.4	0.72			391.69	34.3
61	Ethylene	C_2H_4	28.05	283.1	50.5		103.97	3.351	169.45	13.54
62	Ethylene glycol	$C_2H_6O_2$	62.07			1.11	260	11.23	470.4	56.9
63	Ferric oxide	Fe_2O_3	159.70			5.12	1833			
64	Ferric sulfide	Fe_2S_3	207.90			4.30				
65	Ferrous sulfide	FeS	87.92			4.84	1466			
66	Formaldehyde	H_2CO	30.03			0.81	154.9		253.9	24.5
67	Formic acid	CH_2O_2	46.03			1.22	281.46	12.7	373.7	22.3
68	Glycerol	$C_3H_8O_3$	92.09			1.26	291.36	18.30	563.2	
69	Helium	He	4.00	5.26	2.26		3.5	0.02	4.216	0.084
70	Heptane	C_7H_{16}	100.20	540.2	27.0	0.68	182.57	14.03	371.59	31.69
71	Hexane	C_6H_{14}	86.17	507.9	29.9	0.66	177.84	13.03	341.90	28.85

No.	Substance	Formula								
72	Hydrogen	H_2	2.016	12.8	33.3		13.96	0.12	20.39	0.904
73	Hydrogen chloride	HCl	36.47	81.5	324.6		158.94	1.99	188.11	16.15
74	Hydrogen fluoride	HF	20.01	—	503.2	1.15	238		293	
75	Hydrogen sulfide	H_2S	34.08	88.9	373.6		187.63	2.38	212.82	18.67
76	Iodine	I_2	253.8	—	826.0	4.93	386.5	15	457.4	
77	Iron	Fe	55.85	—	—	7.70	1808	138	3073	353
78	Iron oxide	Fe_3O_4	231.55			5.20	1867			
79	Lead	Pb	207.21			11.34	600.6	5.10	2023	180
80	Lead oxide	PbO	223.21			9.50	1159	11.7	1745	213
81	Magnesium	Mg	24.32			1.74	923	9.2	1393	132
82	Magnesium chloride	$MgCl_2$	95.23			2.33	987	43.1	1691	137
83	Magnesium hydroxide	$Mg(OH)_2$	58.34			2.40				
84	Magnesium oxide	MgO	40.32			3.65	3173	77.4	3873	
85	Mercury	Hg	200.61	45.8	190.70	13.54				
86	Methane	CH_4	16.04				90.68	0.941	111.67	8.180
87	Methanol	CH_3OH	32.04			0.79	175.26	3.17	337.9	35.3
88	Methyl acetate	$C_3H_6O_2$	74.08	46.30	506.7	0.93	174.3		330.3	
89	Methyl amine	CH_5N	31.06	73.60	429.9	0.70	180.5		266.3	
90	Methyl chloride	CH_3Cl	50.49	65.80	416.1		175.3		249	
91	Methyl ethyl ketone	C_4H_8O	72.10			0.81	186.1		352.6	
92	Methyl cyclohexane	C_7H_{14}	98.18			0.77	146.58	6.751	374.10	31.7
93	Molybdenum	Mo	95.95			10.2				
94	Napthalene	$C_{10}H_8$	128.16			1.15	353.2		491.0	
95	Nickel	Ni	58.69			8.90	1725	10.47	3173	
96	Nitric acid	HNO_3	63.02			1.50	231.56		359	30.30
97	Nitrobenzene	$C_6H_5O_2N$	123.11			1.20	278.7		483.9	
98	Nitrogen	N_2	28.02	33.5	126.20		63.15	0.720	77.34	5.577
99	Nitrogen dioxide	NO_2	46.01	100.0	431.0		263.86	7.334	294.46	14.73

(Continued)

TABLE A.1 (*Continued*)

Physical Properties of Various Organic and Inorganic Substances

No.	Compound	Formula	Molecular Weight	T_c K	P_c atm	Sp. Gr.	T_m K	$\Delta \bar{h}_m$ KJ/mol	T_b K	$\Delta \bar{h}_v$ KJ/mol
100	Nitrogen (nitric) oxide	NO	30.01	179.20	65.0		109.51	2.301	121.39	13.78
101	Nitrogen pentoxide	N_2O_5	108.02			1.63	303		320	
102	Nitrogen tetraoxide	N_2O_4	92	431.0	99.0	1.45	263.7		294.3	
103	Nitrogen trioxide	N_2O_3	76.02			1.45	171		276.5	
104	Nitrous oxide	N_2O	44.02	309.5	71.70	1.23	182.1		184.4	
105	n-Nonane	C_9H_{20}	128.25	595	23.0	0.72	219.4		423.8	
106	n-Octane	C_8H_{18}	114.22	595.0	22.5	0.70	216.2		398.7	
107	Oxalic acid	$C_2H_2O_4$	90.04			1.90				
108	Oxygen	O_2	32.00	154.4	49.7		54.40	0.443	90.19	6.820
109	n-Pentane	C_5H_{12}	72.15	469.80	33.3	0.63	143.49	8.393	309.23	25.77
110	iso-Pentane	$iso\text{-}C_5H_{12}$	72.15	461.0	32.9	0.62	113.1		300.9	
111	1-Pentane	C_5H_{10}	70.13	474	39.9	0.64	107.96	4.937	303.13	
112	Phenol	C_6H_5OH	94.11	692.1	60.5	1.07	315.66	11.43	454.56	
113	Phenyl hydrazine	$C_6H_8N_2$	108.14			1.10	292.76	16.43	51.66	
114	Phosphoric acid	H_3PO_4	98.00			1.83	315.51	10.5		
115	Phosphorus (red)	P_4	123.90			2.20	863	81.17	863	41.84
116	Phosphorus (white)	P_4	123.90			1.82	317.4	2.5	553	49.71
117	Phosphorus pentoxide	P_2O_5	141.95			2.39				
118	Propane	C_3H_8	44.09	369.9	42.0		85.47	3.524	231.09	18.77
119	Propene	C_3H_6	42.08	365.1	45.4		87.91	3.002	255.46	18.42
120	Propionic acid	$C_3H_6O_2$	74.08			0.99	252.2		414.4	
121	n-Propyl alcohol	C_3H_5OH	60.09	536.7	49.95	0.80	146		370.2	

122	iso-Propyl alcohol	C$_3$H$_8$O	60.09	508.8	53.0	0.79	183.5		355.4	
123	n-Propyl benzene	C$_9$H$_{12}$	120.19	638.7	31.3	0.86	173.660	8.54	432.38	38.2
124	Silicon dioxide	SiO$_2$	60.09			2.25	1883	8.54	2503	
125	Sodium bisulfate	NaHSO$_4$	120.07			2.74	455			
126	Sodium carbonate	Na$_2$CO$_3$·10H$_2$O	286.15			1.46	306.5			
127	Sodium carbonate	Na$_2$CO$_3$	105.99			2.53	1127	33.4		
128	Sodium chloride	NaCl	58.45			2.16	1081	28.5	1738	171
129	Sodium cyanide	NaCN	49.01				835	16.7	1770	155
130	Sodium hydroxide	NaOH	40.00			2.13	592	8.4	1663	
131	Sodium nitrate	NaNO$_3$	85.00			2.26	583	15.9		
132	Sodium nitrite	NaNO$_2$	69.00			2.17	544			
133	Sodium sulfate	Na$_2$SO$_4$	142.05			2.70	1163	24.3		
134	Sodium sulfide	Na$_2$S	78.05			1.86	1223	6.7		
135	Sodium sulfite	Na$_2$SO$_3$	126.05			2.63				
136	Sodium thiosulfate	Na$_2$S$_2$O$_3$	158.11			1.67				
137	Sulfur (rhombic)	S$_8$	256.53			2.07	386	10.0	717.76	84
138	Sulfur (monoclinic)	S$_8$	256.53			1.96	392	14.17	717.76	84
139	Sulfur chloride (mono)	S$_2$Cl$_2$	135.05			1.69	193.0		411.2	36.0
140	Sulfur dioxide	SO$_2$	64.07	430.7	77.8		197.68	7.402	263.14	24.92
141	Sulfur trioxide	SO$_3$	80.07	491.4	83.8		290.0	24.5	316.5	41.8
142	Sulfuric acid	H$_2$SO$_4$	98.08			1.83	283.51	9.87		
143	Toluene	C$_6$H$_5$CH$_3$	92.13	593.9	40.3	0.87	178.169	6.619	383.78	33.5
144	Water	H$_2$O	18.016	647.4	218.3	1.00	273.16	6.009	373.16	40.65
145	m-Xylene	C$_8$H$_{10}$	106.16	619	34.6	0.86	225.288	11.57	412.26	34.4
146	o-Xylene	C$_8$H$_{10}$	106.16	631.5	35.7	0.88	247.978	13.60	417.58	36.8
147	p-Xylene	C$_8$H$_{10}$	106.16	619	33.9	0.86	286.423	17.11	411.51	36.1
148	Zinc	Zn	65.38			7.14	692.7	6.673	1180	114.8
149	Zinc sulfate	ZnSO$_4$	161.44			3.74				

TABLE A.2

Heats of Formation and Combustion

No.	Compound	Formula	Molecular Weight	State	Δh_f^o KJ/mol	Δh_c^o KJ/mol
1	Acetic acid	CH₃COOH	60.05	l	−486.2	−871.69
				g		−919.73
2	Acetaldehyde	CH₃CHO	40.052	g	−166.4	−1192.36
3	Acetone	C₃H₆O	58.08	aq	−410.03	
				g	−216.69	−1821.38
4	Acetylene	C₂H₂	26.04	g	226.75	−1299.61
5	Ammonia	NH₃	17.032	l	−67.20	
				g	−46.191	−382.58
6	Ammonium carbonate	(NH₄)₂CO₃	96.09	c		
				aq	−941.86	
7	Ammonium chloride	NH₄Cl	53.50	c	−315.4	
8	Ammonium hydroxide	NH₄OH	35.05	aq	−366.5	
9	Ammonium nitrate	NH₄NO₃	80.05	c	−366.1	
				aq	−339.4	
10	Ammonium sulfate	(NH₄)SO₄	132.15	c	−1179.3	
				aq	−1173.1	
11	Benzaldehyde	C₆H₅CHO	106.12	l	−88.83	
				g	−40.0	
12	Benzene	C₆H₆	78.11	l	48.66	−3267.6
				g	82.927	−3301.5
13	Boron oxide	B₂O₃	69.64	c	−1263	
					−1245.2	
14	Bromine	Br₂	159.832	l	0	
				g	30.7	
15	*n*-Butane	C₄H₁₀	58.12	l	−147.6	−2855.6
				g	−124.73	−2878.52
16	Isobutane	C₄H₁₀	58.12	l	−158.5	−2849.0
				g	−134.5	−2868.8
17	1-Butene	C₄H₈	56.104	g	1.172	−2718.58
18	Calcium arsenate	Ca₃(AsO₄)₂	398.06	c	−3330.5	
19	Calcium carbide	CaC₂	64.10	c	−62.7	
20	Calcium carbonate	CaCO₃	100.09	c	−1206.9	
21	Calcium chloride	CaCl₂	110.99	c	−794.9	
22	Calcium cyanamide	CaCN₂	80.11	c	−352	
23	Calcium hydroxide	Ca(OH)₂	74.10	c	−986.56	
24	Calcium oxide	CaO	56.08	c	−635.6	
25	Calcium phosphate	Ca₃(PO₄)₂	310.19	c	−4137.6	
26	Calcium silicate	CaSiO₃	116.17	c	−1584	

(*Continued*)

TABLE A.2 (*Continued*)

Heats of Formation and Combustion

No.	Compound	Formula	Molecular Weight	State	Δh_f^o KJ/mol	Δh_c^o KJ/mol
27	Calcium sulfate	$CaSO_4$	136.15	c	−1432.7	
				aq	−1450.5	
28	Calcium sulfate (gypsum)	$CaSO_4 \cdot 2H_2O$	172.18	c	−2021.1	
29	Carbon	C	12.01	c	0	−393.51
30	Carbon dioxide	CO_2	44.01	g	−393.51	
				l	−412.92	
31	Carbon disulfide	CS_2	76.14	l	87.86	−1075.2
				g	115.3	−1102.6
32	Carbon monoxide	CO	28.01	g	−110.52	−282.99
33	Carbon tetrachloride	CCl_4	153.838	l	−139.5	−352.2
				g	−106.69	−384.9
34	Chloroethane	C_2H_5Cl	64.52	g	−105.0	−1421.1
				l	−41.20	−5215.44
35	Cumene (isopropylbenzene)	$C_6H_5CH(CH_3)_2$	120.19	g	3.93	−5260.59
				g	−769.86	
36	Cupric sulfate	$CuSO_4$	159.61	aq	−843.12	
				c	−751.4	
37	Cyclohexane	C_6H_{12}	84.16	g	−123.1	−3953.0
38	Cyclopentane	C_5H_{10}	70.130	l	−105.8	−3290.9
				g	−77.23	−3319.5
39	Ethane	C_2H_6	30.07	g	−84.667	−1559.9
40	Ethyl acetate	$CH_3CO_2C_2H_5$	88.10	l	−442.92	−2274.48
41	Ethyl alcohol	C_2H_5OH	46.068	l	−277.63	−1366.91
				g	−235.31	−1409.25
42	Ethyl benzene	$C_6H_5 \cdot C_2H_5$	106.16	l	−12.46	−4564.87
				g	29.79	−4607.13
43	Ethyl chloride	C_2H_5Cl	64.52	g	−105	
44	Ethylene	C_2H_4	28.052	g	52.283	−1410.99
45	Ethylene chloride	C_2H_3Cl	62.50	g	31.38	−1271.5
46	3-Ethyl hexane	C_8H_{18}	114.22	l	−250.5	−5470.12
				g	−210.9	−5509.78
47	Ferric chloride	$FeCl_3$		g	−403.34	
48	Ferric oxide	Fe_2O_3	159.70	g	−822.156	
49	Ferric sulfide	FeS_2	119.98	g	−177.9	
50	Ferrosoferric oxide	Fe_3O_4	231.55	c	−1116.7	
51	Ferrous chloride	$FeCl_2$		c	−342.67	−303.76

(*Continued*)

TABLE A.2 (*Continued*)

Heats of Formation and Combustion

No.	Compound	Formula	Molecular Weight	State	Δh_f^o KJ/mol	Δh_c^o KJ/mol
52	Ferrous oxide	FeO	71.85	c	−267	
53	Ferrous sulfide	FeS	87.92	c	−95.06	
54	Formaldehyde	H_2CO	30.026	g	−115.89	−563.46
55	*n*-Heptane	C_7H_{16}	100.20	l	−224.4	−4816.91
				g	−187.8	−4853.48
56	*n*-Hexane	C_6H_{14}	86.17	l	−198.8	−4163.1
				g	−167.2	−4194.753
57	Hydrogen	H_2	2.016	g	0	−285.84
58	Hydrogen bromide	HBr	80.924	g	−36.23	
59	Hydrogen chloride	HCl	36.465	g	−92.311	
60	Hydrogen cyanide	HCN	27.026	g	130.54	
61	Hydrogen sulfide	H_2S	34.082	g	−20.15	−562.589
62	Iron sulfide	FeS_2	119.98	c	−177.9	
63	Lead oxide	PbO	223.21	c	−219.2	
64	Magnesium chloride	$MgCl_2$	95.23	c	−641.83	
65	Magnesium hydroxide	$Mg(OH)_2$	58.34	c	−924.66	
66	Magnesium oxide	MgO	40.32	c	−601.83	
67	Methane	CH_4	16.041	g	−74.84	−890.4
68	Methyl alcohol	CH_3OH	32.042	l	−238.64	−726.55
				g	−201.25	−763.96
69	Methyl chloride	CH_3Cl	50.49	g	−81.923	−766.63
70	Methyl cyclohexane	C_7H_{14}	98.182	l	−190.2	−4565.29
				g	−154.8	−4600.68
71	Methyl cyclopentane	C_6H_{12}	84.156	l	−138.4	−3937.7
				g	−106.7	−3969.4
72	Nitric acid	HNO_3	63.02	l	−173.23	
				aq	−206.57	
73	Nitric oxide	NO	30.01	g	90.374	
74	Nitrogen dioxide	NO_2	46.01	g	33.85	
75	Nitrous oxide	N_2O	44.02	g	81.55	
76	*n*-Pentane	C_5H_{12}	72.15	l	−173.1	−3509.5
				g	−146.4	−3536.15
77	Phosphoric acid	H_3PO_4	98.00	c	−1281	
				aq	−1278	
78	Phosphorus	P_4	123.90	c	0	
79	Phosphorus pentoxide	P_2O_5	141.95	c	−1506	

(*Continued*)

TABLE A.2 (*Continued*)

Heats of Formation and Combustion

No.	Compound	Formula	Molecular Weight	State	Δh_f^o KJ/mol	Δh_c^o KJ/mol
80	Propane	C_3H_8	44.09	l	−119.84	−2204.0
				g	−103.85	−2220.0
81	Propene	C_3H_6	42.078	g	20.41	−2058.47
82	*n*-Propyl alcohol	C_3H_8O	60.09	g	−255	−2068.6
83	*n*-Propylbenzene	$C_6H_5 \cdot CH_2 \cdot C_2H_5$	120.19	l	−38.40	−5218.2
				g	7.824	−5264.5
84	Silicon dioxide	SiO_2	60.09	c	−851.0	
85	Sodium bicarbonate	$NaHCO_3$	84.01	c	−945.6	
86	Sodium bisulfate	$NaHSO_4$	120.07	c	−1126	
87	Sodium carbonate	Na_2CO_3	105.99	c	−1130	
88	Sodium chloride	NaCl	58.45	c	−411.00	
89	Sodium cyanide	NaCN	49.01	c	−89.79	
90	Sodium nitrate	$NaNO_3$	85.00	c	−466.68	
91	Sodium nitrite	$NaNO_2$	69.00	c	−359	
92	Sodium sulfate	Na_2SO_4	142.05	c	−1384.5	
93	Sodium sulfide	Na_2S	78.05	c	−373	
94	Sodium sulfite	Na_2SO_3	126.05	c	−1090	
95	Sodium thiosulfate	$Na_2S_2O_3$	158.11	c	−1117	
96	Sulfur	S	32.07	c	0	
97	Sulfur chloride	S_2Cl_2	135.05	l	−60.3	
98	Sulfur dioxide	SO_2	64.066	g	−296.90	
99	Sulfur trioxide	SO_3	80.066	g	−395.18	
100	Sulfuric acid	H_2SO_4	98.08	l	−811.32	
				aq	−907.51	
101	Toluene	$C_6H_5CH_3$	92.13	l	11.99	−3909.9
				g	50.000	−3947.9
102	Water	H_2O	18.016	l	−285.840	
				g	−241.826	
103	*m*-Xylene	$C_6H_4(CH_3)_2$	106.16	l	−25.42	−4551.86
				g	17.24	−4594.53
104	*o*-Xylene	$C_6H_4(CH_3)_2$	106.16	l	−24.44	−4552.86
				g	19.00	−4596.29
105	*p*-Xylene	$C_6H_4(CH_3)_2$	106.16	l	−24.43	−4552.86
				g	17.95	−4595.25
106	Zinc sulfate	$ZnSO_4$	161.45	c	−978.55	
				aq	−1059.93	

Note: Heats of formation and combustion of compounds at 25°C. Standard states of products for $\Delta \hat{H}_c^0$ are CO_2 (g), H_2O (l), N_2 (g), SO_2 (g), and HCl (aq).

Appendix A.2

Table A.3 contains heat capacity equations for organic and inorganic compounds as a function of temperature.

Forms	Heat Capacity
1.	$C_p \, J/(mol)(K \text{ or } {}^{\circ}C) = a + bT + cT^2 + dT^3$
2.	$C_p \, J/(mol)(K \text{ or } {}^{\circ}C) = a + bT + cT^{-2}$

TABLE A.3

Heat Capacity Equations for Organic and Inorganic Compounds (at Low Pressures)

No.	Compound	Formula	Mol. Wt.	State	Form	T	a	b	c	d	Temp. Range
1	Acetone	CH_3COCH_3	58.08	l	1	°C	123.0	0.186			−30 to 60
			58.08	g	1	°C	71.96	0.201	-1.278×10^{-4}	3.476×10^{-8}	0–1200
2	Acetylene	C_2H_2	26.04	g	1	°C	42.43	0.06053	-5.033×10^{-5}	1.820×10^{-8}	0–1200
3	Air		29.0	g	1	°C	28.94	0.004147	0.3191×10^{-5}	-1.965×10^{-9}	0–1500
				g	1	K	28.09	0.001965	0.4799×10^{-5}	-1.965×10^{-9}	273–1800
4	Ammonia	NH_3	17.03	g	1	°C	35.15	0.02954	0.4421×10^{-5}	-6.686×10^{-9}	0–1200
5	Ammonium sulfate	$(NH_4)_2SO_4$	132.15	c	1	K	215.9		0		275–328
6	Benzene	C_6H_6	78.11	l	1	°C	126.5	0.234			6–67
				g	1	°C	74.06	0.3295	-2.520×10^{-4}	7.757×10^{-8}	0–1200
7	Isobutane	C_4H_{10}	58.12	g	1	°C	89.46	0.3013	-1.891×10^{-4}	4.987×10^{-8}	0–1200
8	n-Butane	C_4H_{10}	58.12	g	1	°C	92.30	0.2788	-1.547×10^{-4}	3.498×10^{-8}	0–1200
9	Isobutene	C_4H_8	56.10	g	1	°C	82.88	0.2564	-1.727×10^{-4}	5.050×10^{-8}	0–1200
10	Calcium carbide	CaC_2	64.10	c	2	K	68.62	0.0119	-8.66×10^{-5}	—	298–720
11	Calcium carbonate	$CaCO_3$	100.09	c	2	K	82.34	0.04975	-1.287×10^{-4}	—	273–1033
12	Calcium hydroxide	$Ca(OH)_2$	74.10	c	1	K	89.5				276–373
13	Calcium oxide	CaO	56.08	c	2	K	41.84	0.0203	-4.52×10^{-5}		273–1173
14	Carbon	C	12.01	c[a]	2	K	11.18	0.01095	-4.891×10^{-5}		273–1373
15	Carbon dioxide	CO_2	44.01	g	1	°C	36.11	0.04233	-2.887×10^{-5}	7.464×10^{-9}	0–1500
16	Carbon monoxide	CO	28.01	g	1	°C	28.95	0.00411	0.3548×10^{-5}	-2.220×10^{-9}	0–1500
17	Carbon tetrachloride	CCl_4	153.84	l	1	K	12.285	0.0001095	-3.1826×10^{-3}	3.4252×10^{-6}	273–343
18	Chlorine	Cl_2	70.91	g	1	°C	33.60	0.01367	-1.607×10^{-5}	6.473×10^{-9}	0–1200
19	Copper	Cu	63.54	c	1	K	22.76	0.0006117			273–1357
20	Cumene	C_9H_{12}	120.19	g	1	°C	139.2	0.5376	-3.979×10^{-4}	1.205×10^{-7}	0–1200

(Continued)

TABLE A.3 (*Continued*)

Heat Capacity Equations for Organic and Inorganic Compounds (at Low Pressures)

No.	Compound	Formula	Mol. Wt.	State	Form	T	a	b	c	d	Temp. Range
21	Cyclohexane	C_6H_{12}	84.16	g	1	°C	94.140	0.4962	-3.190×10^{-4}	8.063×10^{-8}	0–1200
22	Cyclopentane	C_5H_{10}	70.13	g	1	°C	73.39	0.3928	-2.554×10^{-4}	6.866×10^{-8}	0–1200
23	Ethane	C_2H_6	30.07	g	1	°C	49.37	0.1392	-5.816×10^{-5}	7.280×10^{-9}	0–1200
24	Ethyl alcohol	C_2H_6O	46.07	l	1	°C	158.8				100
				g	1	°C	61.34	0.1572	-8.749×10^{-5}	1.983×10^{-8}	0–1200
25	Ethylene	C_2H_4	28.05	g	1	°C	40.75	0.1147	-6.891×10^{-5}	1.766×10^{-8}	0–1200
26	Ethylbenzene	$C_2H_4Cl_2$	98.96	l	1	°C	182.0				
				g	1	°C	118	0.30			
27	Ethylene oxide	C_2H_4O	44.05	g	1	K	−4.69	0.206	-9.995×10^{-5}		
28	Ferric oxide	Fe_2O_3	159.70	g	2	K	103.4	0.06711	-17.72×10^{-5}		273–1097
29	Formaldehyde	CH_2O	30.03	g	1	°C	34.28	0.04268	0.0000	-8.694×10^{-9}	0–1200
30	Helium	He	4.00	g	1	°C	20.8				All
31	n-Hexane	C_6H_{14}	86.17	l	1	K	31.421	0.0097606	-2.3537×10^{-3}	3.0927×10^{-6}	273–400
				g	1	°C	137.44	0.4085	-2.392×10^{-4}	5.766×10^{-8}	0–1200
32	Hydrogen	H_2	2.016	g	1	°C	28.84	0.0000765	0.3288×10^{-5}	-0.8698×10^{-9}	0–1500
33	Hydrogen bromide	HBr	80.92	g	1	°C	29.10	−0.000227	0.9887×10^{-5}	-4.858×10^{-9}	0–1200
34	Hydrogen chloride	HCl	36.47	g	1	°C	29.13	−0.001341	0.9715×10^{-5}	-4.335×10^{-9}	0–1200
35	Hydrogen cyanide	HCN	27.03	g	1	°C	35.3	0.02908	1.092×10^{-5}		0–1200

No.	Name	Formula	MW			Unit					Range
36	Hydrogen sulfide	H$_2$S	34.08	g	1	°C	33.51	0.01547	0.3012×10^{-5}	-3.292×10^{-9}	0–1500
37	Magnesium chloride	MgCl$_2$	95.23	c	1	K	72.4	0.0158	-8.732×10^{-5}		273–991
38	Magnesium oxide	MgO	40.32	c	2	K	45.44	0.005008			273–2073
39	Methane	CH$_4$	16.04	g	1	°C	34.31	0.05469	0.3661×10^{-5}	-1.100×10^{-8}	0–1200
				g	1	K	19.87	0.05021	1.268×10^{-5}	-1.100×10^{-8}	273–1500
40	Methyl alcohol	CH$_3$OH	32.04	l	1	°C	75.86	0.1683	-1.87×10^{-5}	-8.03×10^{-9}	0–65
				g		°C	42.93	0.08301			0–700
41	Methyl cyclohexane	C$_7$H$_{14}$	98.18	g	1	°C	121.3	0.5653	-3.772×10^{-4}	1.008×10^{-7}	0–1200
42	Methyl cyclopentane	C$_6$H$_{12}$	84.16	g	1	°C	98.83	0.45857	-3.044×10^{-4}	8.381×10^{-8}	0–1200
43	Nitric acid	HNO$_3$	63.02	l	1	°C	110.0				25
44	Nitric oxide	NO	30.01	g	1	°C	29.50	0.008188	-0.2925×10^{-5}	0.3652×10^{-9}	0–3500
45	Nitrogen	N$_2$	28.02	g	1	°C	29.00	0.002199	0.5723×10^{-5}	-2.871×10^{-9}	0–1500
46	Nitrogen dioxide	NO$_2$	46.01	g	1	°C	36.07	0.0397	-2.88×10^{-5}	7.87×10^{-9}	0–1200
47	Nitrogen tetraoxide	N$_2$O$_4$	92.02	g	1	°C	75.7	0.125	-1.13×10^{-4}		0–300
48	Nitrous oxide	N$_2$O	44.02	g	1	°C	37.66	0.04151	-2.694×10^{-5}	1.057×10^{-8}	0–1200
49	Oxygen	O$_2$	32.00	g	1	°C	29.10	0.01158	-0.6076×10^{-5}	1.311×10^{-9}	0–1500
50	n-Pentane	C$_5$H$_{12}$	72.15	l	1	K	33.24	1.9241	-2.3687×10^{-3}	1.7944×10^{-5}	270–350
				g		°C	114.8	0.3409	-1.899×10^{-4}	4.226×10^{-8}	0–1200
51	Propane	C$_3$H$_8$	44.09	g	1	°C	68.032	0.2259	-1.311×10^{-4}	3.171×10^{-8}	0–1200
52	Propylene	C$_3$H$_6$	42.08	g	1	°C	59.580	0.1771	-1.017×10^{-4}	2.460×10^{-8}	0–1200
53	Sodium carbonate	Na$_2$CO$_3$	105.99	c	1	K	121				288–371

(Continued)

TABLE A.3 (*Continued*)

Heat Capacity Equations for Organic and Inorganic Compounds (at Low Pressures)

No.	Compound	Formula	Mol. Wt.	State	Form	T	a	b	c	d	Temp. Range
54	Sodium carbonate	$Na_2CO_3 \cdot 10H_2O$	286.15	c	1	K	535.6				298
55	Styrene	C_8H_8	104.2	l	1	°C	209.0	0.270			
56	Sulfur	S	32.07	g	1	°C	115.0				
				c[b]		K	15.2	0.0268			273–368
				c[c]		K	18.5	0.0184			368–392
57	Sulfuric acid	H_2SO_4	98.08	l	1	°C	139.1	0.1559			10–45
58	Sulfur dioxide	SO_2	64.07	g	1	°C	38.91	0.03904	-3.104×10^{-5}	8.606×10^{-9}	0–1500
59	Sulfur trioxide	SO_3	80.07	g	1	°C	48.50	0.09188	-8.540×10^{-5}	3.240×10^{-8}	0–1000
60	Toluene	C_7H_8	92.13	l	1	K	1.8083	0.81222	-151.27×10^{-5}	1.630×10^{-6}	270–370
				g	1	°C	94.18	0.3800	-27.86×10^{-5}	8.033×10^{-8}	0–1200
61	Water	H_2O	18.016	l	1	K	18.2964	0.47212	-133.88×10^{-5}	1.3142×10^{-6}	273–373
				g	1	°C	33.46	0.00688	0.7604×10^{-5}	-3.593×10^{-9}	0–1500

a Graphite.

b Rhombic.

c Monoclinic (at 1 atm).

Appendix A.3: Steam Table

Table A.4 contains properties of saturated steam, specific volume, specific internal energy, and specific enthalpy. Table A.5 contains properties of superheated steam.

TABLE A.4

Properties of Saturated Steam

P (bar)	T (°C)	v (m³/kg) Water, v_f	Steam, v_g	u (kJ/kg) Water, u_f	Steam, u_g	h (kJ/kg) Water, h_f	Evap. h_{fg}	Steam, h_g
0.00611	0.01	0.001000	206.2	0.0	2375.6	0.0	2501.6	2501.6
0.008	3.8	0.001000	159.7	15.8	2380.7	15.8	2492.6	2508.5
0.010	7.0	0.001000	129.2	29.3	2385.2	29.3	2485.0	2514.4
0.012	9.7	0.001000	108.7	40.6	2388.9	40.6	2478.7	2519.3
0.014	12.0	0.001000	93.9	50.3	2392.0	50.3	2473.2	2523.5
0.016	14.0	0.001001	82.8	58.9	2394.8	58.9	2468.4	2527.3
0.018	15.9	0.001001	74.0	66.5	2397.4	66.5	2464.1	2530.6
0.020	17.5	0.001001	67.0	73.5	2399.6	73.5	2460.2	2533.6
0.022	19.0	0.001002	61.2	79.8	2401.7	79.8	2456.6	2536.4
0.024	20.4	0.001002	56.4	85.7	2403.6	85.7	2453.3	2539.0
0.026	21.7	0.001002	52.3	91.1	2405.4	91.1	2450.2	2541.3
0.028	23.0	0.001002	48.7	96.2	2407.1	96.2	2447.3	2543.6
0.030	24.1	0.001003	45.7	101.0	2408.6	101.0	2444.6	2545.6
0.035	26.7	0.001003	39.5	111.8	2412.2	111.8	2438.5	2550.4
0.040	29.0	0.001004	34.8	121.4	2415.3	121.4	2433.1	2554.5
0.045	31.0	0.001005	31.1	130.0	2418.1	130.0	2428.2	2558.2
0.050	32.9	0.001005	28.2	137.8	2420.6	137.8	2423.8	2561.6
0.060	36.2	0.001006	23.74	151.5	2425.1	151.5	2416.0	2567.5
0.070	39.0	0.001007	20.53	163.4	2428.9	163.4	2409.2	2572.6
0.080	41.5	0.001008	18.10	173.9	2432.3	173.9	2403.2	2577.1

0.090	43.8	0.001009	16.20	183.3	2435.3	183.3	2397.9	2581.1
0.10	45.8	0.001010	14.67	191.8	2438.0	191.8	2392.9	2584.8
0.11	47.7	0.001011	13.42	199.7	2440.5	199.7	2388.4	2588.1
0.12	49.4	0.001012	12.36	206.9	2442.8	206.9	2384.3	2591.2
0.13	51.1	0.001013	11.47	213.7	2445.0	213.7	2380.4	2594.0
0.14	52.6	0.001013	10.69	220.0	2447.0	220.0	2376.7	2596.7
0.15	54.0	0.001014	10.02	226.0	2448.9	226.0	2373.2	2599.2
0.16	55.3	0.001015	9.43	231.6	2450.6	231.6	2370.0	2601.6
0.17	56.6	0.001015	8.91	236.9	2452.3	236.9	2366.9	2603.8
0.18	57.8	0.001016	8.45	242.0	2453.9	242.0	2363.9	2605.9
0.19	59.0	0.001017	8.03	246.8	2455.4	246.8	2361.1	2607.9
0.20	60.1	0.001017	7.65	251.5	2456.9	251.5	2358.4	2609.9
0.22	62.2	0.001018	7.00	260.1	2459.6	260.1	2353.3	2613.5
0.24	64.1	0.001019	6.45	268.2	2462.1	268.2	2348.6	2616.8
0.26	65.9	0.001020	5.98	275.6	2464.4	275.7	2344.2	2619.9
0.28	67.5	0.001021	5.58	282.7	2466.5	282.7	2340.0	2622.7
0.30	69.1	0.001022	5.23	289.3	2468.6	289.3	2336.1	2625.4
0.35	72.7	0.001025	4.53	304.3	2473.1	304.3	2327.2	2631.5
0.40	75.9	0.001027	3.99	317.6	2477.1	317.7	2319.2	2636.9
0.45	78.7	0.001028	3.58	329.6	2480.7	329.6	2312.0	2641.7
0.50	81.3	0.001030	3.24	340.5	2484.0	340.6	2305.4	2646.0
0.55	83.7	0.001032	2.96	350.6	2486.9	350.6	2299.3	2649.9
0.60	86.0	0.001033	2.73	359.9	2489.7	359.9	2293.6	2653.6
0.65	88.0	0.001035	2.53	368.5	2492.2	368.6	2288.3	2656.9
0.70	90.0	0.001036	2.36	376.7	2494.5	376.8	2283.3	2660.1
0.75	91.8	0.001037	2.22	384.4	2496.7	384.5	2278.6	2663.0
0.80	93.5	0.001039	2.087	391.6	2498.8	391.7	2274.1	2665.8

(*Continued*)

TABLE A.4 (*Continued*)

Properties of Saturated Steam

P (bar)	T (°C)	v (m³/kg)		u (kJ/kg)		h (kJ/kg)		
		Water, v_f	Steam, v_g	Water, u_f	Steam, u_g	Water, h_f	Evap. h_{fg}	Steam, h_g
0.85	95.2	0.001040	1.972	398.5	2500.8	398.6	2269.8	2668.4
0.90	96.7	0.001041	1.869	405.1	2502.6	405.2	2265.6	2670.9
0.95	98.2	0.001042	1.777	411.4	2504.4	411.5	2261.7	2673.2
1.00	99.6	0.001043	1.694	417.4	2506.1	417.5	2257.9	2675.4
1.01325	100.0	0.001044	1.673	419.0	2506.5	419.1	2256.9	2676.0
1.1	102.3	0.001046	1.549	428.7	2509.2	428.8	2250.8	2679.6
1.2	104.8	0.001048	1.428	439.2	2512.1	439.4	2244.1	2683.4
1.3	107.1	0.001049	1.325	449.1	2514.7	449.2	2237.8	2687.0
1.4	109.3	0.001051	1.236	458.3	2517.2	458.4	2231.9	2690.3
1.5	111.4	0.001053	1.159	467.0	2519.5	467.1	2226.2	2693.4
1.6	113.3	0.001055	1.091	475.2	2521.7	475.4	2220.9	2696.2
1.7	115.2	0.001056	1.031	483.0	2523.7	483.2	2215.7	2699.0
1.8	116.9	0.001058	0.977	490.5	2525.6	490.7	2210.8	2701.5
1.9	118.6	0.001059	0.929	497.6	2527.5	497.8	2206.1	2704.0
2.0	120.2	0.001061	0.885	504.5	2529.2	504.7	2201.6	2706.3
2.2	123.3	0.001064	0.810	517.4	2532.4	517.6	2193.0	2710.6
2.4	126.1	0.001066	0.746	529.4	2535.4	529.6	2184.9	2714.5
2.6	128.7	0.001069	0.693	540.6	2538.1	540.9	2177.3	2718.2
2.8	131.2	0.001071	0.646	551.1	2540.6	551.4	2170.1	2721.5
3.0	133.5	0.001074	0.606	561.1	2543.0	561.4	2163.2	2724.7
3.2	135.8	0.001076	0.570	570.6	2545.2	570.9	2156.7	2727.6
3.4	137.9	0.001078	0.538	579.6	2547.2	579.9	2150.4	2730.3
3.6	139.9	0.001080	0.510	588.1	2549.2	588.5	2144.4	2732.9

3.8	141.8	0.001082	0.485	596.4	2551.0	596.8	2138.6	2735.3
4.0	143.6	0.001084	0.462	604.2	2552.7	604.7	2133.0	2737.6
4.2	145.4	0.001086	0.442	611.8	2554.4	612.3	2127.5	2739.8
4.4	147.1	0.001088	0.423	619.1	2555.9	619.6	2122.3	2741.9
4.6	148.7	0.001089	0.405	626.2	2557.4	626.7	2117.2	2743.9
4.8	150.3	0.001091	0.389	633.0	2558.8	633.5	2112.2	2745.7
5.0	151.8	0.001093	0.375	639.6	2560.2	640.1	2107.4	2747.5
5.5	155.5	0.001097	0.342	655.2	2563.3	655.8	2095.9	2751.7
6.0	158.8	0.001101	0.315	669.8	2566.2	670.4	2085.0	2755.5
6.5	162.0	0.001105	0.292	683.4	2568.7	684.1	2074.7	2758.9
7.0	165.0	0.001108	0.273	696.3	2571.1	697.1	2064.9	2762.0
7.5	167.8	0.001112	0.2554	708.5	2573.3	709.3	2055.5	2764.8
8.0	170.4	0.001115	0.2403	720.0	2575.5	720.9	2046.5	2767.5
8.5	172.9	0.001118	0.2268	731.1	2577.1	732.0	2037.9	2769.9
9.0	175.4	0.001121	0.2148	741.6	2578.8	742.6	2029.5	2772.1
9.5	177.7	0.001124	0.2040	751.8	2580.4	752.8	2021.4	2774.2
10	179.9	0.001127	0.1943	761.5	2581.9	762.6	2013.6	2776.2
11	184.1	0.001133	0.1774	779.9	2584.5	781.1	1998.5	2779.7
12	188.0	0.001139	0.1632	797.1	2586.9	798.4	1984.3	2782.7
13	191.6	0.001144	0.1511	813.2	2589.0	814.7	1970.7	2785.4
14	195.0	0.001149	0.1407	828.5	2590.8	830.1	1957.7	2787.8
15	198.3	0.001154	0.1317	842.9	2592.4	844.7	1945.2	2789.9
16	201.4	0.001159	0.1237	856.7	2593.8	858.6	1933.2	2791.7
17	204.3	0.001163	0.1166	869.9	2595.1	871.8	1921.5	2793.4
18	207.1	0.001168	0.1103	882.5	2596.3	884.6	1910.3	2794.8
19	209.8	0.001172	0.1047	894.6	2597.3	896.8	1899.3	2796.1
20	212.4	0.001177	0.0995	906.2	2598.2	908.6	1888.6	2797.2

(Continued)

TABLE A.4 (*Continued*)

Properties of Saturated Steam

P (bar)	T (°C)	v (m³/kg) Water, v_f	Steam, v_g	u (kJ/kg) Water, u_f	Steam, u_g	h (kJ/kg) Water, h_f	Evap. h_{fg}	Steam, h_g
21	214.9	0.001181	0.0949	917.5	2598.9	920.0	1878.2	2798.2
22	217.2	0.001185	0.0907	928.3	2599.6	931.0	1868.1	2799.1
23	219.6	0.001189	0.0868	938.9	2600.2	941.6	1858.2	2799.8
24	221.8	0.001193	0.0832	949.1	2600.7	951.9	1848.5	2800.4
25	223.9	0.001197	0.0799	959.0	2601.2	962.0	1839.0	2800.9
26	226.0	0.001201	0.0769	968.6	2601.5	971.7	1829.6	2801.4
27	228.1	0.001205	0.0740	978.0	2601.8	981.2	1820.5	2801.7
28	230.0	0.001209	0.0714	987.1	2602.1	990.5	1811.5	2802.0
29	232.0	0.001213	0.0689	996.0	2602.3	999.5	1802.6	2802.2
30	233.8	0.001216	0.0666	1004.7	2602.4	1008.4	1793.9	2802.3
32	237.4	0.001224	0.0624	1021.5	2602.5	1025.4	1776.9	2802.3
34	240.9	0.001231	0.0587	1037.6	2602.5	1041.8	1760.3	2802.1
36	244.2	0.001238	0.0554	1053.1	2602.2	1057.6	1744.2	2801.7
38	247.3	0.001245	0.0524	1068.0	2601.9	1072.7	1728.4	2801.1
40	250.3	0.001252	0.0497	1082.4	2601.3	1087.4	1712.9	2800.3
42	253.2	0.001259	0.0473	1096.3	2600.7	1101.6	1697.8	2799.4
44	256.0	0.001266	0.0451	1109.8	2599.9	1115.4	1682.9	2798.3
46	258.8	0.001272	0.0430	1122.9	2599.1	1128.8	1668.3	2797.1
48	261.4	0.001279	0.0412	1135.6	2598.1	1141.8	1653.9	2795.7
50	263.9	0.001286	0.0394	1148.0	2597.0	1154.5	1639.7	2794.2
52	266.4	0.001292	0.0378	1160.1	2595.9	1166.8	1625.7	2792.6
54	268.8	0.001299	0.0363	1171.9	2594.6	1178.9	1611.9	2790.8
56	271.1	0.001306	0.0349	1183.5	2593.3	1190.8	1598.2	2789.0

58	273.3	0.001312	0.0337	1194.7	2591.9	1202.3	1584.7	2787.0
60	275.6	0.001319	0.0324	1205.8	2590.4	1213.7	1571.3	2785.0
62	277.7	0.001325	0.0313	1216.6	2588.8	1224.8	1558.0	2782.9
64	279.8	0.001332	0.0302	1227.2	2587.2	1235.7	1544.9	2780.6
66	281.8	0.001338	0.0292	1237.6	2585.5	1246.5	1531.9	2778.3
68	283.8	0.001345	0.0283	1247.9	2583.7	1257.0	1518.9	2775.9
70	285.8	0.001351	0.0274	1258.0	2581.8	1267.4	1506.0	2773.5
72	287.7	0.001358	0.0265	1267.9	2579.9	1277.6	1493.3	2770.9
74	289.6	0.001364	0.0257	1277.6	2578.0	1287.7	1480.5	2768.3
76	291.4	0.001371	0.0249	1287.2	2575.9	1297.6	1467.9	2765.5
78	293.2	0.001378	0.0242	1296.7	2573.8	1307.4	1455.3	2762.8
80	295.0	0.001384	0.0235	1306.0	2571.7	1317.1	1442.8	2759.9
82	296.7	0.001391	0.0229	1315.2	2569.5	1326.6	1430.3	2757.0
84	298.4	0.001398	0.0222	1324.3	2567.2	1336.1	1417.9	2754.0
86	300.1	0.001404	0.0216	1333.3	2564.9	1345.4	1405.5	2750.9
88	301.7	0.001411	0.0210	1342.2	2562.6	1354.6	1393.2	2747.8
90	303.3	0.001418	0.02050	1351.0	2560.1	1363.7	1380.9	2744.6
92	304.9	0.001425	0.01996	1359.7	2557.7	1372.8	1368.6	2741.4
94	306.4	0.001432	0.01945	1368.2	2555.2	1381.7	1356.3	2738.0
96	308.0	0.001439	0.01897	1376.7	2552.6	1390.6	1344.1	2734.7
98	309.5	0.001446	0.01849	1385.2	2550.0	1399.3	1331.9	2731.2
100	311.0	0.001453	0.01804	1393.5	2547.3	1408.0	1319.7	2727.7
105	314.6	0.001470	0.01698	1414.1	2540.4	1429.5	1289.2	2718.7
110	318.0	0.001489	0.01601	1434.2	2533.2	1450.6	1258.7	2709.3
115	321.4	0.001507	0.01511	1454.0	2525.7	1471.3	1228.2	2699.5
120	324.6	0.001527	0.01428	1473.4	2517.8	1491.8	1197.4	2689.2
125	327.8	0.001547	0.01351	1492.7	2509.4	1512.0	1166.4	2678.4

(Continued)

TABLE A.4 (*Continued*)

Properties of Saturated Steam

P (bar)	T (°C)	v (m³/kg)		u (kJ/kg)		h (kJ/kg)		
		Water, v_f	Steam, v_g	Water, u_f	Steam, u_g	Water, h_f	Evap. h_{fg}	Steam, h_g
130	330.8	0.001567	0.01280	1511.6	2500.6	1532.0	1135.0	2667.0
135	333.8	0.001588	0.01213	1530.4	2491.3	1551.9	1103.1	2655.0
140	336.6	0.001611	0.01150	1549.1	2481.4	1571.6	1070.7	2642.4
145	339.4	0.001634	0.01090	1567.5	2471.0	1591.3	1037.7	2629.1
150	342.1	0.001658	0.01034	1586.1	2459.9	1611.0	1004.0	2615.0
155	344.8	0.001683	0.00981	1604.6	2448.2	1630.7	969.6	2600.3
160	347.3	0.001710	0.00931	1623.2	2436.0	1650.5	934.3	2584.9
165	349.8	0.001739	0.00883	1641.8	2423.1	1670.5	898.3	2568.8
170	352.3	0.001770	0.00837	1661.6	2409.3	1691.7	859.9	2551.6
175	354.6	0.001803	0.00793	1681.8	2394.6	1713.3	820.0	2533.3
180	357.0	0.001840	0.00750	1701.7	2378.9	1734.8	779.1	2513.9
185	359.2	0.001881	0.00708	1721.7	2362.1	1756.5	736.6	2493.1
190	361.4	0.001926	0.00668	1742.1	2343.8	1778.7	692.0	2470.6
195	363.6	0.001977	0.00628	1763.2	2323.6	1801.8	644.2	2446.0
200	365.7	0.00204	0.00588	1785.7	2300.8	1826.5	591.9	2418.4
205	367.8	0.00211	0.00546	1810.7	2274.4	1853.9	532.5	2386.4
210	369.8	0.00220	0.00502	1840.0	2242.1	1886.3	461.3	2347.6
215	371.8	0.00234	0.00451	1878.6	2198.1	1928.9	366.2	2295.2
220	373.7	0.00267	0.00373	1952	2114	2011	185	2196
221.2	374.15	0.00317	0.00317	2038	2038	2108	0	2108

TABLE A.5

Properties of Superheated Steam

P (bar) (T_{sat} °C)		Saturated Water	Saturated Steam	Temperature (°C)							
				50	75	100	150	200	250	300	350
0.0 (—)	h	—	—	2595	2642	2689	2784	2880	2978	3077	3177
	u	—	—	2446	2481	2517	2589	2662	2736	2812	2890
	v	—	—	—	—	—	—	—	—	—	—
0.1 (45.8)	h	191.8	2584.8	2593	2640	2688	2783	2880	2977	3077	3177
	u	191.8	2438.0	2444	2480	2516	2588	2661	2736	2812	2890
	v	0.00101	14.7	14.8	16.0	17.2	9.5	21.8	24.2	26.5	28.7
0.5 (81.3)	h	340.6	2646.0	209.3	313.9	2683	2780	2878	2979	3076	3177
	u	340.6	2484.0	209.2	313.9	2512	2586	2660	2735	2811	2889
	v	0.00103	3.24	0.00101	0.00103	3.41	3.89	4.35	4.83	5.29	5.75
1.0 (99.6)	h	417.5	2675.4	209.3	314.0	2676	2776	2875	2975	3074	3176
	u	417.5	2506.1	209.2	313.9	2507	2583	2658	2734	2811	2889
	v	0.00104	1.69	0.00101	0.00103	1.69	1.94	2.17	2.40	2.64	2.87
5.0 (151.8)	h	640.1	2747.5	209.7	314.3	419.4	632.2	2855	2961	3065	3168
	u	639.6	2560.2	209.2	313.8	418.8	631.6	2643	2724	2803	2883
	v	0.00109	0.375	0.00101	0.00103	0.00104	0.00109	0.425	0.474	0.522	0.571
10 (179.9)	h	762.6	2776.2	210.1	314.7	419.7	632.5	2827	2943	3052	3159
	u	761.5	2582	209.1	313.7	418.7	631.4	2621	2710	2794	2876
	v	0.00113	0.194	0.00101	0.00103	0.00104	0.00109	0.206	0.233	0.258	0.282
20 (212.4)	h	908.6	2797.2	211.0	315.5	420.5	633.1	852.6	2902	3025	3139
	u	906.2	2598.2	209.0	313.5	418.4	603.9	850.2	2679	2774	2862
	v	0.00118	0.09950	0.00101	0.00102	0.00104	0.00109	0.00116	0.111	0.125	0.139

(*Continued*)

TABLE A.5 (*Continued*)

Properties of Superheated Steam

P (bar) (T_{sat} °C)		Saturated Water	Saturated Steam	Temperature (°C)							
				50	75	100	150	200	250	300	350
40 (250.3)	h	1087.4	2800.3	212.7	317.1	422.0	634.3	853.4	1085.8	2962	3095
	u	1082.4	2601.3	208.6	313.0	417.8	630.0	848.8	1080.8	2727	2829
	v	0.00125	0.04975	0.00101	0.00102	0.00104	0.00109	0.00115	0.00125	0.0588	0.0665
60 (275.6)	h	1213.7	2785.0	214.4	318.7	423.5	635.6	854.2	1085.8	2885	3046
	u	1205.8	2590.4	208.3	312.6	417.3	629.1	847.3	1078.3	2668	2792
	v	0.00132	0.0325	0.00101	0.00103	0.00104	0.00109	0.00115	0.00125	0.0361	0.0422
80 (295.0)	h	1317.1	2759.9	216.1	320.3	425.0	636.8	855.1	1085.8	2787	2990
	u	1306.0	2571.7	208.1	312.3	416.7	628.2	845.9	1075.8	2593	2750
	v	0.00139	0.0235	0.00101	0.00102	0.00104	0.00109	0.00115	0.00124	0.0243	0.0299
100 (311.0)	h	1408.0	2727.7	217.8	322.9	426.5	638.1	855.9	1085.8	1343.4	2926
	u	1393.5	2547.3	207.8	311.7	416.1	627.3	844.4	1073.4	1329.4	2702
	v	0.00145	0.0181	0.00101	0.00102	0.001049	0.00109	0.00115	0.00124	0.00140	0.0224
150 (342.1)	h	1611.0	2615.0	222.1	326.0	430.3	641.3	858.1	1086.2	1338.2	2695
	u	1586.1	2459.9	207.0	310.7	414.7	625.0	841.0	1067.7	1317.6	2523
	v	0.00166	0.0103	0.00101	0.00102	0.00104	0.00108	0.00114	0.00123	0.00138	0.0115
200 (365.7)	h	1826.5	2418.4	226.4	330.0	434.0	644.5	860.4	1086.7	1334.3	1647.1
	u	1785.7	2300.8	206.3	309.7	413.2	622.9	837.7	1062.2	1307.1	1613.7
	v	0.00204	0.005875	0.00100	0.00102	0.00103	0.00108	0.00114	0.00122	0.00136	0.00167
221.2(Pc)	h	2108	2108	228.2	331.7	435.7	645.8	861.4	1087.0	1332.8	1635.5
(374.15)(Tc)	u	2037.8	2037.8	206.0	309.2	412.8	622.0	836.3	1060.0	1302.9	1600.3
	v	0.00317	0.00317	0.00100	0.00102	0.00103	0.00108	0.00114	0.00122	0.00135	0.00163

P (bar) (T_{sat} °C)		Saturated Water	Saturated Steam	400	450	500	550	600	650	700	750
250 (—)	h	—	—	230.7	334.0	437.8	647.7	862.8	1087.5	1331.1	1625.0
	u	—	—	205.7	308.7	412.1	620.8	834.4	1057.0	1297.5	1585.0
	v	—	—	0.00100	0.00101	0.00103	0.00108	0.00113	0.00122	0.00135	0.00160
300 (—)	h	—	—	235.0	338.1	441.6	650.9	865.2	1088.4	1328.7	1609.9
	u	—	—	205.0	307.7	410.8	618.7	831.3	1052.1	1288.7	1563.3
	v	—	—	0.0009990	0.00101	0.00103	0.00107	0.00113	0.00121	0.00133	0.00155
500 (—)	h	—	—	251.9	354.2	456.8	664.1	875.4	1093.6	1323.7	1576.3
	u	—	—	202.4	304.0	405.8	611.0	819.7	1034.3	1259.3	1504.1
	v	—	—	0.0009911	0.00100	0.00102	0.00106	0.00111	0.00119	0.00129	0.00144
1000 (—)	h	—	—	293.9	394.3	495.1	698.0	903.5	1113.0	1328.7	1550.5
	u	—	—	196.5	295.7	395.1	594.4	795.3	999.0	1207.1	1419.0
	v	—	—	0.0009737	0.0009852	0.00100	0.00104	0.00108	0.00114	0.00122	0.00131

P (bar) (T_{sat} °C)		Saturated Water	Saturated Steam	Temperature (°C) 400	450	500	550	600	650	700	750
0.0	h	—	—	3280	3384	3497	3597	3706	3816	3929	4043
	u	—	—	2969	3050	3132	3217	3303	3390	3480	3591
	v	—	—	—	—	—	—	—	—	—	—
0.1 (45.8)	h	191.8	2584.8	3280	3384	3489	3596	3706	3816	3929	4043
	u	191.8	2438.0	2969	3050	3132	3217	3303	3390	3480	3571
	v	0.00101	14.7	21.1	33.3	35.7	38.0	40.3	42.6	44.8	47.2
0.5 (81.3)	h	340.6	2646.0	3279	3383	3489	3596	3705	3816	3929	4043
	u	340.6	2484.0	2969	3049	3132	3216	3302	3390	3480	3571
	v	0.00103	3.24	6.21	6.67	7.14	7.58	8.06	8.55	9.01	9.43
1.0 (99.6)	h	417.5	2675.4	3278	3382	3488	3596	3705	3816	3928	4042
	u	417.5	2506.1	2968	3049	3132	3216	3302	3390	3479	3570
	v	0.00104	1.69	3.11	3.33	3.57	3.80	4.03	4.26	4.48	4.72

(*Continued*)

TABLE A.5 (Continued)

Properties of Superheated Steam

P (bar) (T_sat °C)		Saturated Steam	Saturated Steam	Temperature (°C)							
				400	450	500	550	600	650	700	750
5.0 (151.8)	h	640.1	2747.5	3272	3379	3484	3592	3702	3813	3926	4040
	u	639.6	2560.2	2964	3045	3128	3213	3300	3388	3477	3569
	v	0.00109	0.375	0.617	0.664	0.711	0.758	0.804	0.850	0.897	0.943
10 (179.9)	h	762.6	2776.2	3264	3371	3478	3587	3697	3809	3923	4038
	u	761.5	2582	2958	3041	3124	3210	3296	3385	3475	3567
	v	0.00113	0.194	0.307	0.330	0.353	0.377	0.402	0.424	0.448	0.472
20 (212.4)	h	908.6	2797.2	3249	3358	3467	3578	3689	3802	3916	4032
	u	906.2	2598.2	2946	3031	3115	3202	3290	3379	3470	3562
	v	0.00118	0.09950	0.151	0.163	0.175	0.188	0.200	0.211	0.223	0.235
40 (250.3)	h	1087.4	2800.3	3216	3331	3445	3559	3673	3788	3904	4021
	u	1082.4	2601.3	2922	3011	3100	3188	3278	3368	3460	3554
	v	0.00125	0.04975	0.0734	0.0799	0.0864	0.0926	0.0987	0.105	0.111	0.117
60 (275.6)	h	1213.7	2785.0	3180	3303	3422	3539	3657	3774	3892	4011
	u	1205.8	2590.4	2896	2991	3083	3174	3265	3357	3451	3545
	v	0.00132	0.0325	0.0474	0.0521	0.0566	0.0609	0.0652	0.0693	0.0735	0.0776
80 (295.0)	h	1317.1	2759.9	3142	3274	3399	3520	3640	3759	3879	4000
	u	1306.0	2571.7	2867	2969	3065	3159	3252	3346	3441	3537
	v	0.00139	0.0235	0.0344	0.0382	0.0417	0.0450	0.0483	0.0515	0.0547	0.0578

100	h	1408.0	2727.7	3100	3244	3375	3500	3623	3745	3867	3989
(311.0)	u	1393.5	2547.3	2836	2946	3047	3144	3240	3335	3431	3528
	v	0.00145	0.0181	0.0264	0.0298	0.0328	0.0356	0.0383	0.0410	0.0435	0.0461
150	h	1611.0	2615.0	2975	3160	3311	3448	3580	3708	3835	4962
(342.1)	u	1586.1	2459.9	2744	2883	2999	3105	3207	3307	3407	3507
	v	0.00166	0.0103	0.0157	0.0185	0.0208	0.0229	0.0249	0.0267	0.6286	0.0304
200	h	1826.5	2418.4	2820	3064	3241	3394	3536	3671	3804	3935
(365.7)	u	1785.7	2300.8	2622	2810	2946	3063	3172	3278	3382	3485
	v	0.00204	0.005875	0.009950	0.0127	0.0148	0.0166	0.0182	0.0197	0.0211	0.0225
221.2(Pc)	h	2108	2108	2733	3020	3210	3370	3516	3655	3790	3923
(374.15)(Tc)	u	2037.8	2037.8	2553	2776	2922	3045	3157	3265	3371	3476
	v	0.00317	0.00317	0.008157	0.0110	0.0130	0.0147	0.0162	0.0176	0.0190	0.0202
250	h	—	—	2582	2954	3166	3337	3490	3633	3772	3908
(—)	u	—	—	2432	2725	2888	3019	3137	3248	3356	3463
	v	—	—	0.006013	0.009174	0.0111	0.0127	0.0141	0.0143	0.0166	0.0178
300	h	—	—	2162	2826	3085	3277	3443	3595	3740	3880
(—)	u	—	—	2077	2623	2825	2972	3100	3218	3330	3441
	v	—	—	0.002830	0.006734	0.008680	0.0102	0.0114	0.0126	0.0136	0.0147
500	h	—	—	1878	2293	2723	3021	3248	3439	3610	3771
(—)	u	—	—	1791	2169	2529	2765	2946	3091	3224	3350
	v	—	—	0.001726	0.002491	0.003882	0.005112	0.006112	0.007000	0.007722	0.008418
1000	h	—	—	1798	2051	2316	2594	2857	3105	3324	3526
(—)	u	—	—	1653	1888	2127	2369	2591	2795	2971	3131
	v	—	—	0.001446	0.00162	0.001893	0.00224	0.00266	0.00310	0.003536	0.00395

Index